THIS BOOK
BELONGS TO:

meg Schartgen

Structural Equation Modeling with LISREL

Structural Equation Modeling with LISREL

Essentials and Advances

Leslie A. Hayduk

The Johns Hopkins University Press
Baltimore and London

The Johns Hopkins University Press, 701 West 40th Street, Baltimore, Maryland 21211
The Johns Hopkins Press Ltd., London

The paper used in this publication meets the minimum requirements of American National Standard for Information Sciences—Permanence of Paper for Printed Library Materials, ANSI Z39.48-1984.

Library of Congress Cataloging in Publication Data

Hayduk, Leslie Alec.
    Structural equation modeling with LISREL.

    Bibliography: pp. 371–96
    Includes index.
1. LISREL (Computer program.   2. Path analysis—Data processing.
3. Social sciences—Statistical methods.   I. Title.
QA278.3.H39 1987   519.5'35   87-2844
ISBN 0-8018-3478-3   (alk. paper)

LISREL is a registered trademark of Scientific Software, Inc. The program and manuals can be acquired by writing to
        Scientific Software Inc.
        P.O. Box 536
        Mooresville, Indiana
        U.S.A. 46158
The LISREL program is also available as a USERPROC within SPSSX.
        SPSS Inc.
        Marketing Department
        Suite 3300
        444 North Michigan Avenue
        Chicago, Illinois
        U.S.A. 60611

To Ann, Vincent, and Daniel

LISREL is not an easy program to learn. . . .
It does, however, provide the most complete solution
to the estimation problem of structural models.

D. Kenny, 1979

Contents

ix

Preface

I wrote this book because I could not honestly tell my students they had a solid foundation in structural equation modeling, or even that they were prepared for advancing themselves in structural equation modeling, by the time they finished Duncan's (1975) classic book *Introduction to Structural Equation Models*. Though they understood the modeling methods of the 1970s, this did not provide a solid foundation for learning the procedures of the 1980s and 1990s. The absence of matrix algebra, the failure to consistently distinguish between concepts and indicators, and outdated procedures for testing model fit provided topics my students were unlikely to master on their own. Furthermore, they would have to learn LISREL if they were to implement any of their new-found knowledge.

LISREL (*li*near *s*tructural *rel*ations) is the most general program that is widely available for estimating structural equation models, and it covers a truly amazing variety of models. LISREL can be used to analyze data from surveys, experiments, quasi-experimental designs, and longitudinal studies. LISREL allows one to test the goodness of fit of models, to diagnose problems with models, to fix or constrain model coefficients, to do multiple-group analyses, to estimate means and intercepts as well as slopes, and most importantly, to distinguish consistently between latent concepts and observed indicators. These features have made LISREL a major force ushering in "a new generation of statistical techniques for implementing structural equation methods" in the social sciences (Heise and Simmons, 1985: 429).

LISREL originated in the late 1960s and early 1970s when Karl Joreskog, Dag Sorbom, and others at Educational Testing Services created computer programs for use in educational research and factor analysis (e.g., ACOVS by Joreskog, Gruvaeus, and van Thillo, 1970). The theory of maximum likelihood estimation of structural coefficients had been known for some time, but the estimation of even modest models had proven prohibitively expensive. Joreskog and Sorbom improved the efficiency of estimation by capitalizing on a recently developed and efficient procedure for maximizing multivariate functions (Davidon, 1959; Fletcher and Powell, 1963).

Though the initial work centered on factor analysis, the ability to fix model coefficients and to express some factors as functions of other factors (see Chapter 4) shifted the focal interest to *confirmatory* factor analysis and the structural modeling of causal relations among factors. It soon became apparent that the basic LISREL model amounted to a general procedure for doing structural equation modeling (path analysis) in a way that preserved the distinction between concepts and indicators. This model suited the needs of the social sciences more generally, and LISREL broke loose from its roots in education and factor analysis.

Bielby and Hauser (1977) and Bentler (1980, 1986) clearly document the spread of LISREL and structural equation modeling throughout the social sciences, so we will mention only a few recent LISREL applications: **education** Alwin and Thornton (1984), Entwisle and Hayduk (1982), Marsh and Hocevar (1983), Nelson, Lomax, and Perlman (1984), Parkerson et al. (1984); **intelligence testing** Bynner and Romney (1986), Hultsch, Hertzog, and Dixon (1984), O'Grady (1983); **marketing and administration** Anderson (1985), Bagozzi (1980), Bagozzi and Phillips (1982), Joreskog and Sorbom (1982); **personality studies** Bagozzi (1981), Kohn and Schooler (1981, 1983), Marsh and Hocevar (1985), Miller, Kohn, and Schooler (1985), Newton et. al. (1984); **child development** Crano and Mendoza (1987); **status attainment** Campbell (1983), Hauser, Tsai, and Sewell (1983), Lindsay and Knox (1984); **criminology and deviance** Liska and Reed (1985), McCarthy and Hoge (1984), Piliavin et al. (1986), Smith and Patterson (1984), Silverman and Kennedy (1985), Thornberry and Christenson (1984); **family studies** Thomson and Williams (1982); **sex roles** Bielby and Bielby (1984), Thornton, Alwin, and Camburn, (1983); **organizational analysis** Laumann, Knoke, and Kim (1985); **group differences** Baer and Curtis (1984); **measurement** Bohrnstedt (1983), Bohrnstedt and Borgatta (1981), Corcoran, (1980), Fornell (1982), Mare and Mason (1981); **evaluation research** Bentler and Woodward (1978), Magidson (1977, 1978), Sorbom and Joreskog (1982); **experimentation** Alwin (1985), Alwin and Tessler (1974), Bagozzi (1980), Hayduk (1985), Sorbom and Joreskog (1982); **health studies** Chen and Land (1986), Kennedy, Starrfield, and Baffi (1983); **demography** Beckman et al. (1983), Thomson (1983); **twin studies** Fulker, Baker, and Bock

(1983); **race relations** Stahura (1986); **work studies** Lincoln and Kalleberg (1985), Lorence and Mortimer (1985); **longitudinal research** Baker, Mcdnick, and Brock (1984), Bentler (1984), and McArdle and Epstein (1987).

Even if you do not intend to use LISREL, you must be able to evaluate the use of LISREL by others in order to stay current (cf. Biddle and Marlin, 1987; Bohrnstedt, 1983; Fornell, 1983; Fornell and Larcker, 1981; Kessler and Greenberg, 1981).

The first widely distributed version of LISREL was LISREL III, and each new version added substantial new features. LISREL IV simplified the procedures for model specification, allowed multiple-group analyses and analyses with means and intercepts, and provided the standard errors of the estimates as output. LISREL V unified the handling of the signs of the effect coefficients and added automatic start values, a variety of estimation techniques, modification indices, standardized residuals, and much more. LISREL VI further improved the efficiency of the estimation procedures and was adopted as a USERPROC supplementing SPSSX (SPSS Inc., 1984). LISREL is now available at most major computer facilities.

Two philosophies ground this introduction to structural equation modeling through LISREL: (1) an emphasis on the basics, the whole basics, and nothing but the basics; and (2) that LISREL and structural equation models complement, but do not replace, knowledge of one's data set and theoretical subject area.

Emphasizing the basics leads me to present three chapters of review and preparation before beginning LISREL in Chapter 4. I count among the basics a recognition of the difference between mere program constraints and limitations created by data or theory (cf. Chapter 7 on tricking the LISREL program into doing intelligent things). I also count the following as basic: recognizing signs of troubles, knowing what can be done if such signs appear, and understanding the philosophy and fundamental conceptualizations grounding the various procedures and interpretations (cf. the discussion of models with loops in Chapter 8). The basics go beyond what is "most frequently used." Models containing means and intercepts (Chapter 9), for example, are seen infrequently, but they are basic because they add much to what can be done with structural equation models. Similarly, the discussions of interaction and nonlinearity among concepts (Chapter 7) have never before been presented in LISREL, but they remain basic because they open up new possibilities. Indeed, this represents so fundamental a shift that it even seems to contradict the first two letters in the name LISREL.

Emphasizing nothing but the basics prompted me to strive for the shortest path through the maze of key ideas. Comparisons between different strategies of estimation, for example, are minimized by focusing

xiv

on the most broadly applicable procedure—maximum likelihood estimation. It is safe to assume that LISREL will appear in a seventh version, so this emphasis on the immutable basics provides the reader the most lasting knowledge possible.

Our second philosophy is not new. Joreskog and Wold, for example, characterize LISREL as "theory oriented" and emphasizing "the transition from exploratory to confirmatory analysis" (1982c: 270). LISREL and this introduction will be most beneficial to "scientists who are thoroughly familiar with their subject matter . . . (and) who are also willing and able to spend a portion of each day in quiet, analytical thought, thinking through and anticipating possible challenges to their models as currently defined, and designing new studies or reanalysis of available data to meet those challenges" (Cooley, 1979). The joy of LISREL is that it provides opportunites for an active interplay between theory, modeling, and estimation. This rapprochement forces users to become experts at their theory, rather than mere number crunchers.

The more you put into LISREL, the more you get out of it. If only a covariance matrix goes in, then only an exploratory factor analysis comes out. If a theory, knowledge of the data collection procedures, *and* the covariance matrix go in, substantive findings come out. We emphasize the confirmatory use of LISREL, and in doing so we assume that the researcher has identified several concepts of interest, obtained indicators of those concepts, and is attempting to confirm suspected relations among those concepts.

We have not attempted to embrace or avoid causal terminology. Though statisticians are correct in pointing out that there is no causation in any structural equation, including LISREL's equations (cf. DeLeeuw, 1985), causal thinking is consistent with LISREL and structural equations in general. There is no "causation in any regression equation," but identifying Y as the "causally dependent" variable certainly helps us decide to regress Y on X rather than X on Y.

Nor is it necessary to defend causal forces as existing in the real world. Causal thinking may merely constitute a general and parsimonious way for our brains to grasp and summarize data whose ultimate determinential essence is beyond our current knowledge. The ultimate specification of the mechanisms providing the "causal" forces between variables may or may not involve ever finer specifications of increasingly specific intervening variables. Associations created by what a researcher is temporarily willing to describe by employing that "strange and unknown motivator" called causation may subsequently be resolved into noncausal explanatory components. Causation may not be in the real world or in the equations, but it definitely is in our thinking (cf. Mulaik, 1987, forthcoming). As long as social scientists find it useful to think of one thing as influencing, bringing about, effecting, determining, or causing another, there is no reason to abandon causal statements. Despite the philosopher's unease with statements of causation and the statistician's insistence that there is nothing causal about equations, social scientists are free to judge the usefulness of thinking causally about their data sets and to exploit the consistencies between structural equations and causal thinking.

Some suggestions for the LISREL *novice.* First, do it with a friend! Second, learn LISREL as if you are "learning a forest." Do not memorize every tree (equation), but do try to follow the path through the trees. The logic linking each equation to the next tells you where you have been and where you are going. This logic will remain with you long after the details of particular equations have been forgotten.

To instructors using this as a class text. Do not cover the introductory chapters too quickly. Parts of Chapters 1 and 2 may seem "low level," but I find students really do need the review, and the full context of this material helps them see the connections among the more complex issues. Do not get bogged down in the details of the later sections on matrix algebra; they can be reviewed as necessary. Chapters 4–6 should be reviewed as a set after they have been addressed individually. Encourage students to read the LISREL manual and enter their own models in conjunction with these three chapters. Pay special attention to the first exemplary LISREL model because variations on this model are used throughout the book. The beginning of Chapter 7 and all of Chapter 8 deserve special attention, but feel free to pick and choose among the later sections of Chapter 7 and Chapter 9 to suit your students.

Acknowledgments

Thanks are due to the University of Alberta Department of Sociology for supporting this book's preparation, to the participants in my LISREL classes of 1984–1987 for providing a student's perspective, to Terry Taerum for computer consultation, to Karl Joreskog for his intellectual companionship and copies of various works, and to Victor Thiessen, Bob Silverman, Herb Northcott, Mike Gillespie, Bill Avison, Aaron Pallas, Elisabeth Ten Verget, Janet McDonald, and Efraim Darom for comments on various ideas and drafts. Though we can take no responsibility for the LISREL program, we remain responsible for any deficiencies in its elucidation.

Chapter 1

Getting Started

Structural equation models are models of relationships among variables. Hence, we begin by examining "variables" and their "relationships." Next, we investigate how equations involving variables imply links between the *distributions* of those variables and, in particular, how the *structural coefficients* in equations control the location and spread of the concerned variables. We conclude by showing how a simple system of two equations (specifying a spurious relationship) implies a specific *quantifiable* relationship between the two dependent variables. The overall objective of this chapter is to provide the reader with a solid grasp of *how the structural coefficients in equations coordinate the various aspects of the distributions of the concerned variables.*

A **variable** is a characteristic (or quantity) that varies from person to person (or case to case). A **constant** is a characteristic common to all the cases of interest. An uppercase letter (e.g., X) represents a variable and the corresponding lowercase letter with a subscript (e.g., x_i) represents the different values of X. For data sets containing a large number (N) of cases or individuals, each of the n values of X may appear more than once, because several cases may have the same X value. (If X is religion, several persons may share the same religious affiliation.) Thus the subscript i indexes the possible values of X rather than the individuals possessing those values.

The values of X possessed by some set of cases may also be listed in a column (Figure 1.1A) or plotted as a frequency distribution by counting the frequency with which each value of X appears (Figure 1.1B). We assume that the values of X are measured on an interval scale unless we state otherwise; therefore Figure 1.1 depicts "grouped" values of X.

Figure 1.1   Representations of variable X.

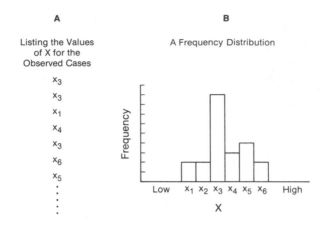

Each of the three representations of a variable has advantages. The x_i notation is easiest to manipulate in equations, the column representation meshes with matrix algebra, and the frequency distribution best reveals the relative placement and clustering of individual values.

1.1 Means, Variances, and Covariances

For a variable X, we consider several possible distributions of scores (values) for 10 cases (individuals). The distribution of values in Figure 1.2A differs from that in B only in location: A is located on lower values of X. Distribution A is located in the same place as C but is less spread out. Distribution D differs from A in location, shape, and spread.

The **location** and **spread** of a distribution are cornerstones of structural equation modeling, so we discuss them in detail now. The **shape** of a distribution is crucial in more advanced issues, so we will postpone this topic. The location of a distribution is determined by the kinds of X values likely to appear. In distribution A, low (small) values of X are likely, whereas in distribution B high (large) values are likely. The likelihood or **probability**, $p(x_i)$, of any particular value x_i appearing is defined and calculated as the frequency with which that value appears (f_i) divided by

Figure 1.2   Distributions.

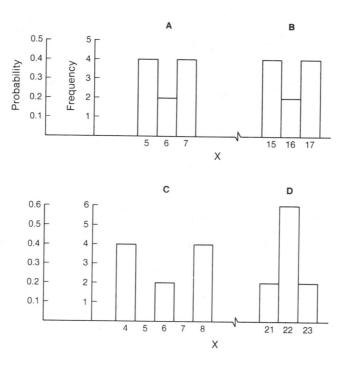

the total frequency (N); hence $p(x_i) = f_i/N$. The probability of a value appearing is the proportion of cases having that value or, equivalently, the proportion of the total area in the distribution that appears in the "bar" above the value.

Probabilities and proportions differ despite being calculated by the same formula. A proportion describes the data at hand; a probability describes the following hypothetical task. Imagine that a case is selected at random from the 10 cases for which we have scores, and you are asked to guess the X value for that case. Since you know the distribution from which the case was drawn, you know something about the values that can appear for the selected case. If a particular value does not appear in the distribution, that value has frequency zero and probability $p(x_i) = 0/N = 0$; hence that value will not appear for the randomly selected case. If there was only one value in the distribution (i.e., if the "variable" was a constant), the probability of that value appearing for the randomly selected case is $p(x_i) = f_i/N = N/N = 1$, and we can guess the value with certainty. Frequencies between 0 and N lead to probabilities between 0 (impossibility) and 1 (certainty) and indicate the greater or lesser likelihood of that value appearing in the randomly selected case. If 9 out of 10 cases had value x_3, then $p(x_3) = .9$, and the selected case would be very *likely* but *not certain* to have value x_3.

Statisticians call the combined set of variable values and accompanying probabilities a **random variable**. "Random" reflects the single hypothetical random selection grounding the probabilities, and "variable" refers to the dimension underlying the particular set of values.

1.1.1 Best Guesses

We might ask what is the single "best" guess of the value of the randomly selected case? If "best" means the guess most likely to be correct, we would guess the value with the highest probability (or frequency), commonly called the **mode** or the **modal value**. If "best" means "coming closest to" the actual value, then we can calculate a better guessed value (see Figure 1.3). We first list the values of X and the associated probabilities. Then we create a column of numbers by multiplying each value by its associated probability, and we sum the entries in this column. We then get 6.2 as the best guess or "expected value." (The final column is discussed later.)

Summing the product of each value with its associated probability (weight) provides the **expected value** of X [$E(X)$ or μ]. The expected value, in equation form, is

$$E(X) = \sum_i x_i p(x_i) \qquad\qquad 1.1$$

and is identical to the mean or average. (Appendix A gives a brief introduction to summation (Σ) notation.) For data listed as individual cases, $p(x_i) = 1/N$ for each individual's value, so

$$E(X) = \sum_i x_i p(x_i) = \sum_i x_i(1/N) = \sum_i x_i/N \qquad\qquad 1.2$$

which is the usual procedure for calculating a mean. For grouped data $p(x_i) = f_i/N$, so

$$E(X) = \sum_i x_i(f_i/N) = \left(\sum_i x_i f_i\right)/N \qquad\qquad 1.3$$

which is a "grouped data" formula for the mean.

In what sense is the expected value (6.2) our best guess for the data in Figure 1.3? The data contain only the values 5, 6, and 7, so our guess will never exactly match the hidden randomly selected value. The two

Figure 1.3   The expected value and variance of X.

Scores or Values of X x_i	Probabilities of those Values $p(x_i) = f_i/N$	The product $x_i p(x_i)$	$(x_i - E(X))^2 p(x_i)$
5	2/10	1.0	$(5-6.2)^2$ 2/10
6	4/10	2.4	$(6-6.2)^2$ 4/10
7	4/10	2.8	$(7-6.2)^2$ 4/10
		6.2	0.56

The sum of this column is the expected value of X since
$$E(X) = \sum_i x_i \, p\,(x_i)$$

The sum of this column is the variance of X since
$$Var(X) = \sum_i (x_i - E(X))^2 \, p(x_i)$$

justifications for why this guess is best require that we consider the spread of cases in the distribution.

1.1.2 The Spread of Cases

How far wrong is our guess? Obviously, cases score above the expected value, others score below it, and some scores are closer to the expected value than others are. Thus, the question, "how far" are the cases from the expected value? must be phrased as, "on average," how far are they? That is, what is a typical, usual, average, or *expected* size for the deviations of the actual values (scores) from the expected value or mean?

The distance between any score and the mean can be written as $x_i - E(X)$. Think of this difference as the single number reporting the size of the error that would have been made if we had "guessed" the mean (expected value) when the true value was x_i. We might calculate the average size of these "errors" in hopes that this would tell us how far the values are from the mean. The average or expected size of these differences can be denoted as $E[X - E(X)]$ and calculated as

$$E[X - E(X)] = \sum_i [x_i - E(X)]p(x_i) = \sum_i [x_i - E(X)]\frac{f_i}{N} \qquad 1.4$$

since the various deviations appear with the same probabilities as the corresponding values of X. [Read $E[X - E(X)]$ as the expected or average value of the single numbers representing the deviation of each value from the mean or expected value. Read $[x_i - E(X)]$ as the single number encapsulating the deviation of the particular value x_i from $E(X)$.]

Unfortunately, $E[X - E(X)]$ is always zero, and so measures nothing about spread, because

$$\sum_i [x_i - E(X)]\frac{f_i}{N} = \sum_i x_i(\frac{f_i}{N}) - \sum_i E(X)(\frac{f_i}{N})$$

$$= E(X) - E(X)(\sum_i f_i)/N \qquad 1.5$$

$$= E(X) - E(X) = 0$$

However, this demonstrates one way in which the mean (expected value) provides the best guess, or the center of the distribution. The expected value is best in that the amount of overestimation is the same as the amount of underestimation; that is, the average size of the guessing errors is zero.

Another way of seeing that the expected value is the best guess requires that we persist in our quest for a measure of the spread of scores in a distribution. Let us examine the **variance** [$Var(X)$ or σ_x^2] of the distribution, which is the *average (or expected value) of the squared deviations from the mean*. Squaring the deviations avoids the problem of positive and negative deviations (under- and overestimates) canceling out. Variance is calculated exactly as it is verbally defined—it is the average or expected value of the squared deviations from the mean $E(X)$:

$$\sigma_x^2 = Var(X) = E[(X - E(X))^2] \qquad 1.6$$

$$Var(X) = \sum_i (x_i - E(X))^2 p(x_i) = \sum_i (x_i - E(X))^2 \frac{f_i}{N} \qquad 1.7$$

Think of the term $(x_i - E(X))^2$ as the single number obtained by squaring the number representing the deviation of one value x_i from $E(X)$. Hence, Eq. 1.7 is merely the average (expected value) of these numbers rather than the average of the basic values as presented in Eq. 1.1. The last column in Figure 1.3 illustrates the calculation of variance.

The second way of seeing that the expected value (mean) is our best guess at the value of the hidden randomly selected case is to learn that the sum of the squared deviations of the values from this guess (the mean) is a minimum. That is, no other guessed value [no other number that could be placed in Eq. 1.7 instead of $E(X)$] provides a smaller sum of squared errors—if we call the difference between what we guess and the actual value of the hidden randomly selected case an error. This "least squares" property of the mean is proven in most introductory statistics texts (e.g., Blalock, 1979b).

A standard deviation σ is the positive square root of the variance. Both the standard deviation and the variance measure the spread of cases around the mean. If the cases cluster closely around the X mean σ_x and σ_x^2 will be small; if they deviate markedly, σ_x and σ_x^2 will be large.

1.1.3 Relationships and Covariance

Having seen that the mean and variance are based on taking expected values, we should not be surprised that the third fundamental statistical quantity, the *covariance*, is also based on expectations. Unlike the mean and variance, however, covariance involves two variables and the relationship between them.

A **relationship** exists between two variables *if the distribution of the cases on the values of one variable differs, depending on which value of the other variable the cases possess.* Consider plotting separate distributions of Y scores (y_j) for various values of X (x_i), as shown in Figure 1.4. Traditionally, we compare these conditional distributions of Y scores by comparing the expected values (means) of the distributions. Figure 1.4A shows that there is a relationship between Y and a dichotomous X because the Y distributions differ, depending on which X value the case has. One would predict a different Y value, depending on which X value a case was known to possess. Similarly, Figure 1.4B illustrates a relationship between Y and a continuous X, where the dashed lines provide some arbitrary categories of X values for which conditional Y distributions are plotted. The means of these conditional distributions *differ*, and hence we say X and Y are related.

The concept *covariance* numerically summarizes the degree of difference between the conditional Y distributions and therefore summarizes the extent of the relationship between X and Y. Equivalently, covariance describes the overall clustering of points if each case is plotted on a pair of X and Y axes using the X and Y scores of the cases.

The formula defining covariance is closely related to the formula for variance, which measures the spread of the scores on the variables individually. The **covariance** between X and Y [Cov(XY) or σ_{xy}] *is defined and calculated as the expected value (average size) of the product of the deviations of the X and Y scores from their respective means*:

$$\text{Cov}(XY) = E[(X - E(X))(Y - E(Y))] \qquad 1.8$$

$$= \sum_i \sum_j (x_i - E(X))(y_j - E(Y))p(x_iy_j) \qquad 1.9$$

Figure 1.4   Relationships.

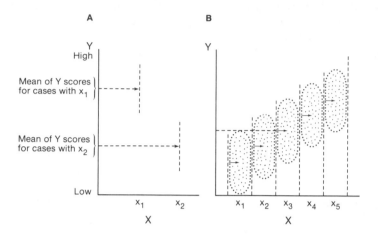

The sum is over all combinations of the values of X and Y, indexed by i and j, respectively, and $p(x_iy_j)$ is the probability of the ith value of X and the jth value of Y appearing together (i.e., the frequency of each possible pair of values divided by the total N of x_iy_j pairs—the total number of cases). If the expected values are replaced with the traditional notation for the population mean (μ) we write

$$\text{Cov}(XY) = \sum_i \sum_j (x_i - \mu_x)(y_j - \mu_y)p(x_iy_j).\qquad 1.10$$

1.2 Thinking about Covariances, Variances, and Means

Figure 1.5 gives a procedure for hand calculating the covariance of two variables; however it is Figure 1.6 that substantially advances our understanding of covariance. Figure 1.6A illustrates the deviations of four cases from the X and Y means by the lengths of the lines connecting the points (cases) to the lines plotted at the means. Case 1 is above the X and Y means (hence the + signs), and case 3 is below both the X and Y means

Figure 1.5   The covariance of X and Y.

① The means of X and Y (E(X) and E(Y)) have been calculated as per Figure 1.3.

② The cases (individuals) have been listed as separate rows of this table. Hence, the index i runs from 1 to N. All existing pairings of X-Y values will be included in some row, and the probability associated with the pair of X-Y values appearing in any one row is 1/N. If several individuals possess the same pair of X-Y values, each individual will contribute a probability of 1/N to the probability of this pairing (and its mean deviations) at the time of summation.

$$= 1/N$$

i	x_i	y_i	$(x_i{-}E(X))(y_i{-}E(Y))p(x_iy_i)$
1			(# — #) (# — #) #
2			
3			
4			
5			
6			
.			
.			
.			
.			
N			

$$\text{Cov}(XY) = \Sigma \underline{\qquad}$$

The sum of the entries in the right column is the covariance of X and Y since

$$\text{Cov}(XY) = \sum_i (x_i{-}E(X))(y_i{-}E(Y))p(x_iy_i)$$

or

$$\text{Cov}(XY) = \sum_i\sum_j (x_i{-}E(X))(y_j{-}E(Y))p(x_iy_j)$$

if i and j subscript all the possible values of X and Y instead of individuals.

Figure 1.6   Covariance.

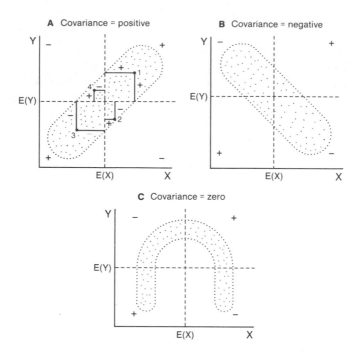

A Covariance = positive

B Covariance = negative

C Covariance = zero

(hence the − signs). Cases 2 and 4 are above only one or the other of the means (hence mixed signs), and the shorter line segments for these cases indicate that their values are relatively close to the X and Y means.

The calculation of covariance (Eqs. 1.9 or 1.10) involves multiplying the lengths of the two lines connecting each point (case) to the means. The signs of these deviations are preserved during multiplication, so a positive product appears if the case is in the upper right or lower left quadrants, and a negative product appears if the case is in the upper left or lower right quadrants. Averaging these product terms for all the cases [summing the products and dividing by N, or multiplying the products by their associated probabilities and then summing (Eqs. 1.9 or 1.10)] gives the expected value of the products of the deviations of the cases from the X and Y means, which is covariance.

If the cases cluster primarily in the upper right and lower left quadrants (Figure 1.6A), the covariance is positive, because most cases contribute positive product terms in the covariance formula. If the cases cluster in the upper left and lower right quadrants (Figure 1.6B), the covariance is negative due to the many negative product cases. Clusterings providing counterbalancing sets of cases in the four quadrants (Figure 1.6C) produce zero covariance, because the positive and negative product terms cancel. This is the basis for the oft-noted caution that covariance and the statistics based on covariance can be misleading if nonlinear clusterings appear in the data. The covariance is zero, but X and Y are still related because the distributions of Y scores differ, depending on the particular X value considered.

We must be able to estimate quickly the approximate numerical magnitude of means, variances, and covariances, and to visualize what these quantities are if we confront them as mere numerical entities. In Figure 1.7 the mean can be visualized as the center of the distribution or as the value at which the distribution would balance if a fulcrum were placed beneath it. Thus, the distribution in Figure 1.7A would come closer to balancing if the fulcrum were placed at 4 rather than at 3 or 5; so visual inspection convinces us that the numerical value of the mean (expected value) must be nearer to 4 than to 3 or 5.

To visualize variance, consider which case (a to e) of the distribution in Figure 1.7B displays a deviation from the mean that is *typical*, usual, average, or expected. Cases a and b would be atypical because they are too close to the mean; d and e are too far from the mean. Thus, c appears to be most typical, and its deviation from the mean is a reasonable estimate of the standard deviation; therefore the square of this deviation is a reasonable estimate of the variance [the average (typical or expected) value of the squared deviations from the mean]. *The squaring in the definition of variance implies that extreme cases contribute disproportionately into variance, so judgments of typical deviations should acknowledge the greater impact of extreme cases.* Deviations below the mean must also be considered, so our decision to count c as typical should recognize that going an equal distance below the mean to c′ provides a deviation typical of cases below the mean.

To obtain a rough estimate of the numerical magnitude of covariance, we again use a typical case to replace the mathematically expected value. In Figure 1.7C, a case describing the typical clustering of the cases falls on the line drawn roughly through the center of the cluster of data points. Cases a and b are properly oriented relative to the intersection of the two means, but they are too close to the means to represent the set of points as a whole. Cases d and e are too far from the means to be typical. Thus case c depicts the typical clustering of the cases, and the product of the deviations of this case from the means approximates the covariance. Case c′ is equally representative of the clustering and will give a second estimate

Figure 1.7   Visual estimation.

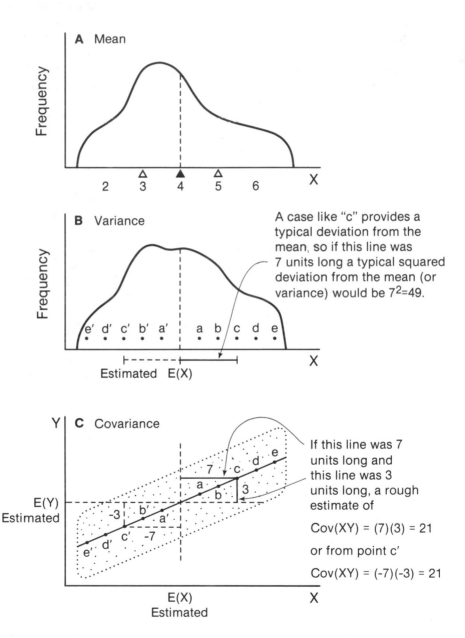

of the covariance, as indicated in the figure.

Whenever we confront variance and covariance as numerical entities, we should feel comfortable visualizing what this implies in terms of distributions of cases. We should feel equally comfortable providing verbal statements for the mean, variance, and covariance as the expected (average) sizes of specific quantities (Eqs. 1.1, 1.6–1.7, and 1.8–1.10).

To test your ability to switch between the verbal, diagrammatic, and equation modes, use the preceding equations to show that *the variance of a variable X is the covariance of that variable with itself.* (*Hint*: Replace the *Y*'s in the covariance formula with *X*'s.) Draw a scatterplot of X against X (itself) to estimate the covariance of X with itself (all the points fall on a 45° line), and consider the parallel between examining the "product of two deviations from means" in covariance and the "squaring of the deviation of the typical case" in obtaining variance.

For another test, show that *the covariance between a variable and a constant must be zero.* (*Hint*: If Y in Eq. 1.9 is a constant, the Y mean would equal that constant, and hence the deviations of the Y scores from the Y mean would always be zero. What would the sum in 1.9 equal?) What does a diagram of this covariance look like?

1.3 Expectations and Equations

Consider creating a new variable (Y) from a variable (X), for which we have existing scores (values) for a set of persons. The new scores (y_i's) will be obtained by multiplying an individual's original x_i score by a number b and then adding another number a. The same numbers a and b will be used to transform each individual's original scores into their new (adjusted) score, so a and b are constants, though the numerical magnitudes of these constants remain unspecified for the moment. Figure 1.8 shows the procedure and makes clear that the probabilities associated with the original X values are precisely the probabilities associated with the new Y values, because the first new score y_1 occurs exactly as often as the old x_1 appears, and so on.

We have seen that we can calculate $E(X)$ and $Var(X)$ as long as we know the actual numerical values of X (the x_i's) and the corresponding probabilities—that is, we can describe the location and spread of the random variable X with the appropriate calculations. We now have a new random variable Y because we have a new set of values (y_i's) and their associated probabilities; therefore we can also describe the location and spread of the distribution of scores on the new variable Y. We could do this by putting the Y values and probabilities into the preceding formulas,

Hence, the variance of the new scores can be expressed as a function of the original X scores and the transformation, without referring to the Y scores, but we will again want to simplify the right side of the expression.

Removing the inner parentheses cancels the a terms.

$$\text{Var}(Y) = \sum_i (a + bx_i - a - bE(X))^2 p(x_i) \qquad 1.19$$

Factoring out the b, squaring, and moving b^2 outside the sum provides

$$\text{Var}(Y) = \sum_i [b(x_i - E(X))]^2 p(x_i) \qquad 1.20$$

$$- \sum_i b^2(x_i - E(X))^2 p(x_i) \qquad 1.21$$

$$= b^2 \sum_i (x_i - E(X))^2 p(x_i) \qquad 1.22$$

Recognizing the sum as $\text{Var}(X)$ (Eq. 1.7) allows us to write

$$\text{Var}(Y) = \text{Var}(a + bX) = b^2 \text{Var}(X) \qquad 1.23$$

Thus, the variance of the new scores is the multiplicative constant squared times the variance of the original scores or, equivalently, the standard deviation of the new scores is b times the standard deviation of the original scores. The a term is absent, and the a's in Eq. 1.19 cancel, because the individual scores and the mean of the scores have both been moved a units by the addition of a during calculation of the new Y scores. Each score is increased by a units, and the mean is increased a units (Eq. 1.16), but the spread of the cases around the mean has not changed. The original X distribution is moved up or down along the axis (depending on whether a is positive or negative), but the curve is neither spread out nor squashed in by the addition of a to each of the original values.

The b^2 term arises from the squaring that is inherent in calculating the variance of Y as the average or expected value of the *squared* deviations of the Y scores from the Y mean (Eq. 1.21). Equation 1.20 demonstrates that each deviation of the Y scores from the Y mean is merely b times as large as the corresponding deviation of the X scores from the X mean. Stretching or shrinking the Y scale relative to the X scale by multiplying each X score by b not only transforms each X score into a new score b times as large, but it also implies that the difference between any two values is b times as large as it was previously.

These observations should not distract us from the main point, which is that the spread of the new scores is predictable or calculable if one knows both the original spread of the scores and that the transformation is of the form $Y = a + bX$. Thus we can describe the spread of cases in the new distribution (the new set of scores) without ever actually calculating the new scores for each case.

As an exercise, use Eqs. 1.16 and 1.23 to express the covariance between the original X and new Y scores as a function of the original scores. You should find that $Cov(XY) = b\,Var(X)$.

1.3.3 Combining Several Original Variables into a New Variable

Consider creating a new Y variable as a linear function of several X variables and another set of constants (c's), as follows:

$$Y = c_1X_1 + c_2X_2 + c_3X_3 + \cdots + c_mX_m \qquad 1.24$$

Thus the new Y scores are created by summing c_1 times the individual's score on X_1 plus c_2 times their score on X_2, and so on, until all the m original variables have been used. Figure 1.9 depicts this process in tabular form (without the constants c_1 to c_m). The means and variances of the original X variables are known, and the mean and variance of the newly created scores are being sought. The equations linking the mean and variance of the new Y variable to the means and variances of the original X variables are derived in Mood, Graybill, and Boes (1974). The mean of the new Y variable is

$$E(Y) = c_1E(X_1) + c_2E(X_2) + c_3E(X_3) + \cdots + c_mE(X_m) \qquad 1.25$$

The formula expressing the variance of the new variable Y as a function of the original variable is more complex and has two equivalent forms.

$$Var(Y) = \sum_{\substack{all \\ j,k}} c_jc_k Cov(X_jX_k)$$

$$= \sum_{j} c_j^2 Var(X_j) + \sum_j \sum_k c_jc_k Cov(X_jX_k) \qquad 1.26$$
$$\quad (j = k) \qquad\qquad (j \neq k)$$

The second version of Eq. 1.26 arises because the covariance of a variable with itself (when $j = k$) is a variance, and hence the implicit variances can be considered separately from the "real" covariances in the equation.

Though initially foreboding, this equation is my candidate for the single most important equation in structural equation modeling. To convince you of this, we first summarize what the equation is, explain the notation, and so on. Then we give simpler versions of the formula appropriate for only two or three original variables. Finally, we discuss a special case of this formula appropriate when the original variables are independent of one another.

Figure 1.9 $Y = c_1X_1 + c_2X_2 + c_3X_3 + \cdots + c_mX_m$

Variables	X_1	X_2	X_3	X_4	........X_j........	X_m	Y
Individual 1	(#)	(#)	(#)	(#)	(#)	(#)	(#)
2							
3							
4							
5							
6							
7							
.							
.							
.							
N							
Calculable	$E(X_1)$	$E(X_2)$	$E(X_3)$	$E(X_4)$	$\cdots E(X_j)\cdots$	$E(X_m)$	$E(Y)=?$
Quantities	$Var(X_1)$	$Var(X_2)$	$Var(X_3)$	$Var(X_4)$	$\cdots Var(X_j)\cdots$	$Var(X_m)$	$Var(Y)=?$

From the equation we note that the variance of Y is expressed in terms of the constants used in creating the new Y scores, the variances of the original variables, and the covariances among the original variables. The sums are not over individual values but are over sets of variables, emphasized by our switch to subscripts j and k, both of which run from 1 to m (the number of original variables). The summing of individual scores and deviations from means is subsumed within the variances and covariances. To see what the j and k subscripts do, consider Figure 1.10. The original variables, indexed by j, head the rows of this table, and the same variables, indexed by k, head the columns. The entries in the cells of the table are created by first considering the covariance between the relevant row and column variables. (Recall that the covariance of a variable with itself is the variance of that variable.) These covariances are systematically multiplied by the constants (c's) used in creating the new variable Y. Each row of the table is multiplied by the constant associated with the appropriate row variable, and each column is also multiplied by the constant appropriate for the column variable. Thus each cell of the table is composed of two weights—one from the row the cell appears in and one from the column the cell appears in—and the covariance between the variables heading the row and column.

The variance of the new set of scores created as the weighted sum of these m original variables and expressed as Eq. 1.26 is the sum of all

Figure 1.10   Variance of $Y = c_1 X_1 + c_2 X_2 + \cdots + c_m X_m$

	X_1	X_2	$X_3 \cdots \cdots \cdots X_k \cdots \cdots X_m$				Vector of c's
X_1	$c_1 c_1 \text{Var}(X_1)$	$c_1 c_2 \text{Cov}(X_1 X_2)$	$c_1 c_3 \text{Cov}(X_1 X_3)$		$c_1 c_k \text{Cov}(X_1 X_k)$	$c_1 c_m \text{Cov}(X_1 X_m)$	c_1
X_2	$c_2 c_1 \text{Cov}(X_2 X_1)$	$c_2 c_2 \text{Var}(X_2)$	$c_2 c_3 \text{Cov}(X_2 X_3)$		$c_2 c_k \text{Cov}(X_2 X_k)$		c_2
X_3	$c_3 c_1 \text{Cov}(X_3 X_1)$	$c_3 c_2 \text{Cov}(X_3 X_2)$	$c_3 c_3 \text{Var}(X_3)$				c_3
X_j	$c_j c_1 \text{Cov}(X_j X_1)$	$c_j c_2 \text{Cov}(X_j X_2)$	$c_j c_3 \text{Cov}(X_j X_3)$		$c_j c_k \text{Cov}(X_j X_k)$		
X_m	$c_m c_1 \text{Cov}(X_m X_1)$					$c_m c_m \text{Var}(X_m)$	c_m

In Chapter 3 we will see that if **c** is the vector on the right **c**'(Covariance Matrix)**c** provides the sum of all the cells in the above matrix, which is the variance of Y created by $Y = c_1 X_1 + c_2 X_2 + c_3 X_3 \ldots c_m X_m$.

entries in this table. The m diagonal elements of this table (where $j = k$) contain variances and the square of the corresponding c constants. These elements appear as the first sum in the second version of Eq. 1.26. The off-diagonal elements (when $j \neq k$) contain covariances and the constants for row j and column k. These elements appear as the second sum in Eq. 1.26.

If Eq. 1.26 has only two original variables, the variance of Y simplifies to

$$\text{Var}(Y) = c_1^2 \text{Var}(X_1) + c_2^2 \text{Var}(X_2)$$
$$+ c_1 c_2 \text{Cov}(X_1 X_2) + c_2 c_1 \text{Cov}(X_2 X_1) \qquad 1.27$$

or, because of the equivalence of the last terms, to

$$\text{Var}(Y) = c_1^2 \text{Var}(X_1) + c_2^2 \text{Var}(X_2) + 2c_1 c_2 \text{Cov}(X_1 X_2) \qquad 1.28$$

Thus the variance of the new variable behaves as in Eq. 1.23, where the constants arc squared in translating the variances of the original variables into the variance of the new variable, but there is an additional concern that the variance of the new variable also reflects the covariance between the variables (the extent to which the original variables vary together). The covariance may be positive or negative, so the variance of the new Y variable may end up above or below the sum of the weighted variances of the X variables (the first two terms).

Within the two-variable case, we might consider only specific values of the constants. If we *select* $c_1 = c_2 = 1$, then Y is the sum of the two X variables, and, according to Eq. 1.28, the variance of the *sum* Y would be the sum of the X variances plus the possibly negative magnitude of the covariance between the original variables. Or, if we had added one variable to 4 times the other ($c_1 = 1$, $c_2 = 4$), the variance of this weighted sum would be dominated by the second variable, it being 1 times the variance of the first variable plus 16 times the variance of the second plus 8 times the covariance.

Next, consider the simplification of Eq. 1.26 that arises *if the original variables are independent of one another*. By definition, two events (*values of variables*) are independent if the probability of the events appearing together equals the product of the probabilities of the events appearing separately.

$$p(x_i y_j) = p(x_i)p(y_j) \qquad \text{if } x_i \text{ and } y_j \text{ independent} \qquad 1.29$$

Two *variables* are independent if each pair of values for those variables is independent.

One implication of the independence between variables is that the covariance between the variables must be zero. This is easily seen if we take the formula for covariance (Eq. 1.9) and replace the joint $x_i y_j$ probability with the separate probabilities (thereby assuming the variables are independent).

$$\text{Cov}(XY) = \sum_i \sum_j (x_i - E(X))(y_j - E(Y))p(x_i)p(y_j) \qquad 1.30$$

Factoring the X terms out of the Y sum provides

$$\text{Cov}(XY) = \sum_i (x_i - E(X))p(x_i) \sum_j (y_j - E(Y))p(y_j) \qquad 1.31$$

in which the second sum is zero because it is the average of the deviation of the Y values from the Y mean (Eq. 1.5).

$$\text{Cov}(XY) = \sum_i (x_i - E(X))p(x_i)(0) = 0 \qquad 1.32$$

Therefore, if we know or assume that we are creating a new variable from original variables that are independent of one another, the covariances in Eq. 1.26 become zero, and the equation simplifies to

$$\text{Var}(Y) = \sum_j c_j^2 \text{Var}(X_j) \qquad 1.33$$

For two original variables

$$\text{Var}(Y) = c_1^2 \text{Var}(X_1) + c_2^2 \text{Var}(X_2). \qquad 1.34$$

For the sum of two independent variables (i.e., the covariance is zero and we choose $c_1 = c_2 = 1$ so that $Y = X_1 + X_2$),

$$\text{Var}(Y) = \text{Var}(X_1) + \text{Var}(X_2) \qquad 1.35$$

Equation 1.35 partitions the variance of the Y variable into two pieces. This partitioning underlies using a squared correlation coefficient to speak of "explained and unexplained variance" (e.g., if X_1 is a predictor of Y and X_2 is an independent error variable). The same equation grounds the "error variance versus explained variance" discussions in analysis of variance. Equation 1.26 also underlies the partitioning of variance in the context of multiple regression, where some variables are independent (the error variable) but others are not (the predictor variables may be correlated). Naturally, this wide range of commonalities arises because all these discussions of explained versus error variance share the same general underlying model, one where some variables give rise to other variables by following a specific structural form (specifically Eq. 1.24).

Equation 1.35 also hints that many of the procedures for estimating structural equation models require only a covariance matrix as input, as opposed to requiring individual scores. In Chapter 5, for example, maximum likelihood estimation of the coefficients (c's) in Eq. 1.24 and partitionings comparable to Eq. 1.35 can be done even if the covariance matrix (and not individual scores) is known. For this reason, some authors claim to be "analyzing covariance structures" when they engage in these types of analyses.

1.3.4 Standardizing Variables

The equations linking the means and variances of new and original random variables are the logical basis for rescaling variables. We have seen that if we restrict ourselves to a transformation of the form $Y_i = a + bX_i$, then Eqs. 1.16 and 1.23 link the means and variances of the new and old variables:

$$E(Y) = E(a + bX) = a + bE(X) \qquad\qquad 1.36$$

$$\text{Var}(Y) = \text{Var}(a + bX) = b^2\text{Var}(X) \qquad\qquad 1.37$$

Earlier we arbitrarily specified values for a and b for illustrative purposes, but we can use whatever criteria we like to specify these constants. In the context of standardizing, we select the values of the constants to provide the new scores with a desired mean and variance (specifically, mean 0.0 and variance 1.0). The above equations thus become

$$0 = a + bE(X) \qquad\qquad 1.38$$

$$1 = b^2\text{Var}(X) \qquad\qquad 1.39$$

$E(X)$ and $\text{Var}(X)$ take on values calculated from the original data, so the numerical value of b required to give the new scores a variance of 1 is

$$b^2 = 1/\text{Var}(X) \qquad\qquad 1.40$$

$$b = 1/\text{std.dev.} = 1/\sigma_x \qquad\qquad 1.41$$

Inserting this known value of b in Eq. 1.38 provides the numerical magnitude of a, the coefficient that shifts the distribution up or down the scale until it is centered on zero:

$$0 = a + (1/\text{std.dev.})E(X) \qquad\qquad 1.42$$

Therefore

$$a = -E(X)/\text{std.dev.} \qquad\qquad 1.43$$

Inserting these a and b values in the equation for obtaining the new Y scores ($Y_i = a + bX_i$) gives

$$Y_i = -(\text{mean}/\text{std.dev.}) + (1/\text{std.dev.})X_i \qquad\qquad 1.44$$

Creating a common denominator and rearranging gives

$$Y_i = \frac{X_i - \text{mean}}{\text{std.dev.}} \qquad\qquad 1.45$$

The new standardized variable is usually called Z, rather than Y, so this is traditionally written as

$$Z_i = \frac{X_i - \mu_x}{\sigma_x} \qquad \text{or} \qquad Z_i = \frac{X_i - \bar{X}}{s} \qquad\qquad 1.46$$

for means and variances arising from a population (first equation) or a sample (second equation).

Thus, standardizing is a special case of creating a new random variable from an old one, and it should be thought of as filling in a table, as illustrated in Figure 1.8. Notice that the probabilities associated with the different values of the old and new random variables are identical. A Z transformation (more precisely, creating a new random variable with mean 0.0 and variance 1.0) does not alter the shape of the distribution. If X was normally distributed, the new variable is normally distributed. If X was nonnormal, it remains nonnormal after standardizing by the calculation of Z scores.

We can think of the standardizing transformation as changing the scale on which the values of the original variable are reported, as shown in Figure 1.11A, with the distribution remaining untouched. Equivalently, we can imagine the transformation as sliding the original distribution up or down until it is centered on zero (the origin) and then spreading or squashing the distribution until one standard deviation on the distribution corresponds to one unit on the scale (Figure 1.11B). Either way, the shape of the distribution remains unchanged.

We can use the foregoing procedure to create new variables with *any* mean or variance. Consider creating new scores scaled like IQ scores to have mean 100 and variance 225. Equations 1.36 and 1.37 become

$$100 = a + bE(X) \qquad\qquad 1.47$$

$$225 = b^2\text{Var}(X) \qquad\qquad 1.48$$

and we could solve as before for the a and b that rescale a set of "raw" X scores into "IQ-standardized" scores.

Another important type of standardization is creating a set of scores with mean zero while the variance of the scores is unchanged. This is done by making $b = 1.0$ (hence no change in variance), and $a = -E(X)$. Thus the new scores are obtained from $y_i = x_i - E(X)$ or as deviations from the mean.

Figure 1.11   Standardizing.

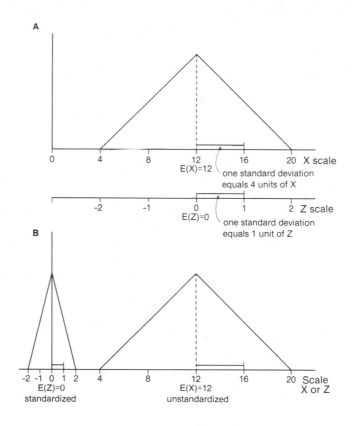

1.3.5 Standardizing Covariances

Consider what happens to the covariance between two variables if one variable (say Y) is transformed as   new-$y_i = a + by_i$ prior to calculating the covariance. The expected value of the transformed Y's is $a + bE(Y)$ (by Eq. 1.16). The covariance of X with new-Y (from Eq. 1.9) is then

$$\text{Cov}(X, (a + bY)) = \sum_i \sum_j (x_i - E(X))(a + by_j - (a + bE(Y)))p(x_i y_j) \quad 1.49$$

Canceling the a's and factoring the b first out of the right parentheses and then outside the sums lets us rewrite this as

$$\text{Cov}(X, (a + bY)) = b \sum_i \sum_j (x_i - E(X))(y_j - E(Y))p(x_i y_j) \qquad 1.50$$

$$= b\,\text{Cov}(XY) \qquad 1.51$$

So the covariance of X with a linear function of some original Y variable is merely b times the covariance of X with the original Y variable.

Similarly, had X been modified with a linear transformation, we would have

$$\text{Cov}((a + bX), Y) = b\,\text{Cov}(XY) \qquad 1.52$$

If both variables had been transformed, we would have

$$\text{Cov}((a + bX), (a' + b'Y)) = bb'\,\text{Cov}(XY) \qquad 1.53$$

where the prime means that the transformation on Y differs from the transformation on X.

The import of these mathematical ramblings becomes clear if we consider the covariance among two standardized variables—the correlation coefficient r. From Eq. 1.41, we know that the multiplicative terms required for standardizing are $b = 1/\sigma_x$, and $b' = 1/\sigma_y$ in Eq. 1.53. (The additive terms are not needed because the a's canceled from the preceding equations.) Hence, from Eq. 1.53,

$$\text{Cov(standardized } X, \text{ standardized } Y) = r_{xy} = \frac{\text{Cov}(XY)}{\sigma_x \sigma_y} \qquad 1.54$$

This informs us that the correlation coefficient, the covariance between the standardized versions of two variables, may be calculated by dividing the covariance between the variables by the standard deviations of both variables.

1.4 Structural Equation Models

We have been discussing structural equations ever since we considered creating new Y scores from some original X scores. The equation $Y = a + bX$ is a **structural equation**, and the constants a and b are **structural coefficients**. Although it is occasionally helpful to consider such single equations as structural equation models, the term "models" is usually not applied until at least two equations simultaneously describe the set of variables under consideration.

1.4.1 How Structural Equation Models Imply Covariances/Correlations

Imagine we have five variables, three of which are important to us and are labeled X_1, X_2, and X_3. The two other variables are called e_2 and e_3 because they are "extraneous" or "error" variables. We may not know the identity of these variables even though we know (we are postulating that) they are connected with X_2 and X_3, respectively. Further imagine that X_1 and e_2 contribute to the production of the values of X_2 by

$$X_2 = b_{21}X_1 + e_2 \qquad\qquad 1.55$$

while X_1 and e_3 contribute to the production of X_3 by

$$X_3 = b_{31}X_1 + e_3 \qquad\qquad 1.56$$

These equations are structural equations of the form of Eq. 1.24 with two variables, where the constants preceding the e variables are 1 and the constants preceding the X variables are b's. The first subscript on each b identifies the newly resultant (or dependent) variable, and the second subscript identifies the original (or independent) variable.

Two representations of this model appear in Figure 1.12. The right portion of the figure emphasizes columns of numbers that are used to obtain other columns of numbers (without presenting the actual coefficients or probabilities). The left portion of the figure is a path diagram. Here the variables are represented as letters (or names), and the structural coefficients appear as appropriately labeled arrows. Equation 1.55 can be recovered from the path diagram by "seeing" that X_2 results from X_1 and e_2, with coefficients b_{21} and 1, respectively. Equation 1.56 can be recovered by "seeing" that X_3 results from X_1 and e_3, with coefficients b_{31} and 1, respectively.

Even this simple model illustrates two fundamental characteristics of structural equation modeling. First, there is always one equation for each dependent (new or resultant) variable among the set of variables considered. Second, some variables can be included in more than one equation.

Consider what happens if we force the common cause X_1 to take on a high value—we write a large value in the first column of the right side of Figure 1.12 or imagine the value of X_1 in the path diagram to be large—and assume for the moment that the constants b_{21} and b_{31} are positive numbers near 1.0. Then X_2 would take on a relatively high value, because one of the values contributing to its calculation is high (right of Figure 1.12 and Eq. 1.55), and X_3 would also become relatively high (Eq. 1.56). Similarly, examining the path diagram convinces us that if X_1 is

Figure 1.12   A common cause model.

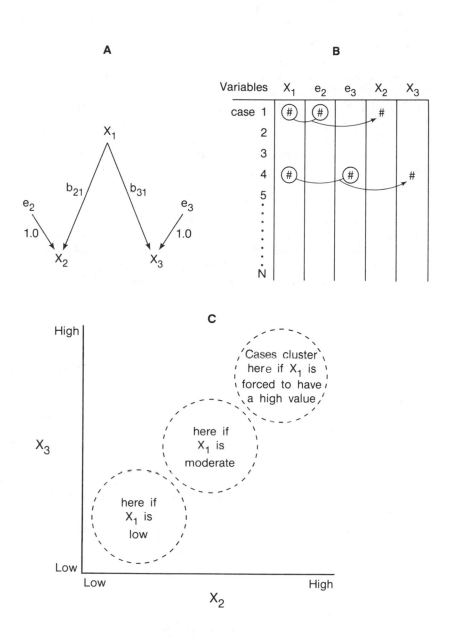

large, the large value is transmitted to both X_2 and X_3 via the causal arrows (paths) b_{21} and b_{31}.

Correspondingly, if X_1 is forced to take on a low value, both X_2 and X_3 become relatively low; if X_1 takes on a moderate value, X_2 and X_3 take on moderate values. Figure 1.12C plots the different values of X_2 and X_3 that are likely to appear if X_1 takes on high, moderate, or low values. The reason Figure 1.12C depicts the regions in which the X_2 and X_3 values cluster rather than depicting precisely the value of X_2 that equals b_{21} times the specific X_1 value, and the value of X_3 that is b_{31} times that same X_1 value, is that we have not specified what values e_2 and e_3 assume. Positive or negative values of e_2 would move the plotted point(s) to the right or left, respectively, whereas positive or negative values of e_3 would move the plotted point(s) up or down, respectively.

According to the model, variables X_2 and X_3 will be related to one another because of their dependence on the common cause X_1. Variables X_2 and X_3 show a definite relationship to one another (high X_2 values are associated with high X_3 values, Figure 1.12C) not because we purposely selected the values for these variables but *because our model* (Figure 1.12A,B or Eqs. 1.55 and 1.56) *implies they must be related due to the presence of X_1 as a common cause of X_2 and X_3.*

1.4.1.1 Quantifying the implied covariance

Earlier, we used the notion of covariance to quantify the magnitude of the relationship between two variables. If we inquire about the "model-implied" *magnitude* of the relationship between X_2 and X_3 (as indicated by the covariance between these variables), we might expect the extent of the relationship to depend on three things: (1) the variance of X_1, because if X_1 had no variance there would be no high, medium, and low X_1 values to force the separate clusterings in Figure 1.12C; (2) e_2 and e_3, because these variables also influence the placement of points in Figure 1.12C; (3) b_{21} and b_{31}, because these coefficients link the variations in the X_1 values to particular scores on the X_2 and X_3 scales.

To develop the equation for $\text{Cov}(X_2 X_3)$, we will need the means of X_2 and X_3. The mean for X_2 can be written as

$$E(X_2) = \mu_{x2} = E(b_{21}X_1 + e_2) \qquad 1.57$$

or, from Eq. 1.16,

$$\mu_{x2} = b_{21}\mu_{x1} + \mu_{e2}. \qquad 1.58$$

Similarly, the mean of X_3 is

$$E(X_3) = \mu_{x3} = b_{31}\mu_{x1} + \mu_{e3}. \qquad 1.59$$

We are now ready to express the covariance between X_2 and X_3 as a function of the three types of entities listed earlier, but it warrants repeating as to *why* we are attacking the forbidding equations that follow. We outlined the logic resulting in Figure 1.12C, which convinces us that if we have a structural equation model like that in Eqs. 1.55 and 1.56 and Figure 1.12A, variables X_2 and X_3 must be correlated (covary) even though there is no direct causal connection between them. The presence of a common cause X_1 implies that some relationship (correlation) between X_2 and X_3 must exist. To inquire about $Cov(X_2X_3)$ is to inquire about the actual numerical magnitude of the relationship implied by the model.

By definition, the covariance between X_2 and X_3 is

$$Cov(X_2X_3) = E[(X_2 - \mu_{x2})(X_3 - \mu_{x3})]$$

$$= \sum_i \sum_j (x_{2i} - \mu_{x2})(x_{3j} - \mu_{x3})p(x_{2i}x_{3j}) \qquad 1.60$$

Using the structural equation model (Eqs. 1.55 and 1.56) and the equations for the means (Eqs. 1.58 and 1.59), we rewrite this as

$$Cov(X_2X_3) = \sum_i [(b_{21}x_{1i} + e_{2i}) - (b_{21}\mu_{x1} + \mu_{e2})]$$

$$\times [(b_{31}x_{1i} + e_{3i}) - (b_{31}\mu_{x1} + \mu_{e3})](1/N) \qquad 1.61$$

In Eq. 1.61 we use a single subscript and a summation over individuals (cases), rather than two subscripts and summations over values of X_2 and X_3. Each individual possesses a potentially unique set of values on X_1, e_2, and e_3 that define that individual's scores on X_2 and X_3 (via Eqs. 1.55 and 1.56). After we have recorded and summed over all the individuals, we will have recorded and summed over all the possible combinations of X_2 and X_3 values as required by Eq. 1.60. With i subscripting individuals (cases) the probability of the particular combination of X_1, e_2, and e_3 scores unique to the ith individual is $1/N$, and this provides this case's (individual's) contribution to the probability of a particular pair of X_2 and X_3 scores appearing. If two or more individuals have the same combination of X_1, e_2, and e_3 scores, this situation will be automatically compensated by the inclusion of two or more identical terms into the summation over i.

To obtain a more understandable way of writing the right side of Eq. 1.61, we factor out $1/N$, rearrange within the square brackets,

$$Cov(X_2X_3) = \frac{1}{N}\sum_i [b_{21}(x_{1i} - \mu_{x1}) + (e_{2i} - \mu_{e2})]$$

$$\times [b_{31}(x_{1i} - \mu_{x1}) + (e_{3i} - \mu_{e3})] \qquad 1.62$$

multiply the square brackets,

$$\text{Cov}(X_2 X_3) = \frac{1}{N} \sum_i [b_{21} b_{31} (x_{1i} - \mu_{x1})(x_{1i} - \mu_{x1})$$
$$+ b_{21}(x_{1i} - \mu_{x1})(e_{3i} - \mu_{e3}) \qquad 1.63$$
$$+ b_{31}(x_{1i} - \mu_{x1})(e_{2i} - \mu_{e2})$$
$$+ (e_{2i} - \mu_{e2})(e_{3i} - \mu_{e3})]$$

and regroup by summing similar types of terms for all the individuals before summing the sums to obtain the overall sum

$$\text{Cov}(X_2 X_3) = \frac{1}{N} [\sum_i b_{21} b_{31} (x_{1i} - \mu_{x1})(x_{1i} - \mu_{x1})$$
$$+ \sum_i b_{21}(x_{1i} - \mu_{x1})(e_{3i} - \mu_{e3}) \qquad 1.64$$
$$+ \sum_i b_{31}(x_{1i} - \mu_{x1})(e_{2i} - \mu_{e2})$$
$$+ \sum_i (e_{2i} - \mu_{e2})(e_{3i} - \mu_{e3})].$$

Factoring the constants outside the sums and dividing each of the terms individually by N gives

$$\text{Cov}(X_2 X_3) = \frac{b_{21} b_{31} \sum_i (x_{1i} - \mu_{x1})^2}{N} + \frac{b_{21} \sum_i (x_{1i} - \mu_{x1})(e_{3i} - \mu_{e3})}{N}$$

$$1.65$$

$$+ \frac{b_{31} \sum_i (x_{1i} - \mu_{x1})(e_{2i} - \mu_{e2})}{N} + \frac{\sum_i (e_{2i} - \mu_{e2})(e_{3i} - \mu_{e3})}{N}$$

in which we can replace the various averages with the corresponding variances and covariances

$$\text{Cov}(X_2 X_3) = b_{21} b_{31} \text{Var}(X_1) + b_{21} \text{Cov}(X_1 e_3)$$
$$+ b_{31} \text{Cov}(X_1 e_2) + \text{Cov}(e_2 e_3) \qquad 1.66$$

Thus, the magnitude of the model-implied covariance between X_2 and X_3 can be obtained if one knows the magnitudes of the terms on the right side of Eq. 1.66.

1.4.1.2 Independent errors and standardized variables

Consider what happens to Eq. 1.66 if we assume that e_2 and e_3 are independent of the substantively important variable X_1 and also independent of one another—that is, assume e_2 and e_3 function as truly random errors. This implies (by Eq. 1.32) that $\text{Cov}(X_1 e_2)$, $\text{Cov}(X_1 e_3)$, and $\text{Cov}(e_2 e_3)$ are zero. Equation 1.66 then becomes

$$\text{Cov}(X_2 X_3) = b_{21} b_{31} \text{Var}(X_1) \qquad 1.67$$

If we further assume that the X variables are all standardized, then $\text{Cov}(X_2 X_3) = r_{23}$, $\text{Var}(X_1) = 1.0$, and the structural coefficients are standardized coefficients, which we denote by β_{21} and β_{31}. (Standardized coefficients are discussed in Chapter 2.) Equation 1.66 is now

$$r_{23} = \beta_{21} \beta_{31} \qquad 1.68$$

With standardization and independent errors, the magnitude of the implied correlation between X_2 and X_3 is a simple multiplicative function of the structural coefficients. With independent errors (but without standardization) the implied covariance between X_2 and X_3 is a simple multiplicative function of the structural coefficients and the variance of X_1 (Eq. 1.67).

1.4.1.3 Models more generally

In Chapter 4 we extend the notion that models imply specific covariances among variables by considering models of a very general form rather than specific models such as the common cause model discussed here. Chapter 5 demonstrates how to capitalize on these implied covariances to obtain estimates of the structural coefficients, and Chapter 6 discusses comparison of model-implied covariances with the actual observed covariances as a method for testing models.

We will confront two major stumbling blocks in extending these ideas: we typically do not know (1) the precise form of the structural equations linking the variables in the model, or (2) the magnitudes of the coefficients in those structural equations. Our theory about the behavior of the variables in question constitutes a guess about the nature of the structural equation model. This dependence on theory for the basic model binds the statistical procedures discussed from Chapter 4 onward so closely to theory that one literally cannot do good structural equation modeling without being a good theorist. Conversely, we fundamentally confront, and hopefully advance, theory by investigating structural equation models. This dynamic interplay between theory and structural equation modeling is the fundamental strength of structural equation modeling and is prototypical of "good science."

Chapter 2

Traditional Basics

This chapter reviews the fundamental conceptual issues grounding regression and the general linear model. It begins by establishing the use of regression lines as descriptive and predictive devices, and then integrates regression with the discussion of structural equations given in Chapter 1. The procedure for standardizing regression slopes provides a direct precursor for standardizing LISREL models (Section 6.6), and the transition from simple to multiple regression clarifies the nature of the statistical controlling implicit in all structural equation modeling. The problem of colinearity discussed in Section 2.4 resurfaces when we discuss the estimation of structural coefficients in Chapter 5. Sections 2.5 and 2.6 on interaction and nonlinearity provide the basic format for the discussions of interaction and nonlinearity among *unobserved concepts* (Sections 7.4 and 7.5).

2.1 Fitting a Line to a Scatterplot

Consider two interval level variables (variables whose values are recorded in units on some scale) where one variable (X) is suspected to cause, influence, or result in the other variable (Y). If the X and Y scores are recorded for a set of persons (cases), these scores can be conveniently summarized in a scatterplot, such as Figure 2.1, where each dot refers to a case, and the placement of each dot is determined by the individual's X and Y scores. The center of the cluster of points is determined by locating the point specified by the mean X and Y scores (the centers of the X and Y distributions individually), and then a line may be drawn through this point using whatever slope makes the line come close to as many of the dots

Figure 2.1   Simple Regression.

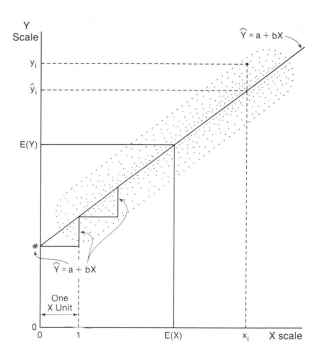

(cases) as possible.

The equation for such a line is $Y = a + bX$, where a is the Y value at which the line intercepts the Y axis, and b is the slope of the line, or the amount Y changes following a unit change in X. Before discussing a specific criterion for determining how close the dots are to the line, or examining how to obtain the numerical estimates of the intercept a and slope b, we address the more fundamental question, Of what use is such a line?

First, the line provides a *parsimonious description* of the clustering of the cases' X and Y values. The cases can be described as clustering around the line, and this description is parsimonious because the description comprises only two numbers (a and b), no matter how many cases were plotted in the scatterplot.

Second, the line provides a *predictive device*. If X causes (produces or is even associated with) Y, information on X should help predict Y. Imagine that we now encounter some newcomer considered to be subject to the same causal forces as the cases plotted in Figure 2.1 and that we know this individual's X score but not the Y score. A prediction for the newcomer's Y score may be obtained by locating the newcomer's X score on the X axis, moving vertically until we intersect the line, and then moving horizontally to locate the score on the Y scale that constitutes our prediction for the newcomer's Y score. But *why* do we use the line to obtain the predicted Y score? If the newcomer's score does not correspond to any of the initially plotted X scores, we would need some mechanism for interpolating between the existing values (and the line satisfies this function admirably), but this is not the fundamental reason. Imagine that the newcomer's score corresponds exactly to the X score of the individual illustrated with the enlarged dot. Could we not take the similarity in X scores between these cases as indicating that the cases are "alike" and thus use the Y score corresponding to the "large dot" case as our prediction, rather than the slightly lower predicted Y value given by the fitted line?

The answer demands an understanding of why the cases do not fall exactly on a line. The foregoing assumes that X is a cause, but not necessarily the only cause, of Y. If we allow Y to be influenced by other variables, the vertical distance between any particular dot and the line can be attributed to "other unspecified causes" boosting or suppressing the individual's Y score compared to the Y score that would arise if it were fully determined by the individual's X score. From this perspective, it is easy to see that although we have claimed our newcomer case has the same X score as the large dot case, we have *not* made any claim about the equivalence of these two cases on all the other unspecified and possibly unknown variables that might influence Y. Hence, the possession of identical X scores should not lead to the expectation of identical Y scores unless we can substantiate an additional claim that X is the only cause of Y. Clearly, any spreading of cases around the line is direct evidence that variables other than X influence Y, and hence it argues against a matching of individuals.

2.2 Regression in the Context of Chapter 1

The preceding discussion of Figure 2.1 introduces four variables, each of which has a mean, variance, and covariances with the other variables. Figure 2.2 represents these variables as columns of numbers. The variables X and Y do not require special comment, but be sure you can visualize the means and variances and covariance between these variables as "additions" to Figure 2.1. The values of e (the error variable that is the net sum of all the extra causes of Y) are obtained by measuring the length of the vertical

Figure 2.2   The Four Variables in "Bivariate" Regression.

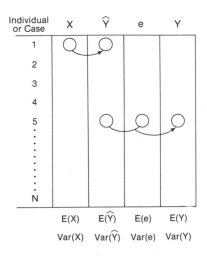

lines connecting each of the points to the regression line. The variable $\hat{Y}$ records the predicted value of Y for each case, and its values are obtained by moving vertically above each individual's X score until we encounter the regression line and then moving horizontally to read the Y value predicted by the line. If a and b are known (e.g., by reading them directly off Figure 2.1 after the scales of X and Y are specified), an equivalent way of obtaining $\hat{Y}$ is to plug the individual's X score into the equation $\hat{Y} = a + bX$. We now consider the means, variances, and covariances of the implicit variables e and $\hat{Y}$.

The variable e, which is $Y - \hat{Y}$, is typically assumed to be independent of X and to have mean zero. The zero mean can be visualized as either the entries in column e in Figure 2.2 summing to (and hence averaging) zero, or as the total length of the lines connecting the points above the line to the line (in Figure 2.1) equaling the total length of the lines connecting the points below the line to the line. Although $E(e) = 0$ is typically stated as an assumption, this is a necessary consequence if we agree to force the regression line to go through the point defined by the means of X and Y. Consider the following equations describing individual i:

$$\hat{Y}_i = a + bX_i \qquad\qquad 2.1$$

$$Y_i = \hat{Y}_i + e_i = a + bX_i + e_i \qquad\qquad 2.2$$

Equation 2.1 is the regression line giving the predicted value of Y. Equation 2.2 tells us that we can recover any individual's Y score by starting from that individual's predicted value $\hat{Y}$ and then moving above or below that value by the distance e_i unique to the individual. From Eq. 2.2, the mean of the actual Y scores can be written as

$$\text{E}(Y) = \text{E}(a + bX + e) = a + b\text{E}(X) + \text{E}(e) \qquad 2.3$$

by using Eq. 1.25.

It is reasonable to demand that the regression line go through the point specified by the means of X and Y. If a case is "near the middle of the heap" in the causal variable's distribution, it should appear near the middle of the heap on the effect, as well. That is, if we plug $\text{E}(X)$ into Eq. 2.1, then $\text{E}(Y)$ should be predicted to appear:

$$\text{E}(Y) = a + b\text{E}(X) \qquad 2.4$$

But how can Eqs. 2.3 and 2.4 both be true? Both equations can be true only if $\text{E}(e) = 0$. Thus forcing the regression line to go through the two means forces $\text{E}(e) = 0$. Any line going through the two means, even a line with an obviously incorrect slope, will have errors averaging 0.

Next consider obtaining the mean of the predicted Y values by taking the expected value of Eq. 2.1. This gives an equation like 2.4 except that $\text{E}(\hat{Y})$ appears on the left side. Since the right sides of these equations are identical, the left sides must also be equal [i.e., $\text{E}(\hat{Y}) = \text{E}(Y)$]. The means of the predicted and actual Y scores must be equal no matter which of the lines passing through the X and Y means are adopted. (Examine this in the context of Figures 2.1 and 2.2 by considering the Y and $\hat{Y}$ values that could appear if the line does not pass through the X and Y means.)

We turn now to the variance of Y. Since Y can be considered as created from $\hat{Y}$ and e (Figure 2.2 or Eq. 2.2), we can apply Eqs. 1.26 or 1.28 for the variance of a new variable created from two other variables. Since the coefficients are 1.0,

$$\text{Var}(Y) = \text{Var}(\hat{Y}) + \text{Var}(e) + 2\text{Cov}(\hat{Y}e) \qquad 2.5$$

Consider $\text{Cov}(\hat{Y}e)$. From Eq. 2.1 we can write this term as $\text{Cov}((a + bX)e)$, which has the form of Eq. 1.51, so $\text{Cov}((a + bX)e) = b\text{Cov}(Xe)$. Inserting this value in Eq. 2.5 gives

$$\text{Var}(Y) = \text{Var}(\hat{Y}) + \text{Var}(e) + 2b\text{Cov}(Xe) \qquad 2.6$$

Consider creating a huge $\text{Cov}(Xe)$ in the context of Figure 2.1. Imagine that we tried to use a horizontal regression line for these data, ensuring that this line still goes though the point specified by the means of X and Y. A horizontal regression line would have very large positive errors (values of e) for cases having high X scores, and very large negative errors for cases having low X scores. The correspondence between high X and large positive e scores and between low X and large negative e scores would provide a substantial covariance (correlation) between X and e.

How could we reduce this covariance between X and e? Clearly, by choosing a line that provides some positive and negative e values for both high and low values of X so that the pattern described in the last sentence of the previous paragraph is disrupted. But the only way to create positive and negative values of e for both high and low X values is if the selected line passes through the center of the cluster of points. Thus, *if we agree to select not only a regression line that passes through the X and Y means but also a line that passes through the center of the cluster, we will be selecting a line that will be forcing the covariance between X and e to be near zero.* The ordinary least squares regression procedure discussed in the next section forces this covariance to be exactly zero, so if we agree to use ordinary least squares, Eq. 2.6 simplifies to

$$\text{Var}(Y) = \text{Var}(\hat{Y}) + \text{Var}(e) \qquad\qquad 2.7$$

Try expressing this equation in words.

Let us now examine $\text{Var}(\hat{Y})$ by noting that Eq. 2.1 expresses $\hat{Y}$ as a linear function of X; hence $\text{Var}(\hat{Y})$ can be obtained by using Eq. 1.23 for the variance of a linear function. Thus,

$$\text{Var}(\hat{Y}) = b^2 \text{Var}(X) \qquad\qquad 2.8$$

Furthermore, substituting Eq. 2.8 in 2.7 gives

$$\text{Var}(Y) = b^2 \text{Var}(X) + \text{Var}(e) \qquad\qquad 2.9$$

That is, the variance in Y can be partitioned into the variance arising from fluctuations in both the cases' X scores and the extraneous variable e. This is the classical partitioning of the variance of a dependent variable into the variance attributed to the independent variable X and the error or unexplained variance arising from unspecified sources.

We conclude this section by highlighting the fundamental conceptual point underlying the preceding discussion. Figure 2.1 contains four, not two, variables. Two variables were observed (X and Y), and two more were implied by the mere decision to fit a prediction line to the scatterplot ($\hat{Y}$ and e). These incidental variables are fundamental to many topics in regression. For example, e is fundamental to understanding regression diagnostics, and $\hat{Y}$ is useful directly as a set of predicted values and in multiple correlations. (If $\hat{Y}$ is the prediction obtained from several predictor variables, the correlation between Y and $\hat{Y}$ is the multiple correlation coefficient.)

2.2.1 Ordinary Least Squares

Several lines (several different values of a and b) may approximately fit a scatter of points such as that in Figure 2.1. Ordinary least squares (OLS) is one criterion for selecting the **best fitting** line—*the line that makes the sum of the squared e values smallest.* The values of a and b for this line are the OLS estimates of the regression coefficients.

Although OLS is a traditional and robust procedure for selecting the best line, it is not the only way. In a later chapter we will consider selecting the values of a and b that maximize the likelihood that the observed variances and covariances of the X and Y variables could arise as sampling fluctuations if a and b were taken as true population coefficients—or **maximum likelihood estimation** (MLE)—but even this does not exhaust the alternatives.

To obtain the formulas for calculating a and b via OLS, we write the sum of the squared errors as the sum of the squared differences between the Y values and the predicted Y values:

$$\sum_i e_i^2 = \sum_i (Y_i - \hat{Y}_i)^2 = \sum_i (Y_i - (a + bX_i))^2 \qquad 2.10$$

The next few steps will probably be opaque to readers unfamiliar with calculus. You might reconsider the following verbal description after reading Section 3.2 on partial derivatives. We take the partial derivative of this sum first with respect to a and then with respect to b and set these derivatives equal to 0. These steps force a and b to take on values that minimize the squared errors. We now have two equations (the two partial derivatives that have been set equal to zero) containing two unknowns (a and b). These equations are solved for a and b to get formulas for calculating the slope and intercept:

$$b = \frac{\text{Cov}(XY)}{\text{Var}(X)} \qquad 2.11$$

$$a = \text{E}(Y) - b\text{E}(X) \qquad 2.12$$

These equations contain only means, variances, and covariances of the observed X and Y variables, so numerical estimates of a and b can be obtained by substituting in the appropriate observed values.

Rearranged, Eq. 2.12 reads $\text{E}(Y) = a + b\text{E}(X)$, which says that the selected line goes through the point defined by the X and Y means, which in turn guarantees that $\text{E}(e) = 0$. Thus, although the OLS criterion of minimizing the squared values of e makes no mention of the average size of the e's, it turns out that the only way to minimize the squared errors is to select a line that makes the mean size of the errors zero.

The coefficient a is the value we would predict Y to have if X takes on value 0 (Eq. 2.1). If an $X = 0$ is outside the usual range of X values, the prediction for Y given by a may be nonsense substantively. (For example, if age is being used to predict attitude toward a government decision, the interpretation of a is totally fictitious, because it requires our imagining that each child forms such attitudes at the instant of birth.)

The coefficient b is the amount our prediction for Y increases [or decreases if $\mathrm{Cov}(XY)$ and, hence, b are negative] for each unit increase in the value of X. It is also commonly interpreted as the amount Y is expected to change if X increases 1 unit. The term "expected" is used technically to describe the expected value (mean) of the Y distributions observed for (conditional on) particular X values. For example, the mean of Y if $X = 2$ can be obtained from Eq. 2.2 as $\mathrm{E}(a + 2b + e)$, which is simply $a + 2b$. Noting that this is precisely the predicted value of Y we would obtain from 2.1 for an X value of 2 demonstrates the connection between predicted values of Y and the expected value of Y conditional on a particular X value. The predicted value wording is usually used when we are referring to a single case, but the expected value wording is used if we are stressing that predictions for several individuals having the same X value are being made and the mean or center of the distribution of the Y values for these cases is being considered.

One problem with using b as the difference or change in expected Y values for cases 1 unit apart on X is that researchers frequently use a nontechnical meaning of the word "expected" while referring to b. They state that one expects (as in anticipates) a change in Y whenever the X score changes. This can lead one to make the often indefensible assumption that an artificial change of 1 unit in the cause X will be responded to in precisely the same way as a "naturally" occurring difference in X scores. In moving from expected values (or differences between expected values or expected difference in values) to *anticipated* values, one moves from the technical meaning of an expectation as a description of the center of some distribution to an unqualified assertion that some result will occur. This change in meaning is equivalent to adding the extra assumption that all changes in X are responded to equivalently. For example, forcing people to increase their X will lead to the same result as naturally occurring increases in X.

Note further that there is a substantial difference between regressing Y on X versus X on Y. Predicting X from Y implies the error terms would have to be plotted horizontally in Figure 2.1, and that in turn leads to a different best fit line and, hence, different estimates for a and b. Naturally, if there is some implicit causal relation between the variables, the cause should be used as the predictor variable and the effect as the predicted variable. In the absence of any suspected direct causal links, both regression lines could be obtained, and the preference for which line to use would depend on which variable one wishes to predict.

2.2.2 Standardizing

Standardized regression slopes, usually denoted by β, are the slopes that would appear if the variables in the regression had been expressed in standardized form (i.e., provided mean = 0, variance = 1.0) before the regression line was obtained as outlined in the previous section. Figure 2.3 depicts an example in which an attitude (Y) is thought to depend on income (X). In real units, $E(X) = \$20,000$, $Var(X) = \$10,000^2$ (so $\sigma_x = \$10,000$), $E(Y) = 6$, $Var(Y) = 4$ ($\sigma_y = 2$), and $Cov(XY) = 16,000$. If X and Y are both standardized (via Eq. 1.46), the scatterplot of the points would not change, but the scales describing the X and Y variables would change as indicated in Figure 2.3. Since the clustering of the points is unaltered by rescaling the variables, the best fitting regression line using the original metric coincides with the line that provides the best fit for the standardized scales.

Though the best fit standardized and unstandardized lines coincide, the regression coefficients describing these lines differ because the lines are to be depicted using different scales for the X and Y variables. This can be demonstrated by first considering that the axes for the standardized scales correspond to the means of the two variables. This implies that a standardized X score of 0 (the mean X) would predict the mean value of Y (namely 0), which we recognize as merely another way of saying that the Y intercept must be 0 for the standardized regression. Hence, for standardized variables, the regression equation corresponding to Eq. 2.1 is

$$Z_{\hat{y}} = 0 + \beta Z_x = \beta Z_x \qquad 2.13$$

We can calculate the slope of the regression line in metric units as

$$b = \frac{Cov(XY)}{Var(X)} = \frac{16,000}{(10,000)^2} = .00016 \qquad 2.14$$

Thus an individual who makes \$1 more than another is predicted to have an attitude score that is .00016 units higher than the other's attitude. Equivalently, an individual who makes \$10,000 more (one standard deviation more) than another is predicted to have an attitude score that is 10,000(.00016) or 1.6 attitude units higher than the other's attitude. Noting that 1.6 attitudinal units is .8 of a standard deviation gives a third equivalent interpretation: an individual whose income is one standard deviation above another's income is predicted to have an attitude score that

Figure 2.3   Standardized Regression.

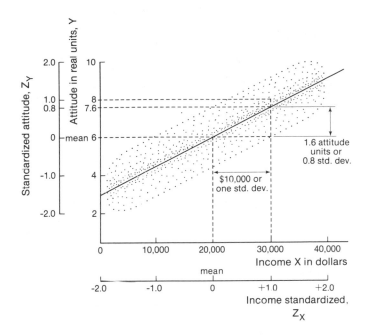

is .8 standard deviations (1.6 units) above that other individual's attitude. (Locate each of these three verbalizations in Figure 2.3.)

Having examined standardizing in diagrammatic and verbal form, we turn now to standardization in equation form. The next few paragraphs begin from the metric form of a regression equation and gradually convert it into the corresponding standardized form of the equation. This demonstrates the mathematics of why the standardized intercept becomes zero and, ultimately, the mathematics of how to transform unstandardized regression slopes into standardized slopes. It also prepares us for some comments on the correlation coefficient that conclude this section. To compact the following equations, we denote the means of X and Y as μ_x and μ_y, respectively, and the respective standard deviations as σ_x and σ_y.

We start with the basic metric regression equation $Y = a + bX + e$. Our strategy is to alter this equation in ways that preserve the equality yet simultaneously transform X and Y into the corresponding standardized variables. We begin by subtracting μ_y from both sides of the equation and adding zero $(+b\mu_x - b\mu_x)$ into the right side

$$Y - \mu_y = a + b\mu_x - \mu_y + bX - b\mu_x + e \qquad 2.15$$

Inserting parentheses and factoring out b gives

$$Y - \mu_y = (a + b\mu_x - \mu_y) + b(X - \mu_x) + e \qquad 2.16$$

The X and Y variables have now been written as deviations from their means, and a simplification is possible. We have from Eq. 2.4 that $a + b\mu_x - \mu_y = 0$. This cancellation removes a from the equation, showing that the "intercept" must be zero in regressions with variables expressed as deviations from their means, and since standardizing involves expressing variables as deviations from their means, standardized variables give regression equations having zero as their intercepts.

Next we multiply the term containing b by 1.0 in the form σ_x/σ_x and divide both sides of the equation by σ_y.

$$\frac{Y - \mu_y}{\sigma_y} = \frac{b\sigma_x(X - \mu_x)}{\sigma_y \; \sigma_x} + \frac{e}{\sigma_y} \qquad 2.17$$

Rewriting the standardized form of Y on the left side and the standardized form of X on the right give

$$Z_y = (b\sigma_x/\sigma_y)Z_x + e/\sigma_y \qquad 2.18$$

Emphasizing the predicted standardized Y (parallel to moving from Eq. 2.2 to 2.1) gives

$$Z_{\hat{y}} = (b\sigma_x/\sigma_y)Z_x \qquad 2.19$$

Comparing this equation to Eq. 2.13 shows that the standardized regression slope is

$$\beta = b\sigma_x/\sigma_y \qquad 2.20$$

Thus we can obtain the standardized regression slope from the unstandardized slope if we know the standard deviations of the variables. Similarly, if we know the standardized regression slope and the standard deviations, we can obtain the unstandardized slope by rearranging this equation:

$$b = \beta\sigma_y/\sigma_x \qquad 2.21$$

A close link between β and r is apparent if we replace b in Eq. 2.20 with its calculation formula (Eq. 2.11):

$$\beta = \frac{b\sigma_x}{\sigma_y} = \frac{\text{Cov}(XY)\sigma_x}{\sigma_x^2 \;\; \sigma_y} = \frac{\text{Cov}(XY)}{\sigma_x\sigma_y} = r \qquad 2.22$$

The last equality arises from the definition of r (Eq. 1.54). The equality of β and r informs us that the correlation coefficient can be interpreted as the slope of the regression line linking two standardized variables. If we are interested in comparing slopes from different groups, then we should use nonstandardized slopes because the variances of the variables might differ between the groups, and this would give different β's even if the b's were the same (cf. Schoenberg, 1972).

2.3 Multiple Regression

Simple regression uses a single predictor of the dependent variable Y, whereas multiple regression uses two or more predictors. The general equation for obtaining the predicted values ($\hat{Y}$'s) is

$$\hat{Y} = a + b_1 X_1 + b_2 X_2 + b_3 X_3 + \cdots + b_k X_k \qquad 2.23$$

The equation exactly reproducing the observed Y scores is obtained as this predicted value plus an error term unique to each individual.

$$Y = a + b_1 X_1 + b_2 X_2 + b_3 X_3 + \cdots + b_k X_k + e \qquad 2.24$$

Equations 2.23 and 2.24 parallel Eqs. 2.1 and 2.2 with the subscript i omitted. We can obtain the predicted Y score for any individual by taking that individual's scores on X_1 to X_k, multiplying these by the corresponding constants b_1 to b_k, and summing these products together with the constant a. (Note the parallels to Figures 1.9 and 2.2.)

In Figure 2.4 we plot a clustering of points for a multiple regression having two predictor variables.

$$\hat{Y} = a + b_1 X_1 + b_2 X_2 \qquad 2.25$$

The n observed cases appear as n points clustering in a three-dimensional cube. As in Figure 2.1, Y scores are plotted vertically and X_1 scores are plotted to the right, but we now plot scores on the additional X_2 variable to the left. Thus each case's Y score governs how high that case is plotted above the floor of zero, its X_1 score governs how far to the right it is plotted, and its X_2 score governs how far forward (toward the left front of the cube) it is plotted. The highlighted case has an X_1 score of 9, so it is plotted 9 units to the right of the origin; an X_2 score of 1, so it is 1 unit forward (or to the left); and a Y score of 50, so it appears 50 units above the origin or at the top of the cube.

Figure 2.4   Multiple Regression.

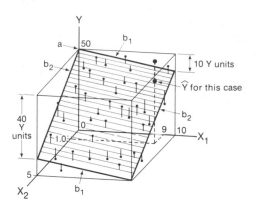

Specifying the value of any one of the three variables specifies a plane in which the dot representing the case must be plotted. If a case has an X_2 score of 0, that case must be plotted somewhere on the back right surface of the cube, with the exact placement of the point in that plane depending on the values of Y and X_1 as in Figure 2.1. If the case has an X_2 score of 5, it must be plotted on the plane providing the front left of the cube, with Y again governing its vertical placement and X_1 governing placement to the right. An X_2 value of 1 places the case somewhere on a vertical plane parallel to but 1 unit in front of the back right of the cube. A Y score of 25 specifies a plane cutting the cube in half horizontally. X_1 scores of 0, 10, and 5 represent, respectively, the back left of the cube, the front right of the cube, and a vertical plane slicing the cube in half midway between the 0 and 10 planes.

You should now be able to visualize the dots as a cloud that clusters generally between the top back left corner and the bottom front right of the cube, with a general absence of points near the top front left and bottom back right of the cube. The regression "plane" is a sloped plane slicing through the clustering from the top back left to the bottom front right of the cube. The precise placement of the regression plane is determined by the criterion of minimizing the squared errors (minimizing the squares of the vertical distance to the points from the plane). The plane is completely determined by the coefficients a, b_1, and b_2 in Eq. 2.25 (an intercept and two slopes), just as an intercept and one slope completely determined the best fit line in Figure 2.1.

This plane provides a set of predicted Y values. Any pair of X_1 and X_2 values specifies a spot on the bottom of the cube and the height of the regression plane directly above that spot gives the predicted Y value for a case having this particular combination of X_1 and X_2 values. Predicted (expected) Y values can be obtained for any combination of X_1 and X_2 values of interest to the researcher. These values may be extreme (such as those at the corners of the cube), or sets of values depicting commonly occurring patterns, but the predicted value is always obtained by inserting the values of interest into the regression equations 2.23 or 2.25.

As before, the value of a in the regression equation is the value Y predicted to appear when the predictor variables have value zero. Convince yourself of this by examining Eq. 2.25. Since the regression plane in Figure 2.4 cuts the vertical Y axis at the value 50 (the top back left corner of the cube), this implies $a = 50$ for the diagrammed multiple regression.

The regression slope b_1 is the change in Y that would be predicted to accompany a unit increase in X_1 no matter what value X_2 has, as long as that X_2 value remains unchanged. (Can you see this in Eq. 2.25?) If X_2 is stable at 0, then b_1 is the negative slope (decline) appearing on the back right of the cube; if $X_2 = 5$, b_1 is the same slope appearing on the front left of the cube, and if X_2 takes on any value between 0 and 5, b_1 is the same negative slope on any one of the numerous vertical planes between the back right and front left of the cube. Similarly, the b_2 slope describes the clustering of points on Y and X_2 in any particular plane defined by a single value of X_1—the back left of the cube if $X_1 = 0$, the front right if $X_1 = 10$, or parallel planes anywhere between if X_1 has any value in between.

For example, if Y is marital happiness, X_1 the number of bills per month, and X_2 the number of arguments per month, Figure 2.4 indicates marital happiness declines with increasing numbers of bills and more frequent arguments. The highest predicted happiness score (50) appears when a family has zero bills and zero arguments, and happiness is predicted to decline to zero with a combination of 10 bills and 5 arguments. Since 10 bills decrease happiness by 10 units, the predicted change in happiness accompanying each additional bill, with any specific number of arguments, is $-10/10 = -1.0$. Thus b_1 is -1.0 in Figure 2.4. Since 5 arguments decrease happiness by 40 units, the change in happiness predicted to accompany each additional argument (i.e., b_2) is $-40/5 = -8$. Thus one additional argument decreases predicted happiness more radically than an additional bill, which is equivalent to saying b_2 or -8 is larger in absolute magnitude than b_1 or -1, and both of these are equivalent to noting that the regression plane is more steeply sloped in the X_2 direction than in the X_1 direction in Figure 2.4.

Be sure you can see that since b_1 and b_2 in Eq. 2.25 each take on only a single value, this forces the slopes at the back right and the front left of the cube to be identical and the slopes at the back left and front right to be identical. It also implies that the effects of the predictor variables are additive in that the overall effect resulting from changing both predictor variables can be calculated as the sum of the effects of the changes in each predictor individually. Section 2.5 shows how a regression plane can be "twisted" to allow different slopes along the parallel sides of the cube.

If we have multiple predictors (as in Eq. 2.23), we cannot keep drawing more and more dimensions because we cannot graph more than three dimensions. It is traditional, however, to maintain the vocabulary of a "regression surface," where that surface represents the predicted values and has a dimensionality one less than the space in which the raw data points are represented.

2.3.1 Contrasting Simple and Multiple Regression with Dummy Variables

The preceding emphasizes several similarities between simple and multiple regression, but it fails to highlight the key differences. We discuss these differences in the context of introducing a dummy variable, because this simplifies the diagrams and discussion, even though the points we make are equally valid for continuous variables. We begin by contrasting the simple regression equation

$$Y = a_0 + b_{01}X_1 + e_0 \qquad\qquad 2.26$$

with the multiple regression equation

$$Y = a + b_1X_1 + b_2X_2 + e \qquad\qquad 2.27$$

where the subscript 0 distinguishes the coefficients in 2.26 from those in 2.27, and where X_2 is a dummy variable. That is, X_2 will be allowed only two values (0 or 1); hence the cases must cluster on either the back right or front left of the cube, as in Figure 2.5.

Consider first the possible correlation between X_1 and X_2 in Eq. 2.27 for which there is no parallel in Eq. 2.26, because there is only a single predictor. The implications of correlated predictor variables are most easily seen by starting out from a multiple regression in which there are correlated predictors and then examining what happens to the simple regression if one of the predictors is omitted.

Before we can illustrate the effects of correlated predictors, we must become familiar with the details of Figure 2.5. The right side of Figure 2.5 is similar to Figure 2.4 in that both X_1 and X_2 display negative slopes to

Figure 2.5   Contrasting Simple and Multiple Regression.

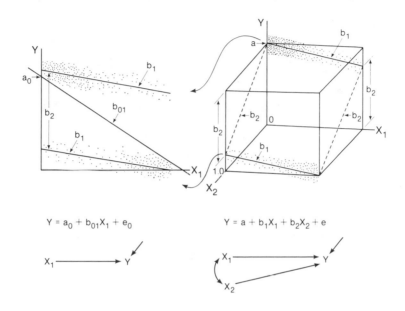

$$Y = a_0 + b_{01}X_1 + e_0$$

$$Y = a + b_1X_1 + b_2X_2 + e$$

the clustering of data points, but it is unlike Figure 2.4 in that variable X_2 is a dummy variable only taking on values 0 or 1. Although there is a slope b_2 in the X_2 direction, it is determined by the clustering of points at X_2 values of 0 and 1 with no cases in between. This implies that the only possible kind of difference or change in X_2 is a unit change, and therefore the magnitude of b_2 can be interpreted as *the difference in Y scores predicted for individuals with similar X_1 scores but differing on X_2.* Thus the magnitude of b_2 indicates the slope in the X_2 direction and how many units of Y the X_2 group coded 1.0 is above (or below) the X_2 group coded 0.0 (sometimes referred to as the reference group or the omitted group).

Notice the clustering of data points near the back top left of the cube and the front bottom right. This clustering is the correlation between X_1 and X_2. It implies that the low value of X_2 often accompanies low values of X_1, and the high value of X_2 most often accompanies high values of X_1. We recognize this as a classic statement of a relationship, and note that had we wished a plot of the clustering for just this relationship we could ignore Y and simply squash the cube from above until all the cases were squashed onto the bottom of the cube (in mathematical jargon, we look at the projection of the points in the X_1-X_2 plane). The pattern of the points on the bottom of the cube (a line of dots at the front left and back right of the square in the X_1-X_2 plane with the indicated clustering of points along these lines) is a diagram of the relationship between X_1 and X_2.

Now we can examine the implications of the correlated predictor variables. Consider what happens if we regress Y on X_1 and ignore X_2, as suggested by Eq. 2.26. The result would be the selection of the best fit line in the two-dimensional figure illustrated on the left side of Figure 2.5. (Only the axes of Y and X_1 appear in the figure.) Imagine the left side of Figure 2.5 as being the clustering of the points resulting from squashing the cube along the X_2 axis (put one hand on the front left surface of the cube, the other on the back right, and press together until the front and back are squashed together). The result is the left side of Figure 2.5. We have plotted the slope b_1 from the multiple regression as going through the centers of each of the two clusterings to be consistent with the slopes in the three-dimensional figure on the right. The slope b_2 is plotted as the difference between the groups (again as in the right figure), and b_{01} is the slope of the line that would appear as the best fit between Y and X_1 in the left figure and corresponds to the slope for the simple regression of Y on X_1 (Eq. 2.26).

The key conceptual issue is this: Why is the slope for X_1 from the simple regression (b_{01}) not the same as the slope for X_1 from the multiple regression (b_1)? The reason is that the pronounced clustering of the points near the top left of the upper blob and the bottom right of the lower blob (on the left of Figure 2.5) pulls b_{01} away from the slope b_1 that would have appeared if there had been an even spreading of the cases throughout the blobs. We recognize these clusterings as the same clusterings underlying the correlation between X_1 and X_2 in the right side of the figure. The vertical displacement of the blobs in the left figure implies there is also a relationship between the dependent variable and the omitted variable X_2. In short, the b_{01} slope from the simple regression is biased toward too steep a slope because this regression omits a variable X_2 that is *both* a cause of Y and related to the included predictor X_1.

Though b_{01} provides a reasonable statement of the *predictive power* of X_1 in the left figure, it provides a poor statement of the *causal effectiveness* of X_1. The fact that b_{01} subsumes information on the related and effective X_2 is reasonable if we are seeking mere prediction and only have information on X_1, but it is unreasonable if we are seeking a statement of the causal effectiveness of X_1. (An omitted cause that is *un*correlated with X_1 is of no concern because this is another way of stating e, which is already in Eq. 2.26.)

The bias in b_{01} results from a misspecified model—Eq. 2.26 omits a causal variable that is correlated with X_1. Bias will not always produce an overly steep slope. Had a positive b_1 been combined with a negative b_2, the b_{01} from the simple regression would have been less than the corresponding b_1 from the multiple regression. Nor does this phenomena depend on X_2 being a dichotomy. Imagine scores spread throughout the X_2 direction, with a similar relationship between X_1 and X_2, and you will see that the same type of shifting of b_{01} away from b_1 appears.

2.3.2 Multiple Dummy Categories

As a check on your understanding and as a minor extension of the foregoing discussion, consider the case where X_2 has three categories, which we might label "none," "some," and "many." Lacking evidence that a score of some is midway between none and many, and hence doubting the scale is interval, we decide to use dummy variables to represent the values of X_2. We create two dummy variables. The first, DS_2 ("DummySome for variable 2") is scored 1.0 for those reporting some on X_2 and 0 for everyone else. The second dummy, DM_2, is scored 1.0 for all cases reporting many on X_2 and 0 for everyone else. Cases reporting none on X_2 are scored 0 on DS_2 and DM_2, and this group serves as the reference group if we interpret the resulting regression slopes for the dummy variables as differences.

The regression equation we would estimate using OLS is

$$\hat{Y} = a + b_1 X_1 + b_{2s} DS_2 + b_{2m} DM_2 \qquad\qquad 2.28$$

and the slopes we would obtain are as diagrammed on the right of Figure 2.6. Although we normally cannot draw a four-dimensional figure, Figure 2.6 takes advantage of the fact that DS_2 and DM_2 originate from the same variable X_2 and plots both these variables on a single axis with a scale break (zigzag) reminding us of this fact. The regression slopes for the dummy variables are plotted as both slopes and differences to emphasize the implicit use of the comparison group none appearing at the top back of the cube. The clumping of cases within particular portions of the three overall clusters of points again indicates a relationship between X_2 and X_1.

The important point concerns the regression that would have been found if X_2 had been ignored and Y had simply been regressed on X_1 alone. The left portion of Figure 2.6 illustrates the slope that would appear as b_{01}, which again differs from b_1, the corresponding slope for X_1 from the multiple regression. Again, we see that the slopes for X_1 in the simple and multiple regressions differ because the omitted variable X_2 is both a cause of Y (so b_{2s} and b_{2m} are nonzero) and correlated with X_1 (so omission of X_2 information allows X_1 to capitalize on its association with X_2 to increase its own predictive power in the simple regression on the left). Slope b_{01} is biased as an estimate of the causal effectiveness of X_1 because b_{01} reflects

Figure 2.6   Multiple Dummy Categories.

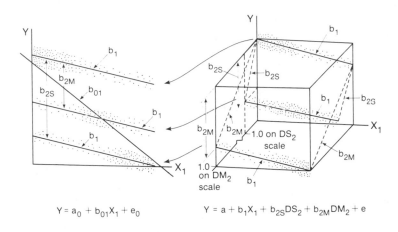

$$Y = a_0 + b_{01}X_1 + e_0$$

$$Y = a + b_1X_1 + b_{2S}DS_2 + b_{2M}DM_2 + e$$

not only the effectiveness of X_1 but also the causal effectiveness of the related variable X_2.

2.4 Colinearity

The problem of colinearity (or multicolinearity) occurs when two (or more) of the predictor variables in a multiple regression are highly correlated. Consider the regression equation

$$\hat{Y} = a + b_1X_1 + b_2X_2 \qquad 2.29$$

if X_1 and X_2 are perfectly correlated ($r = 1.0$). The perfect correlation makes X_1 the same as X_2 (if they are measured on the same scales), and we could then rewrite the equation and apply the common name X to the variables.

$$\hat{Y} = a + (b_1 + b_2)X \qquad 2.30$$

Regressing Y on the single variable X would provide some specific slope such as .5. Thus $.5 = b_1 + b_2$, but any values of b_1 and b_2 that sum to .5 could appear in Eqs. 2.29 or 2.30. That is, there is *no unique solution for*

Figure 2.7   Colinearity.

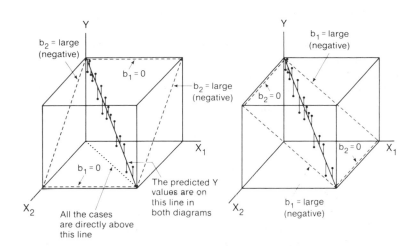

the values of b_1 and b_2 if X_1 and X_2 are perfectly correlated. In other words, the coefficients b_1 and b_2 are *underidentified* in that our model (Eq. 2.29) and our data (the appropriate variances and covariances that provide the hypothetical .5 slope) are insufficient to restrict us to any particular numerical estimates of b_1 and b_2.

A diagrammatic representation of colinearity appears in Figure 2.7. Restricting the correlation between X_1 and X_2 to 1.0 restricts all the data points to fall above a line from the origin to the bottom front right of the cube. The *two* indicated regression planes provide *equally good fits to the data* because they provide identical sets of predicted values—namely, the values appearing on a line from the top back left to the bottom front right of the cube. The left diagram depicts $b_1 = 0$ and $b_2 = $ large (which claims X_2 is an important predictor while X_1 is unimportant). The right diagram depicts $b_1 = $ large and $b_2 = 0$ (which claims X_1 is important and X_2 is unimportant). (Can you visualize a moderate effect of both of the variables that provides the same set of predicted values?) Hence, regression slopes may be inaccurate or unstable if the predictor variables are highly correlated.

We detect colinearity by observing large simple or multiple correlations among the independent variables or by observing unusually large standard errors for the slopes (see Section 6.3). As solutions, we can eliminate one or the other of the colinear variables or create a scale combining the colinear variables. In the context of LISREL we might use "nearly" colinear items as multiple indicators of a common concept. There is no known way to get estimates of both b_1 and b_2 in Eq. 2.29 if X_1 and X_2 are colinear.

If X_1 and X_2 are highly (but not perfectly) correlated, this implies a few cases appear above points to either the left or the right of the line connecting the origin to the bottom front right of the cube. This implies these few cases may provide information making either of the planes in Figure 2.6 (or any of the intermediary planes) preferable, but this preferability is based on a few deviant cases and hence is unreliable.

2.5 Interaction

An **interaction** *exists if the magnitude of the effect of one variable on another differs, depending on the particular value possessed by some third variable* (often some special condition describing the situation or environment). For example, if X_1 has a negative effect on Y when X_2 is low and no effect when X_2 is high, then X_1 and X_2 interact in the production of Y. As we hinted earlier, modeling interactions in multiple regression amounts to relaxing the requirement that the regression plane be flat, as guaranteed by Eq. 2.25 and illustrated in Figure 2.4.

Figure 2.8 illustrates how a "curved" regression surface can be created by including an additional product variable into a multiple regression equation. The extra variable is X_1X_2, created as the product of each individual's X_1 and X_2 scores. Thus, the regression equation is

$$\hat{Y} = a + b_1X_1 + b_2X_2 + b_3X_1X_2. \qquad 2.31$$

The way this additional product variable bends the regression surface becomes obvious if we consider two alternative representations of this equation:

$$\hat{Y} = a + (b_1 + b_3X_2)X_1 + b_2X_2 \qquad 2.32$$

and

$$\hat{Y} = a + b_1X_1 + (b_2 + b_3X_1)X_2 \qquad 2.33$$

The first equation emphasizes that the effect of X_1 in Eq. 2.31 depends on the value of X_2 that we wish to consider—a classic statement of interaction. Similarly, from the second regrouping we see that the effect of X_2 depends on the X_1 value considered.

Figure 2.8   Interaction.

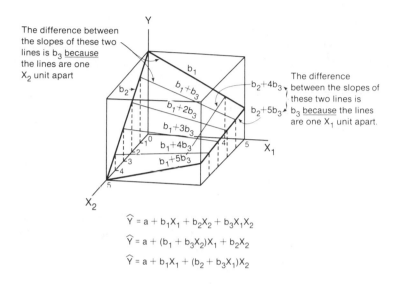

$$\widehat{Y} = a + b_1 X_1 + b_2 X_2 + b_3 X_1 X_2$$

$$\widehat{Y} = a + (b_1 + b_3 X_2) X_1 + b_2 X_2$$

$$\widehat{Y} = a + b_1 X_1 + (b_2 + b_3 X_1) X_2$$

The way these equations provide for a curved regression is easily
illustrated. Consider how large an effect X_1 in Eq. 2.32 has on Y if X_2 is
zero. Clearly the slope characterizing the effect is $b_1 + b_3 0$, or b_1. The
slope encapsulating the effect of X_1 on Y if X_2 is 1.0 is $b_1 + b_3 1.0$, or
$b_1 + b_3$. By direct extension, the slopes describing the effect of X_1 on Y
when X_2 takes on values 2.0, 3.0, and 4.0 are $b_1 + 2b_3$, $b_1 + 3b_3$, and
$b_1 + 4b_3$, respectively. From this it is obvious that b_3 encapsulates how
much the slope linking Y to X_1 changes for each unit increment in X_2.
That is, b_3 is the differences between the slopes of the regression lines in
any two Y-X_1 planes that are 1 unit apart (see Figure 2.8).

Note that each of these regression lines is "straight" even though the
progressive changes in slope imply that the regression surface itself is no
longer flat. You should convince yourself that steps parallel to the
foregoing are also appropriate for Eq. 2.33 where X_1 conditions the effect
of X_2 on Y. The coefficient b_3 is now the difference between the slopes of
the regressions of Y on X_2 appearing for sets of cases differing by 1 unit in
terms of X_1 values.

If X_2 happened to be a dummy variable, only the regression lines at the
front left and rear right of the cube in Figure 2.8 would appear, and we

would observe the classic crossed lines usually employed in introductory expositions of interactions. If a "progressive" interaction is suspected such that unit changes in the region of larger values of X_2 make stronger differences in the Y-X_1 slope, this can be accommodated by using the product variable $X_1 X_2^2$ instead of $X_1 X_2$.

Clearly, only interactions providing relatively smooth curvatures in the regression surface are modelable with this strategy, though combining dummy variables with multiplicative variables does provide some ability to model more radical breaks in the regression surface. Hayduk and Wonnacott (1980) provide further discussion of interactions in regression. We encounter interactions in LISREL in Sections 7.5, 9.1, and 9.2.

2.6 Nonlinearity

One way of dealing with nonlinearity in multiple regression is a variant of the procedure discussed for interaction. If Y is suspected of being nonlinearly related to X, Y is regressed on both X and X^2, where X^2 is a new variable created by squaring each case's value on X. We thus estimate the regression

$$\hat{Y} = a + b_1 X + b_2 X^2 \qquad 2.34$$

The effect of X on Y is now determined by both b_1 and b_2, and the effect of a unit change in X will vary, depending on which two X values are being considered. For example, the dashed curve in Figure 2.9 illustrates the predicted values of Y if the coefficients in Eq. 2.34 had been estimated as $\hat{Y} = 7 - 3X + 1.0X^2$. The coefficient preceding the X^2 variable controls the sharpness of the curve. This can be seen by comparing the relatively sharp curve produced when b_2 is 1.0 to the milder dotted curve that would appear had the b_2 coefficient been estimated as .5 (thereby implying $\hat{Y} = 7 - 3X + .5X^2$) and to the straight line that would appear had b_2 been zero. If b_2 is negative, the ends of the curve point downward.

If the X values are large, then the X^2 values are huge. Thus a very small value of b_2 will be required to keep the predictions for Y in a reasonable range. Since even small values of b_2 can provide substantial curvature, we should routinely graph nonlinear relationships as in Figure 2.9. The partial derivative of the regression equation with respect to X (see Section 3.2) is another useful way of describing nonlinear relations.

Another procedure for investigating nonlinear relations involves regressing Y on some nonlinear transformation of X, such as log X, $1/X$, or $\sqrt{X}$. Estimating a linear relation with a transformed variable implies a nonlinear relation between Y and the original X. Since there are a variety

Figure 2.9   Nonlinearity.

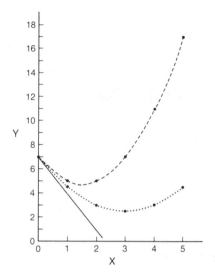

X	X^2	$\widehat{Y}$ from $\widehat{Y}=7-3X+1.0X^2$	$\widehat{Y}$ from $\widehat{Y}=7-3X+0.5X^2$
0	0	7	7
1	1	5	4.5
2	4	5	3
3	9	7	2.5
4	16	11	3
5	25	17	4.5

of nonlinear transformations that could be made to X, we should be guided by theory when choosing a transformation. In the absence of specific guidance as to functional form, the X^2 procedure discussed earlier is a reasonable first step to investigating nonlinearity.

Nonlinearity within LISREL is discussed in Section 7.4, and Wonnacott and Wonnacott (1970) is recommended to those wishing to pursue nonlinearity within the context of regression. Stolzenberg and Land (1983:657) report a simple procedure for calculating the standard error of nonlinear and interactive effects.

Chapter 3

The New Basics

Much recent literature in structural equation modeling, and all of LISREL, uses matrix algebra. The discussion of matrix algebra in Sections 3.1.1 and 3.1.2 is absolutely fundamental because practically everything we say about LISREL assumes that the reader is comfortable with matrix addition, subtraction, multiplication, and inversion. The aspects of matrix algebra in Section 3.1.3 are necessary but are referred to less frequently. This section should be read to obtain a grasp of the basic logic and then reviewed as necessary. Section 3.1.4 summarizes numerous useful identities in matrix algebra. Davis (1973), Fuller (1962), Searle (1982), and Van de Geer (1971) provide additional introductions to matrix algebra. A clear discussion of eigenvalues and eigenvectors appears in the first chapter of Hammarling (1970). This chapter concludes with a brief but mandatory section on derivatives, which completes the arsenal of knowledge required to master LISREL.

3.1 A Touch of Matrix Algebra

A **matrix** is a collection of numbers written in rows and columns. Matrices are represented by bold letters, such as $\mathbf{X}$, or occasionally as $\underset{\sim}{X}$. An example of a matrix is

$$\mathbf{X} = \begin{bmatrix} 1 & 3 & 5 \\ 2 & 4 & 6 \end{bmatrix}$$ 3.1

The numbers in a matrix are called **elements** of the matrix and are denoted by the corresponding lowercase letter subscripted with row and column indices specifying the location of the element within the matrix. Thus, $x_{13} = 5$ since the element in row 1 and column 3 of matrix $\mathbf{X}$ is 5. Two matrices are equal only if all of the corresponding elements in the matrices are equal.

Learning matrix algebra amounts to learning the language describing various aspects of matrices and to memorizing the rules specifying how to manipulate the elements of matrices to accomplish matrix addition, subtraction, multipliction, and so on. The power and elegance of matrix algebra originates in the compactness of the notation, but the joy of matrix algebra is our ability to manipulate matrices containing easily observable entities into other matrices containing desired elements—in particular, the "effect" coefficients comprising structural equation models. We begin with the basic definitions and rules.

A matrix with m rows and n columns is of **order $m \times n$**, and this information is sometimes reported in parentheses below the matrix. For example, the notation $\underset{(r \,\times\, s)}{\mathbf{C}}$ indicates one is considering a matrix $\mathbf{C}$ with elements (c's) arranged in r rows and s columns. Stipulating values for r and s determines how many c values there are and the overall shape of the matrix $\mathbf{C}$. If $\mathbf{C}$ is of order 20×2 it is a tall, skinny matrix containing 40 elements, written in 20 rows of 2 elements.

A matrix with as many rows as columns is a **square** matrix, and the elements on the diagonal from the upper left to lower right are **diagonal** elements. The sum of the diagonal elements of a matrix is called the **trace** of the matrix.

If the elements of a matrix $\mathbf{X}$ are rearranged so that the successive rows of $\mathbf{X}$ become columns of a new matrix, then the new matrix is denoted $\mathbf{X}'$ and is called the **transpose** of $\mathbf{X}$. That is, the elements in the first row of $\mathbf{X}$ are written as the first column of $\mathbf{X}'$, the second row of $\mathbf{X}$ becomes the second column of $\mathbf{X}'$, and so on. Hence, our original matrix $\mathbf{X}$ has transpose

$$\mathbf{X}' = \begin{bmatrix} 1 & 2 \\ 3 & 4 \\ 5 & 6 \end{bmatrix} \qquad 3.2$$

The transpose of a matrix contains the same elements as the original matrix but the original rows of elements become columns and the columns become rows. In other words, $x_{ij} = x'_{ji}$.

A matrix is **symmetric** if the matrix and its transpose are equal: $\mathbf{X} = \mathbf{X}'$. This happens if each row of the matrix is identical to its corresponding column or, equivalently, if $x_{ij} = x_{ji}$. [This equation compares two elements in a single matrix, whereas the equation ending the preceding paragraph compares two elements in two different matrices ($\mathbf{X}$ and $\mathbf{X}'$).] For example, the matrix

$$\begin{bmatrix} \# & 1 & 2 \\ 1 & \# & 3 \\ 2 & 3 & \# \end{bmatrix} \qquad 3.3$$

is symmetric no matter what values are inserted in place of the # symbol. Correlation and covariance matrices are symmetric because the correlation (covariance) between variables i and j is the same as the correlation (covariance) between j and i.

A **diagonal matrix** is a symmetric matrix in which all the off-diagonal elements equal zero.

$$\begin{bmatrix} \# & 0 & 0 & 0 \\ 0 & \# & 0 & 0 \\ 0 & 0 & \# & 0 \\ 0 & 0 & 0 & \# \end{bmatrix} \qquad 3.4$$

If the diagonal elements are all 1's, the matrix is called an **identity matrix** and is denoted **I**. Here is the 3 × 3 identity matrix

$$\begin{bmatrix} 1 & 0 & 0 \\ 0 & 1 & 0 \\ 0 & 0 & 1 \end{bmatrix} \qquad 3.5$$

A matrix containing only 0's either above or below the diagonal is called a **triangular** matrix.

$$\begin{bmatrix} \# & 0 & 0 & 0 \\ \# & \# & 0 & 0 \\ \# & \# & \# & 0 \\ \# & \# & \# & \# \end{bmatrix} \qquad 3.6$$

A matrix having m rows but only one column is of order $m \times 1$ and is called a **column vector**. Similarly, a matrix with one row and n columns is a **row vector** and has order $1 \times n$. Vectors are usually denoted by bold lowercase letters to indicate they are matrices rather than single or scalar numbers and to allow us to link the vector to some larger-order matrix. For example, our **X** is composed of three column vectors:

$$\mathbf{X} = \begin{bmatrix} 1 & 3 & 5 \\ 2 & 4 & 6 \end{bmatrix} \qquad \mathbf{x}_1 = \begin{bmatrix} 1 \\ 2 \end{bmatrix} \qquad \mathbf{x}_2 = \begin{bmatrix} 3 \\ 4 \end{bmatrix} \qquad \mathbf{x}_3 = \begin{bmatrix} 5 \\ 6 \end{bmatrix} \qquad 3.7$$

Alternatively, **X** might be considered as being composed of the row vectors [1, 3, 5] and [2, 4, 6].

3.1.1 Doing Algebra with Matrices

Matrix addition and subtraction parallel the addition and subtraction of ordinary scalar numbers. To add matrices **A** and **B** to get matrix **C**, we add the corresponding elements of **A** and **B** to obtain the elements of **C**. Thus, if **A** and **B** are the matrices on the left of Eq. 3.8, the sum of these matrices (**C**) is as illustrated.

$$\mathbf{A} + \mathbf{B} = \mathbf{C}$$
$$\begin{bmatrix} 1 & 3 \\ 2 & 4 \end{bmatrix} + \begin{bmatrix} 5 & 7 \\ 6 & 8 \end{bmatrix} = \begin{bmatrix} 6 & 10 \\ 8 & 12 \end{bmatrix} \qquad 3.8$$

Only matrices of the same size or order can be added because the procedure of adding matrices requires that there be "corresponding elements" in the matrices being added ($c_{ij} = a_{ij} + b_{ij}$). If **A** contained another row, it would not conform to **B** and the matrices could not be added.

Similarly, subtraction of **B** from **A** is defined as the subtraction of corresponding elements to obtain the elements in resultant matrix **C** (i.e., $c_{ij} = a_{ij} - b_{ij}$).

$$\mathbf{A} - \mathbf{B} = \mathbf{C}$$
$$\begin{bmatrix} 1 & 3 \\ 2 & 4 \end{bmatrix} - \begin{bmatrix} 5 & 7 \\ 6 & 8 \end{bmatrix} = \begin{bmatrix} -4 & -4 \\ -4 & -4 \end{bmatrix} \qquad 3.9$$

Again, **A** and **B** must be of the same order (size), and their order determines the order of **C**.

Matrix addition is commutative (**A** + **B** = **B** + **A**) and associative [**A** + (**B** + **C**) = (**A** + **B**) + **C**].

Matrices may be multiplied by scalars (ordinary numbers) or other matrices. Multiplication by a scalar c multiplies each element of **A** by c. Here are two examples.

$$3\begin{bmatrix} 1 & 3 \\ 2 & 4 \end{bmatrix} = \begin{bmatrix} 3 & 9 \\ 6 & 12 \end{bmatrix}$$

$$3.10$$

$$-1\begin{bmatrix} 1 & 3 \\ 2 & 4 \end{bmatrix} = \begin{bmatrix} -1 & -3 \\ -2 & -4 \end{bmatrix}$$

The second example in Eq. 3.10 shows that "adding -1 times a matrix" is the same as "subtracting that matrix" from some other matrix:

$$\mathbf{A} + (-\mathbf{B}) = \mathbf{A} - \mathbf{B} \qquad 3.11$$

Multiplication of a matrix by another matrix (**AB** = **C**) is demonstrated by

$$\mathbf{AB} = \mathbf{C}$$

$$\begin{bmatrix} 1 & 3 & 5 \\ 2 & 4 & 6 \end{bmatrix} \begin{bmatrix} 1 & 2 \\ 1 & 0 \\ 1 & 1 \end{bmatrix} = \begin{bmatrix} 9 & 7 \\ 12 & 10 \end{bmatrix}$$ 3.12

(2 × 3)(3 × 2)   (2 × 2)

The element 9 appearing in row 1 and column 1 of **C** is obtained by locating row 1 of the left matrix (**A**) and column 1 of the right matrix (**B**) and then summing the products of the corresponding elements in this row and this column [$1(1) + 3(1) + 5(1) = 9$]. In general, the element in row i and column j of the product matrix is obtained by locating the ith row of the left matrix and the jth column of the right matrix and summing the products of the corresponding elements in this row and this column. Hence, the elements of the resultant product matrix are the sums of the products of corresponding elements from one row of **A** and one column of **B**.

Matrix multiplication is possible only for pairs of matrices where *the number of columns in the left matrix is the same as the number of rows in the right matrix*. This is another conformability requirement created by the need for "corresponding elements." The resultant product matrix always has as many rows as there are rows in the first matrix, and as many columns as there are columns in the second matrix. If an $m \times n$ matrix is multiplied by an $n \times k$ matrix, the product matrix will be of order $(m \times \not{n})(\not{n} \times k) = m \times k$. The adjacent canceled n's indicate that the matrices are conformable in that the numbers of columns of the left matrix correspond to the number of rows in the right matrix, and also indicate that n product terms are summed to obtain each element of the product matrix **C**.

The product matrix may be smaller than the multiplied matrices or larger. For example, if **B** from Eq. 3.12 is placed in front of **A** before multiplying, then we get

$$\begin{bmatrix} 1 & 2 \\ 1 & 0 \\ 1 & 1 \end{bmatrix} \begin{bmatrix} 1 & 3 & 5 \\ 2 & 4 & 6 \end{bmatrix} = \begin{bmatrix} 5 & 11 & 17 \\ 1 & 3 & 5 \\ 3 & 7 & 11 \end{bmatrix}$$ 3.13

(3 × 2)(2 × 3)   (3 × 3)

This demonstrates that in matrix multiplication *the order in which the matrices are multiplied is important*. Scalar multiplication is commutative ($ab = ba$), but matrix multiplication is not commutative (**AB** may not equal **BA**). Statisticians say **A** is *postmultiplied* by **B** for **AB**, and **A** is *premultiplied* by **B** for **BA**.

To check your understanding of matrix multiplication, use the following matrices **A** and **B** to verify **AB** and **BA**.

$$\mathbf{A} = \begin{bmatrix} 1 & 1 & 1 \\ 0 & 0 & 1 \end{bmatrix} \qquad \mathbf{B} = \begin{bmatrix} 1 & 4 & 7 \\ 2 & 5 & 8 \\ 3 & 6 & 9 \end{bmatrix}$$

$$\mathbf{AB} = \begin{bmatrix} 6 & 15 & 24 \\ 3 & 6 & 9 \end{bmatrix} \qquad\qquad 3.14$$

$$\mathbf{BA} \underset{(3 \times 3)(2 \times 3)}{=} \left\{ \begin{array}{l} \text{not defined because} \\ \text{the adjacent dimensions} \\ \text{do not conform} \end{array} \right\}$$

Matrix multiplication is associative:

$$(\mathbf{AB})\mathbf{C} = \mathbf{A}(\mathbf{BC}) \qquad\qquad 3.15$$

and distributive with respect to addition:

$$\mathbf{A}(\mathbf{B} + \mathbf{C}) = \mathbf{AB} + \mathbf{AC}$$
$$(\mathbf{B} + \mathbf{C})\mathbf{A} = \mathbf{BA} + \mathbf{CA} \qquad\qquad 3.16$$

Multiplying a matrix **X** by a matrix containing all zero elements gives

$$\mathbf{0X} = \mathbf{0} = \mathbf{X0} \qquad\qquad 3.17$$

and pre- or postmultiplication by a corresponding sized identity matrix leaves the matrix unchanged:

$$\mathbf{IX} = \mathbf{X} = \mathbf{XI} \qquad\qquad 3.18$$

For example,

$$\begin{bmatrix} 1 & 0 \\ 0 & 1 \end{bmatrix}\begin{bmatrix} 1 & 3 & 5 \\ 2 & 4 & 6 \end{bmatrix} = \begin{bmatrix} 1 & 3 & 5 \\ 2 & 4 & 6 \end{bmatrix} = \begin{bmatrix} 1 & 3 & 5 \\ 2 & 4 & 6 \end{bmatrix}\begin{bmatrix} 1 & 0 & 0 \\ 0 & 1 & 0 \\ 0 & 0 & 1 \end{bmatrix} \qquad 3.19$$

If c is a scalar, then

$$c(\mathbf{AB}) = (c\mathbf{A})\mathbf{B} = \mathbf{A}(c\mathbf{B}) \qquad\qquad 3.20$$

This says that multiplying a product matrix by a scalar (Eq. 3.10) accomplishes the same thing as if either of the matrices composing that product had been multiplied by the scalar before matrix multiplication. You can see why this is true if you multiply the left equality in Eq. 3.19 by the scalar number 2 and group the 2 first with $\mathbf{I}$ to get $(2\mathbf{I})\mathbf{X} = 2\mathbf{X}$, and then with the $\mathbf{X}$ to get $\mathbf{I}(2\mathbf{X}) = 2\mathbf{X}$. This exercise demands that you understand the difference between multiplying a matrix by another matrix, and multiplying a matrix by a scalar (Eqs. 3.10 and 3.12).

3.1.1.1 Covariance matrices in matrix algebra

Imagine a data matrix $\mathbf{X}$ containing information on m variables for each of n individuals, with the data recorded so that each row contains the information on a particular variable and each column refers to a particular individual.

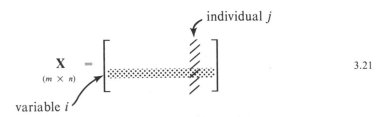

$$\underset{(m \times n)}{\mathbf{X}} = \left[\cdots \cdots \cdots \right] \qquad 3.21$$

individual j

variable i

Imagine further that the data on the individuals are recorded as deviations from the mean, so the element x_{ij} contains the number of units individual j is above or below the mean of variable i. That is, x_{ij} is individual j's score on variable i minus the mean of variable i. [A raw data matrix $(\mathbf{R})$ can be converted to a data matrix containing deviations from the means $(\mathbf{X})$ by creating a column vector of means $\mathbf{m}$ and a row vector composed of as many 1 elements as there are individuals (call it $\mathbf{1}$) and then calculating $\mathbf{X} = \mathbf{R} - \mathbf{m}\mathbf{1}$.]

The transpose of the data matrix containing deviations from the means $(\mathbf{X}')$ would record each individual's deviation scores in a row, with the columns referring to the successive variables. Multiplying $\mathbf{X}$ by $\mathbf{X}'$ gives an $m \times m$ matrix containing the sums of squares of the deviations of the individuals' scores from the mean of each of the m variables as the diagonal elements, whereas the off-diagonal elements contain the sums of the products of the mean deviations for the variables specifying the particular row and column considered.

$$
\mathbf{XX'} = \begin{bmatrix} \text{variable 1} \end{bmatrix}\begin{bmatrix} \text{v} \\ \text{a} \\ \text{r} \\ \text{i} \\ \text{a} \\ \text{b} \\ \text{l} \\ \text{e} \\ \text{1} \end{bmatrix} = \begin{bmatrix} \text{matrix of sum} \\ \text{of squares and} \\ \text{cross products} \end{bmatrix} \qquad 3.22
$$

$(m \times n) \qquad\qquad (n \times m) \qquad\qquad (m \times m)$

Dividing each element of this product matrix by n (multiplying by the scalar $1/n$) gives the expected values of the squared deviations from the mean, namely the variances of the variables, as the diagonal elements. The off-diagonal elements contain the expected values of the products of the deviations of the individuals' scores from two variables' means, or the covariances between the variables specifying the rows and columns of this matrix.

$$
\begin{bmatrix} \text{covariance matrix} \\ \text{for the variables} \\ \text{in } \mathbf{X} \end{bmatrix} = E(\mathbf{XX'}) = \frac{1}{n}\,\mathbf{XX'} \qquad 3.23
$$

Thus the variance/covariance matrix for any given data set is easily calculated as $(1/n)\mathbf{XX'}$ if the data is recorded as deviations from the mean and with the rows referring to variables and columns to individuals. If the rows referred to individuals and columns to variables, the data matrix would be the transpose of the data matrix in 3.21 and the covariance matrix would equal $(1/n)\mathbf{X'X}$.

By a direct parallel to the preceding, if $\mathbf{X}$ is a data matrix recording the deviations of individuals' scores from the means, and if $\mathbf{Y}$ is a matrix containing the deviations of the *same n individuals'* scores from the means of *another set* of variables, then the covariances between the variables in $\mathbf{X}$ and in $\mathbf{Y}$ may be obtained as

$$
\begin{bmatrix} \text{covariances between} \\ \text{the variables} \\ \text{in } \mathbf{X} \text{ and } \mathbf{Y} \end{bmatrix} = E(\mathbf{XY'}) = \frac{1}{n}\,\mathbf{XY'} \qquad 3.24
$$

Equation 3.24 corresponds to examining a rectangularly shaped "chunk" out of the off-diagonal elements of 3.23, because no variable appears twice to give a variance as the covariance of a variable with itself.

3.1.1.2 Elementary matrix operations

Matrix multiplication allows us to make three fundamental types of changes in matrices by three allowable *elementary operations*:

1. Multiplying one row (or column) of a matrix by a constant
2. Interchanging the position of two rows (or columns)
3. Replacing a row (or column) with itself plus k times the value of another row (or column).

These operations are the only three operations allowed in solving sets of equations (multiplying both sides of an equation by a constant, and so on). The matrices that accomplish the elementary operations are called **elementary matrices**. Premultiplication by an elementary matrix operates on the rows of the matrix, and postmultiplication by an elementary matrix operates on the columns of the matrix. For example:

multiplying row 2 by constant k:

$$\begin{bmatrix} 1 & 0 & 0 \\ 0 & k & 0 \\ 0 & 0 & 1 \end{bmatrix} \begin{bmatrix} 1 & 4 \\ 2 & 5 \\ 3 & 6 \end{bmatrix} = \begin{bmatrix} 1 & 4 \\ 2k & 5k \\ 3 & 6 \end{bmatrix} \qquad 3.25$$

interchanging rows 2 and 3:

$$\begin{bmatrix} 1 & 0 & 0 \\ 0 & 0 & 1 \\ 0 & 1 & 0 \end{bmatrix} \begin{bmatrix} 1 & 4 \\ 2 & 5 \\ 3 & 6 \end{bmatrix} = \begin{bmatrix} 1 & 4 \\ 3 & 6 \\ 2 & 5 \end{bmatrix} \qquad 3.26$$

replacing row 1 with itself plus k times row 3:

$$\begin{bmatrix} 1 & 0 & k \\ 0 & 1 & 0 \\ 0 & 0 & 1 \end{bmatrix} \begin{bmatrix} 1 & 4 \\ 2 & 5 \\ 3 & 6 \end{bmatrix} = \begin{bmatrix} 1+3k & 4+6k \\ 2 & 5 \\ 3 & 6 \end{bmatrix} \qquad 3.27$$

multiplying column 2 by a constant k:

$$\begin{bmatrix} 1 & 4 \\ 2 & 5 \\ 3 & 6 \end{bmatrix} \begin{bmatrix} 1 & 0 \\ 0 & k \end{bmatrix} = \begin{bmatrix} 1 & 4k \\ 2 & 5k \\ 3 & 6k \end{bmatrix} \qquad 3.28$$

interchanging the two columns:

$$\begin{bmatrix} 1 & 4 \\ 2 & 5 \\ 3 & 6 \end{bmatrix} \begin{bmatrix} 0 & 1 \\ 1 & 0 \end{bmatrix} = \begin{bmatrix} 4 & 1 \\ 5 & 2 \\ 6 & 3 \end{bmatrix} \qquad 3.29$$

replacing column 1 with itself plus k times column 2:

$$\begin{bmatrix} 1 & 4 \\ 2 & 5 \\ 3 & 6 \end{bmatrix} \begin{bmatrix} 1 & 0 \\ k & 1 \end{bmatrix} = \begin{bmatrix} 1+4k & 4 \\ 2+5k & 5 \\ 3+6k & 6 \end{bmatrix} \qquad 3.30$$

3.1.2 Matrix Inversion

Just as dividing 6 by 3 can be done by multiplying 6 by $1/3$, matrix division can be done by multiplying one matrix by the **inverse** of the other matrix. In algebra a number multiplied by its inverse $[b(1/b)]$ produces 1 (the identity element). In matrix algebra a matrix X multiplied by its inverse X^{-1} gives the identity matrix I:

$$XX^{-1} = I = X^{-1}X \qquad 3.31$$

Only square matrices have inverses. Here are two matrices that are inverses of each other.

$$\begin{bmatrix} 1 & 2 & 3 \\ 0 & 1 & 0 \\ 0 & 0 & 1 \end{bmatrix} \begin{bmatrix} 1 & -2 & -3 \\ 0 & 1 & 0 \\ 0 & 0 & 1 \end{bmatrix} = \begin{bmatrix} 1 & 0 & 0 \\ 0 & 1 & 0 \\ 0 & 0 & 1 \end{bmatrix}$$

$$\qquad 3.32$$

$$\begin{bmatrix} 1 & -2 & -3 \\ 0 & 1 & 0 \\ 0 & 0 & 1 \end{bmatrix} \begin{bmatrix} 1 & 2 & 3 \\ 0 & 1 & 0 \\ 0 & 0 & 1 \end{bmatrix} = \begin{bmatrix} 1 & 0 & 0 \\ 0 & 1 & 0 \\ 0 & 0 & 1 \end{bmatrix}$$

Matrix division, namely, multiplication by an inverse, is easily accomplished by using the steps for multiplication as specified above. The difficult task is to obtain the inverse. This is addressed next.

3.1.2.1 Inverses via diagonalization

Computers have eliminated the need for hand-calculating inverses, but a conceptual understanding of the procedures remains important to serious investigation of LISREL. One procedure for calculating the inverse of any matrix A depends on the elementary operations already discussed. Imagine P_1, P_2, P_3, ... are matrices specifying elementary row operations on A (so these matrices premultiply A). Suppose we select these multiple row operations such that their combined effect is to progressively convert A into an identity matrix I. We could represent this as

$$\cdots P_4P_3P_2P_1A = I. \qquad 3.33$$

We can rewrite this more compactly by multiplying the row transformation matrices (the elementary matrices) together to get an overall row transformation matrix $\mathbf{P} = \cdots \mathbf{P}_4\mathbf{P}_3\mathbf{P}_2\mathbf{P}_1$:

$$\mathbf{PA} = \mathbf{I} \qquad\qquad 3.34$$

In this form we see that $\mathbf{P}$ is the inverse of $\mathbf{A}$ because multiplying $\mathbf{P}$ and $\mathbf{A}$ produces $\mathbf{I}$. The question is, is it possible to specify a series of elementary row transformations $\mathbf{P}$ that convert $\mathbf{A}$ into $\mathbf{I}$? If so, this sequence of elementary transformations accomplishes precisely what an inverse accomplishes, and hence the matrix product of the appropriate sequence is the inverse of the matrix $\mathbf{A}$.

An example will convince us that it is sometimes possible to convert a matrix $\mathbf{A}$ into an identity matrix using only elementary row operations. Consider

$$\mathbf{A} = \begin{bmatrix} 2 & 4 & 6 \\ 4 & 5 & 6 \\ 7 & 8 & 10 \end{bmatrix} \qquad\qquad 3.35$$

We begin converting $\mathbf{A}$ to $\mathbf{I}$ by dividing row 1 by 2 to obtain a 1 as the upper left element by using the elementary row operation:

$$\begin{bmatrix} \frac{1}{2} & 0 & 0 \\ 0 & 1 & 0 \\ 0 & 0 & 1 \end{bmatrix}\begin{bmatrix} 2 & 4 & 6 \\ 4 & 5 & 6 \\ 7 & 8 & 10 \end{bmatrix} = \begin{bmatrix} 1 & 2 & 3 \\ 4 & 5 & 6 \\ 7 & 8 & 10 \end{bmatrix} \qquad 3.36$$

Replacing row 2 (of the matrix resulting from the previous step) with itself minus 4 times row 1 gives a 0:

$$\begin{bmatrix} 1 & 0 & 0 \\ -4 & 1 & 0 \\ 0 & 0 & 1 \end{bmatrix}\begin{bmatrix} 1 & 2 & 3 \\ 4 & 5 & 6 \\ 7 & 8 & 10 \end{bmatrix} = \begin{bmatrix} 1 & 2 & 3 \\ 0 & -3 & -6 \\ 7 & 8 & 10 \end{bmatrix} \qquad 3.37$$

Replacing row 3 (of the matrix resulting from the previous step) with itself minus 7 times row 1 gives another 0:

$$\begin{bmatrix} 1 & 0 & 0 \\ 0 & 1 & 0 \\ -7 & 0 & 1 \end{bmatrix}\begin{bmatrix} 1 & 2 & 3 \\ 0 & -3 & -6 \\ 7 & 8 & 10 \end{bmatrix} = \begin{bmatrix} 1 & 2 & 3 \\ 0 & -3 & -6 \\ 0 & -6 & -11 \end{bmatrix} \qquad 3.38$$

Now $-\frac{1}{3}$ times row 2 gives a second diagonal 1:

$$\begin{bmatrix} 1 & 0 & 0 \\ 0 & -\frac{1}{3} & 0 \\ 0 & 0 & 1 \end{bmatrix}\begin{bmatrix} 1 & 2 & 3 \\ 0 & -3 & -6 \\ 0 & -6 & -11 \end{bmatrix} = \begin{bmatrix} 1 & 2 & 3 \\ 0 & 1 & 2 \\ 0 & -6 & -11 \end{bmatrix} \qquad 3.39$$

Replace row 1 by itself minus 2 times row 2:

$$\begin{bmatrix} 1 & -2 & 0 \\ 0 & 1 & 0 \\ 0 & 0 & 1 \end{bmatrix}\begin{bmatrix} 1 & 2 & 3 \\ 0 & 1 & 2 \\ 0 & -6 & -11 \end{bmatrix} = \begin{bmatrix} 1 & 0 & -1 \\ 0 & 1 & 2 \\ 0 & -6 & -11 \end{bmatrix} \qquad 3.40$$

Replacing row 3 with itself plus 6 times row 2 gives another zero and a gratuitous diagonal 1:

$$\begin{bmatrix} 1 & 0 & 0 \\ 0 & 1 & 0 \\ 0 & 6 & 1 \end{bmatrix}\begin{bmatrix} 1 & 0 & -1 \\ 0 & 1 & 2 \\ 0 & -6 & -11 \end{bmatrix} = \begin{bmatrix} 1 & 0 & -1 \\ 0 & 1 & 2 \\ 0 & 0 & 1 \end{bmatrix} \qquad 3.41$$

Replace row 1 with itself plus row 3:

$$\begin{bmatrix} 1 & 0 & 1 \\ 0 & 1 & 0 \\ 0 & 0 & 1 \end{bmatrix}\begin{bmatrix} 1 & 0 & -1 \\ 0 & 1 & 2 \\ 0 & 0 & 1 \end{bmatrix} = \begin{bmatrix} 1 & 0 & 0 \\ 0 & 1 & 2 \\ 0 & 0 & 1 \end{bmatrix} \qquad 3.42$$

The process is completed by replacing row 2 with itself minus 2 times row 3:

$$\begin{bmatrix} 1 & 0 & 0 \\ 0 & 1 & -2 \\ 0 & 0 & 1 \end{bmatrix}\begin{bmatrix} 1 & 0 & 0 \\ 0 & 1 & 2 \\ 0 & 0 & 1 \end{bmatrix} = \begin{bmatrix} 1 & 0 & 0 \\ 0 & 1 & 0 \\ 0 & 0 & 1 \end{bmatrix} \qquad 3.43$$

The strategy illustrated here is to place a 1 at the appropriate spot in a column and then 0's elsewhere in that column, before moving to the next column to create another diagonal 1 and using this 1 to place 0's elsewhere. Observe that the precise elementary operations required would change if we sought the inverse of a different matrix, and that the elementary row operations are all accomplished by premultiplication by matrices differing only slightly from I. These matrices are P_1, P_2, P_3, . . ., and their product is $P = \cdots P_3P_2P_1$. The elementary matrix providing the last step of the process appears on the extreme left because each step requires a row operation and hence **premultiplication** of the matrix resulting from the previous step by an elementary matrix.

$$P = \begin{bmatrix} 1 & 0 & 0 \\ 0 & 1 & -2 \\ 0 & 0 & 1 \end{bmatrix}\begin{bmatrix} 1 & 0 & 1 \\ 0 & 1 & 0 \\ 0 & 0 & 1 \end{bmatrix}\begin{bmatrix} 1 & 0 & 0 \\ 0 & 1 & 0 \\ 0 & 6 & 1 \end{bmatrix}\begin{bmatrix} 1 & -2 & 0 \\ 0 & 1 & 0 \\ 0 & 0 & 1 \end{bmatrix}$$

$$\times \begin{bmatrix} 1 & 0 & 0 \\ 0 & -\frac{1}{3} & 0 \\ 0 & 0 & 1 \end{bmatrix}\begin{bmatrix} 1 & 0 & 0 \\ 0 & 1 & 0 \\ -7 & 0 & 1 \end{bmatrix}\begin{bmatrix} 1 & 0 & 0 \\ -4 & 1 & 0 \\ 0 & 0 & 1 \end{bmatrix}\begin{bmatrix} \frac{1}{2} & 0 & 0 \\ 0 & 1 & 0 \\ 0 & 0 & 1 \end{bmatrix} \qquad 3.44$$

$$\mathbf{P} = \begin{bmatrix} -1/3 & -4/3 & 1 \\ -1/3 & 11/3 & -2 \\ 1/2 & -2 & 1 \end{bmatrix}$$

You should check that $\mathbf{PA} = \mathbf{I} = \mathbf{AP}$.

This convinces us that the matrix embodying the row operations converting $\mathbf{A}$ into $\mathbf{I}$ is the inverse of $\mathbf{A}$ and could be denoted $\mathbf{A}^{-1}$. The fact that pre- or postmultiplying $\mathbf{A}$ by $\mathbf{P}$ produces an identity matrix indicates that column operations could have been used to transform $\mathbf{A}$ into $\mathbf{I}$ and that one adequate set of column operations is provided by the same elementary matrices required to accomplish this by row transformations.

But can every possible $\mathbf{A}$ be transformed into an identity matrix using elementary row transformations? Unfortunately, some matrices cannot be converted to $\mathbf{I}$ and hence have no inverse. One can begin creating a diagonal matrix by placing 1's on the diagonal and 0's elsewhere, but at some point the process is halted because one row (or column) becomes nothing but 0's, due to it being a linear function of the other rows (or columns), and one is unable to complete the transformation into the identity matrix. This happens if we take a new $\mathbf{A}$ that differs from the $\mathbf{A}$ just used in that element a_{33} equals 9 instead of 10:

$$\mathbf{A} = \begin{bmatrix} 2 & 4 & 6 \\ 4 & 5 & 6 \\ 7 & 8 & 9 \end{bmatrix} \qquad\qquad 3.45$$

Check that using the same first five elementary row transformations (paralleling Eqs. 3.36–3.40) produces

$$\begin{bmatrix} 1 & 0 & -1 \\ 0 & 1 & 2 \\ 0 & -6 & -12 \end{bmatrix} \qquad\qquad 3.46$$

At the next step the process breaks down. To obtain a zero in row 3, column 2, we use the same row transformation as in 3.41 but find

$$\begin{bmatrix} 1 & 0 & 0 \\ 0 & 1 & 0 \\ 0 & 6 & 1 \end{bmatrix} \begin{bmatrix} 1 & 0 & -1 \\ 0 & 1 & 2 \\ 0 & -6 & -12 \end{bmatrix} = \begin{bmatrix} 1 & 0 & -1 \\ 0 & 1 & 2 \\ 0 & 0 & 0 \end{bmatrix} \qquad 3.47$$

in which both the remaining elements in the third row become 0's. Although nonzero elements might be placed in this third row again (e.g., by adding row 1 to row 3), this always adds at least two nonzero elements into the third row; eliminating these excess elements always results in both the desired element (a 1 in row 3, column 3) and the excess elements vanishing simultaneously.

Nor would resorting to column transformations help. Since all the elements in the third row are 0, there are no combinations of columns that can introduce nonzero elements into this row. The only thing the column transformations can do is eliminate the −1 and 2 in the third column (i.e., replace column three with itself plus column 1; then replace the new column 3 with itself minus twice column 2). This gives a matrix like an identity matrix, but the diagonal contains a 0. This matrix is in canonical form, and the number of diagonal 1's indicates the **rank** of the matrix. That is, although the matrix **A** is of order 3 × 3, it is only of rank 2 because the linear dependencies among the elements of the matrix imply that elementary transformations reduce the matrix to a form in which there are only two diagonal unity elements. Thus the rank of a matrix indicates the number of linearly independent rows (or columns) of information in the original matrix **A**, and this equals the order (size) of the largest identity matrix that can be created using elementary transformations on the original matrix. A matrix is of **full rank** if none of the rows (columns) become nothing but 0's—that is, if using the elementary transformations produces a matrix with as many diagonal 1's as there are rows/columns in the original matrix. Matrices that are of full rank must have inverses because there exists a set of row transformations that converts the matrix into **I**.

Row operations alone are sufficient to obtain **I** if the matrix is of full rank [that is the size (order) of the resultant identity matrix is the same as that of the original matrix], but it might be easier to use both row and column elementary operations to transform **A** into **I**. In this case

$$\cdots \mathbf{P}_4\mathbf{P}_3\mathbf{P}_2\mathbf{P}_1\mathbf{A}\mathbf{Q}_1\mathbf{Q}_2\mathbf{Q}_3\mathbf{Q}_4 \cdots - \mathbf{I} \qquad 3.48$$

This equation can be simplified by multiplying the row transformations together to get **P** and the column transformations together to get **Q** so that

$$\mathbf{PAQ} = \mathbf{I} \qquad 3.49$$

If row (**P**) and column (**Q**) transformations result in the form

$$\mathbf{PAQ} = \begin{bmatrix} 1 & 0 & . & . & 0 & 0 & . & . \\ 0 & 1 & . & . & 0 & 0 & . & . \\ . & . & 1 & . & . & . & . & . \\ . & . & . & 1 & . & . & . & . \\ \hline 0 & 0 & . & . & 0 & 0 & . & . \\ 0 & 0 & . & . & 0 & 0 & . & . \\ . & . & . & . & . & . & . & . \\ . & . & . & . & . & . & . & . \end{bmatrix} \qquad 3.50$$

the matrix **A** is of less than full rank, or contains linear dependencies, or is singular. If a complete identity matrix results, the matrix must have an inverse.[1]

Diagonal matrices have simple inverses. The inverse of a diagonal matrix is another diagonal matrix whose elements are the reciprocals of the original diagonal elements. Thus,

$$\begin{bmatrix} a & 0 & 0 \\ 0 & b & 0 \\ 0 & 0 & c \end{bmatrix} \begin{bmatrix} 1/a & 0 & 0 \\ 0 & 1/b & 0 \\ 0 & 0 & 1/c \end{bmatrix} = \begin{bmatrix} 1 & 0 & 0 \\ 0 & 1 & 0 \\ 0 & 0 & 1 \end{bmatrix} \qquad 3.51$$

3.1.3 Determinants

The quantity denoted |A| is a scalar number that is uniquely defined for any matrix A and is called the **determinant** of A. The mathematically sophisticated will feel satisfied knowing it can be proven (e.g., Cullen, 1972) that there exists one and only one function of the elements of a matrix (i.e., the determinant) possessing the following properties:

  a. The determinant of a matrix that is the product of two other matrices must equal the products of the determinants of those other matrices (|AB| = |A||B|).

  b. The determinant of an identity matrix with one diagonal element replaced by k equals k.

From these two properties and the rules of multiplication, we can show that the following must also be true of the determinants for matrices to which the elementary operations are applied:

  1. Multiplying a row (or column) of A by k produces a matrix whose determinant is k times the determinant of A.

  2. Interchanging two rows of A makes the determinant of the new matrix the negative of the determinant of the original matrix A.

  3. The determinant of a matrix remains unchanged if k times the elements in one row is added to another row.

The mathematically unadulterated are likely to exclaim, "So you know you can prove this thing called the determinant exists and behaves in certain ways, but what *is* a determinant?" A mathematician would say that it *is* what its properties are (as listed)—nothing more, nothing less—and display no further interest in one's plight.

We, like the mathematician, can do no *better* than point to a proof (Cullen, 1972) of these properties. A less rigorous feel for what a determinant is, however, can be developed by considering several of the ways for calculating determinants, all of which are less than rigorous in that they either invoke as yet undefined entities or they are a mere recounting of steps in calculations, no matter how familiar those steps may be. For example, the determinant of a 2 × 2 matrix can always be calculated as the product of the main diagonal elements minus the product of the other diagonal's elements.

One way to obtain the determinant of a matrix of any size (order) begins from the equation PAQ=I, which uses row and column transformations to convert a matrix into the identity matrix. If we write out the row and column transformations individually and use the fact that the determinant of a product is the product of the determinants, we can write

$$\cdots |\mathbf{P}_3||\mathbf{P}_2||\mathbf{P}_1||\mathbf{A}||\mathbf{Q}_1||\mathbf{Q}_2||\mathbf{Q}_3| \cdots = |\mathbf{I}| = 1 \qquad 3.52$$

Since the determinants of the elementary operations (the **P**'s and **Q**'s) are simple (any elementary matrix interchanging two rows/columns has determinant -1; any elementary matrix multiplying a row/column by k has determinant k; and any elementary matrix adding k times one row/column to another has determinant 1), we could calculate a lot of these simple determinants and then divide both sides of 3.52 by these many simple determinants to obtain the determinant of the complex matrix **A** as the only quantity that had not canceled out of the left side of Eq. 3.52. If there are no **P**'s and **Q**'s that converted **A** into **I**, which would be the case if **A** were not of full rank, this procedure for obtaining |A| fails.

Three ways of describing matrices where one row of the matrix can be exactly expressed as some linear combination of the other rows are that the matrix will have no inverse, a rank less than n (as we saw in Section 3.1.2.1), and a determinant of zero (by 3.52 with a diagonal zero in **I** and property b of determinants).[2]

A second way to obtain the determinant of a matrix is by expanding the determinant according to any one row or column of the original matrix. This involves expressing the determinant of a matrix as a function of the elements in one row or column of the matrix and the determinants of the slightly smaller "minor" matrices obtained by deleting one row and one column from the original matrix. Repeatedly creating slightly smaller matrices reduces the size of matrices whose determinants are required until we eventually encounter 2 × 2 matrices, whose determinants are easily obtained as the product of the main diagonal elements minus the product of the other diagonal's elements, or where the process could be carried one step further to use the fact that the determinant of a scalar is simply that scalar's value. In short, the second procedure replaces the difficult task of calculating the determinant of a complex matrix with the relatively trivial task of keeping track of the determinants of the many smaller matrices created by successive row/column deletions.

The formula for obtaining a determinant by row expansion is

$$|A| = \sum_i a_{ij}(-1)^{i+j}|Minor_{ij}| \qquad 3.53$$

This formula tells us that the determinant $|A|$ equals the sum (over all entries in any one row) of the values of the elements in that row each multiplied by the signed determinants of the minor (smaller) matrices created by eliminating that row and column from the original matrix.

For example, the determinant of the matrix

$$\begin{bmatrix} 2 & 4 & 6 \\ 4 & 5 & 6 \\ 7 & 8 & 10 \end{bmatrix} \qquad 3.54$$

obtained by expanding according to the first row is

$$|A| = 2(-1)^{1+1}\begin{vmatrix} 5 & 6 \\ 8 & 10 \end{vmatrix} + 4(-1)^{1+2}\begin{vmatrix} 4 & 6 \\ 7 & 10 \end{vmatrix} + 6(-1)^{1+3}\begin{vmatrix} 4 & 5 \\ 7 & 8 \end{vmatrix} \qquad 3.55$$

Expanding each of the remaining determinants according to their first rows produces the same result as the product of the main diagonal elements minus the product of the other diagonal's elements for 2×2 matrices. Thus

$$\begin{aligned} |A| = \; & 2[5(-1)^{1+1}10 + 6(-1)^{1+2}8] \\ & - 4[4(-1)^{1+1}10 + 6(-1)^{1+2}7] \\ & + 6[4(-1)^{1+1}8 + 5(-1)^{1+2}7] \end{aligned} \qquad 3.56$$

$$= 2(50 - 48) - 4(40 - 42) + 6(32 - 35)$$
$$= -6$$

The same determinant is obtained by expanding according to *any* row or column. For large matrices it can be difficult to keep track of the determinants of the progressively shrinking matrices. The process can be simplified by selecting a row or column containing zeros and capitalizing on the resultant dropping of terms due to multiplication by zero. Indeed, the next procedure for obtaining the determinant carries this notion to the extreme.

A third way to calculate the determinant combines three facts: (1) All square matrices can be reduced to triangular matrices by repeated use of the elementary operation of replacing one row (or column) with itself plus k times another row (or column). (2) The determinants of the triangularized and original matrices are the same. (3) The determinant of a triangular matrix is the product of the diagonal elements. This last fact is easily demonstrated because it involves a special case of the preceding procedure—expansion of a matrix according to a row or column where all but one of the entries in that row or column are zero. For example, the following triangular matrix results from the matrix in 3.54 if we replace column 2 by itself minus twice column 1, then replace column 3 with itself minus 3 times column 1, and finally replace the new column 3 by itself minus twice the new column 2.

$$\mathbf{T} = \begin{bmatrix} 2 & 0 & 0 \\ 4 & -3 & 0 \\ 7 & -6 & 1 \end{bmatrix} \qquad 3.57$$

Expanding this according to the first row gives

$$|\mathbf{T}| = 2(-1)^{1+1}\begin{vmatrix} -3 & 0 \\ -6 & 1 \end{vmatrix} + 0(\quad)\begin{vmatrix} \quad \end{vmatrix} + 0(\quad)\begin{vmatrix} \quad \end{vmatrix} \qquad 3.58$$

Expanding the minor matrix by its first row to get its determinant gives

$$|\mathbf{T}| = 2[-3(-1)^{1+1}|1|] + 0(\quad)|\quad| \qquad 3.59$$

$$= 2(-3)1 = -6$$

This demonstrates that the determinant of the overall matrix is the product of the diagonal elements of the triangular matrix, and that the determinants of the triangular and original matrices are the same (by comparing this to 3.56). Thus we can calculate a determinant by triangularizing a matrix with the elementary operation of replacing a row (or column) by itself plus k times another row (or column) (which leaves the determinant unchanged) and then using the fact that the determinant of a triangular matrix is the product of the diagonal elements.

Determinants can also be calculated by the Cayley-Hamilton theorem (Fuller, 1962) or by using the fact that the determinant equals the product of the eigenvalues of the matrix, but these procedures require a knowledge of eigenvalues and eigenvectors, which we do not discuss.

3.1.4 Summary of the Mathematics of Inverses, Determinants, and Transposes

The following describe the mathematics of handling inverses, if the indicated inverses exist—that is, if no row or column is an exact linear combination of any other row(s) or column(s).

$$(\mathbf{A}^{-1})^{-1} = \mathbf{A} \qquad 3.60$$

$$(\mathbf{AB})^{-1} = \mathbf{B}^{-1}\mathbf{A}^{-1} \qquad 3.61$$

$$(\mathbf{ABC})^{-1} = \mathbf{C}^{-1}\mathbf{B}^{-1}\mathbf{A}^{-1} \qquad\qquad 3.62$$

The inverse of a product of four or more matrices equals the product of the inverses in reverse order.

The mathematics of determinants allows

$$|\mathbf{AB}| = |\mathbf{A}||\mathbf{B}| \qquad\qquad 3.63$$

$$|\mathbf{A} + \mathbf{B}| = |\mathbf{A}| + |\mathbf{B}| \qquad\qquad 3.64$$

$$|\mathbf{A}^{-1}| = 1/|\mathbf{A}| \qquad\qquad 3.65$$

If $\mathbf{B}$ is obtained by multiplying one row (or column) of $\mathbf{A}$ by the scalar k, then

$$|\mathbf{B}| = k|\mathbf{A}| \qquad\qquad 3.66$$

If $\mathbf{B}$ is obtained by interchanging two rows (or columns) of $\mathbf{A}$, then

$$|\mathbf{B}| = -|\mathbf{A}| \qquad\qquad 3.67$$

If $\mathbf{B}$ is obtained from $\mathbf{A}$ by replacing one row of $\mathbf{A}$ by that row plus k times another row, then the determinant remains unchanged.

If two rows (or columns) of $\mathbf{A}$ are proportional, then the matrix is said to be "singular" and $|\mathbf{A}| = 0$. The matrix has no inverse, and the matrix has a rank less than its order (size)—that is, using row (or column) transformations to attempt to convert the matrix into an identity matrix results in a matrix containing at least one row (or column) of zero elements. A similar set of statements is true if one row (or column) is equivalent to the weighted sum of any other rows (or columns).

For transposes

$$(\mathbf{A}')' = \mathbf{A} \qquad\qquad 3.68$$

$$(\mathbf{A} + \mathbf{B})' = \mathbf{A}' + \mathbf{B}' \qquad\qquad 3.69$$

$$(k\mathbf{A})' = k\mathbf{A}' \qquad\qquad 3.70$$

$$(\mathbf{AB})' = \mathbf{B}'\mathbf{A}' \qquad\qquad 3.71$$

$$|\mathbf{A}'| = |\mathbf{A}| \qquad\qquad 3.72$$

$$(\mathbf{A}')^{-1} = (\mathbf{A}^{-1})' \qquad\qquad 3.73$$

If a matrix is multiplied by itself so that $\mathbf{AA} = \mathbf{A}^2$ or $\mathbf{AAA} = \mathbf{A}^3$, and so on, then

$$\mathbf{A}^m\mathbf{A}^n = \mathbf{A}^{m+n} \qquad\qquad 3.74$$

$$(\mathbf{A}^m)^n = (\mathbf{A}^n)^m = \mathbf{A}^{mn} \qquad\qquad 3.75$$

and, by definition, $\mathbf{A}^0 = \mathbf{I}$.

3.1.5 Symmetric and Partitioned Matrices

At the beginning of this chapter we used the term "symmetric" to describe matrices that satisfy $\mathbf{A} = \mathbf{A}'$, which is another way of saying that the elements falling above the main diagonal are the mirror image of those falling below the diagonal, or that the first row and column contain the same elements, and so on. Only square matrices can be symmetric.

Since we will encounter numerous symmetric matrices, we should mention how such matrices arise. Symmetric matrices are created by taking any matrix $\mathbf{A}$ and obtaining $\mathbf{AA}'$ or $\mathbf{A}'\mathbf{A}$. For example,

$$\begin{bmatrix} 1 & 3 & 5 \\ 2 & 4 & 6 \end{bmatrix}\begin{bmatrix} 1 & 2 \\ 3 & 4 \\ 5 & 6 \end{bmatrix} = \begin{bmatrix} 35 & 44 \\ 44 & 56 \end{bmatrix} \qquad\qquad 3.76$$

$$\begin{bmatrix} 1 & 2 \\ 3 & 4 \\ 5 & 6 \end{bmatrix}\begin{bmatrix} 1 & 3 & 5 \\ 2 & 4 & 6 \end{bmatrix} = \begin{bmatrix} 5 & 11 & 17 \\ 11 & 25 & 39 \\ 17 & 39 & 61 \end{bmatrix}$$

Since Eq. 3.23 for the covariance matrix is of this form, covariance matrices are symmetric.

In the following chapters we encounter several partitioned matrices. Partitioning matrices allows us to keep track of groups of rows or columns while proceeding with various matrix manipulations. Dotted lines are used to indicate partitioning if the full matrix is presented, and the notation for denoting partitions is as follows:

$$\mathbf{A} = \begin{bmatrix} 1 & 4 & 7 & \vdots & 10 \\ 2 & 5 & 8 & \vdots & 11 \\ \cdots & \cdots & \cdots & + & \cdots \\ 3 & 6 & 9 & \vdots & 12 \end{bmatrix} = \begin{bmatrix} \mathbf{A}_{11} & \vdots & \mathbf{A}_{12} \\ \cdots & + & \cdots \\ \mathbf{A}_{21} & \vdots & \mathbf{A}_{22} \end{bmatrix} \qquad\qquad 3.77$$

$$\mathbf{A}_{11} = \begin{bmatrix} 1 & 4 & 7 \\ 2 & 5 & 8 \end{bmatrix} \qquad \mathbf{A}_{12} = \begin{bmatrix} 10 \\ 11 \end{bmatrix}$$

$$\mathbf{A}_{21} = \begin{bmatrix} 3 & 6 & 9 \end{bmatrix} \qquad \mathbf{A}_{22} = \begin{bmatrix} 12 \end{bmatrix}$$

If two partitioned matrices are to be added, not only must the matrices be of the same overall size, but the partitions must be identical. For example, adding the following $\mathbf{B}$ to the previous $\mathbf{A}$ produces

$$\mathbf{B} = \left[\begin{array}{ccc:c} 1 & 1 & 1 & 0 \\ 1 & 1 & 1 & 0 \\ \hdashline 0 & 0 & 0 & 0 \end{array}\right] = \left[\begin{array}{c:c} \mathbf{B}_{11} & \mathbf{B}_{12} \\ \hdashline \mathbf{B}_{21} & \mathbf{B}_{22} \end{array}\right]$$

$$\mathbf{A} + \mathbf{B} = \left[\begin{array}{cc} \mathbf{A}_{11} & \mathbf{A}_{12} \\ \mathbf{A}_{21} & \mathbf{A}_{22} \end{array}\right] + \left[\begin{array}{cc} \mathbf{B}_{11} & \mathbf{B}_{12} \\ \mathbf{B}_{21} & \mathbf{B}_{22} \end{array}\right] \qquad \text{3.78}$$

$$\mathbf{A} + \mathbf{B} = \left[\begin{array}{c:c} \mathbf{A}_{11} + \mathbf{B}_{11} & \mathbf{A}_{12} + \mathbf{B}_{12} \\ \hdashline \mathbf{A}_{21} + \mathbf{B}_{21} & \mathbf{A}_{22} + \mathbf{B}_{22} \end{array}\right] = \left[\begin{array}{ccc:c} 2 & 5 & 8 & 10 \\ 3 & 6 & 9 & 11 \\ \hdashline 3 & 6 & 9 & 12 \end{array}\right]$$

If two partitioned matrices are to be multiplied, the partitionings must give conformable matrices, after which matrix multiplication proceeds exactly as if the partitions are treated as elements rather than as imbedded matrices.

$$\mathbf{AB} = \left[\begin{array}{c:c} \mathbf{A}_{11} & \mathbf{A}_{12} \\ \hdashline \mathbf{A}_{21} & \mathbf{A}_{22} \end{array}\right] \left[\begin{array}{c} \mathbf{B}_{1} \\ \hdashline \mathbf{B}_{2} \end{array}\right] = \left[\begin{array}{c} \mathbf{A}_{11}\mathbf{B}_{1} + \mathbf{A}_{12}\mathbf{B}_{2} \\ \hdashline \mathbf{A}_{21}\mathbf{B}_{1} + \mathbf{A}_{22}\mathbf{B}_{2} \end{array}\right] \qquad \text{3.79}$$

For example,

$$\left[\begin{array}{cc:c} 1 & 0 & 0 \\ 0 & 1 & 0 \\ \hdashline 0 & 0 & 0 \end{array}\right] \left[\begin{array}{cccc} 1 & 1 & 1 & 1 \\ 1 & 1 & 1 & 1 \\ \hdashline 1 & 1 & 1 & 1 \end{array}\right]$$

$$= \left[\begin{array}{c} \left[\begin{array}{cc} 1 & 0 \\ 0 & 1 \end{array}\right]\left[\begin{array}{cccc} 1 & 1 & 1 & 1 \\ 1 & 1 & 1 & 1 \end{array}\right] + \left[\begin{array}{c} 0 \\ 0 \end{array}\right][1 \quad 1 \quad 1 \quad 1] \\ \hdashline [0 \quad 0]\left[\begin{array}{cccc} 1 & 1 & 1 & 1 \\ 1 & 1 & 1 & 1 \end{array}\right] + [0][1 \quad 1 \quad 1 \quad 1] \end{array}\right] \qquad \text{3.80}$$

$$= \left[\begin{array}{cccc} 1 & 1 & 1 & 1 \\ 1 & 1 & 1 & 1 \\ \hdashline 0 & 0 & 0 & 0 \end{array}\right]$$

Partitioning is especially useful when portions of a matrix have special properties, such as in Eq. 3.50 where certain groups of elements are all zeros, and in Chapters 7 and 9 where we subsume some matrices within others.

3.2 Derivatives in a Few Easy Pages

Two aspects of LISREL modeling require a basic conceptual understanding of derivatives. Derivatives can be used to obtain the maximum or minimum of functions and hence are an essential component of the *maximum likelihood* procedure we will be using to obtain estimates of the coefficients in structural equation models. Furthermore, derivatives and rescaled derivatives called *modification indices* provide useful diagnostic information about model deficiencies and potential model improvements. Though a detailed knowledge of derivatives is not demanded, a clear understanding of what derivatives are and how they can be used is fundamental. Fortunately, the requisite ideas are relatively simple and are easily developed as minor extensions to the concept of the slope of a line.

3.2.1 Derivatives as Limiting Slopes

A **derivative** *is an equation for the slope of a line.* Chapter 2 discussed the slope of a straight line, such as that summarizing the linear relationship between variables X and Y. We continue to focus on two variables but inquire about the slope of a curved line, such as that in Figure 3.1. Clearly, the slope depends on the particular value of X considered. At some values of X the slope is steep, but at others it is gradual.

The slope of a straight line can be represented by a single number b, because it is constant for all values of X. The slope of a curved line changes continuously, depending on the value of X considered; hence it is a function of X and not a constant. *The derivative of Y with respect to X* (dY/dX) is the equation giving the slope of the line for any given value of X:

$$\frac{dY}{dX} = \text{some function containing } X \qquad 3.81$$

The slope of the curve at any value of X can be obtained by inserting that value into the function on the right of the equal sign. The quantity dY/dX is a *single symbol* and should be read as "the derivative of Y with respect to X" or "the slope of Y with respect to X." It does *not* mean dY divided by dX.

The function on the right side of Eq. 3.81 is obtained by a simple extension of the procedure for obtaining the slope of a straight line. The slope of a straight line can be obtained by calculating the amount of change in the value of the Y variable that accompanies a certain amount of change (conveniently 1 unit) in the X variable. That is, we locate any two points on the straight line, measure how much the Y values for these points differ, measure how much the X values for these points differ, and divide the Y difference by the X difference to obtain the amount of change in Y per unit change in X—namely, the slope.

Figure 3.1 shows a variable Y as some continuous nonlinear function of X. The precise nature of the curve (i.e., the equation linking Y to X) is irrelevant for our current purposes, so we merely denote this as $Y = f(X)$. We want the slope of the curve at the value x, so we proceed as in the preceding paragraph. We locate two points on the curve and record the lower X value as x and the upper value as Δx units larger than this, namely $x + \Delta x$. The Y value for the lower point is obtained by inserting the value of x into the function linking Y and X [namely as $f(x)$], and the higher Y value is obtained as the value the function takes on if a value of $x + \Delta x$ is inserted into the function [namely $f(x + \Delta x)$]. Still following the procedure for calculating a slope, we take the difference in Y scores and divide this by the difference in X scores to obtain

$$\text{slope} = \frac{f(x + \Delta x) - f(x)}{(x + \Delta x) - x} = \frac{f(x + \Delta x) - f(x)}{\Delta x} \qquad 3.82$$

But is this the slope of the curve at the value x? Visual inspection of Figure 3.1 indicates this slope might be a good approximation to the average slope of the curve in the region between x and $x + \Delta x$, but it is a little too steep to be the slope right at the point x. The slope depicted with a heavy solid line actually crosses the curved line at x, and what we really desire is the slope of a straight line that just touches (is tangent to) the curve at x, as illustrated by the light solid line.

The solution to this problem is clear: make the quantity Δx smaller. This moves the upper point closer to the lower point and provides a better approximation to the true slope at x. Making Δx half its original size produces the dashed slope in Figure 3.1, and making Δx even smaller gives still better approximations to the true slope at x. To obtain the true slope at x (the derivative at x), we need only obtain the value of the slope presented in 3.82 as Δx approaches zero—that is, the limit of the quantity

Figure 3.1   The Derivative as a Limiting Slope.

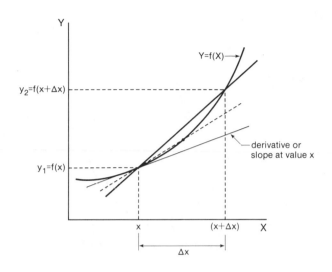

in Eq. 3.82 as Δx approaches zero.

$$\frac{dY}{dX} = \lim_{\Delta x \to 0} \frac{f(x + \Delta x) - f(x)}{\Delta x} \qquad 3.83$$

Note that although Δx is allowed to get as close to zero as we want, it is *not* allowed to equal zero, because this would result in dividing by zero in Eq. 3.83.

To illustrate how this can be done, we focus on one of the many possible continuous nonlinear functions relating Y to X. Figure 3.2 illustrates the function $Y = f(X) = X^2 + 1$. To obtain *the slope of this curved function for any desired value of X* (the derivative of Y with respect to X), we use the $X^2 + 1$ function to specify the difference or change in Y values in the numerator of Eq. 3.83. One value of Y is calculated at the specific but unspecified value x and the other value is Δx units higher. This provides

$$\frac{dY}{dX} = \lim_{\Delta x \to 0} \frac{((x + \Delta x)^2 + 1) - (x^2 + 1)}{\Delta x} \qquad 3.84$$

The only task that remains is to simplify the expression so that the implications of the "limit as $\Delta x \to 0$" becomes apparent. Canceling the 1's

Figure 3.2   The Slopes for $Y = X^2 + 1$.

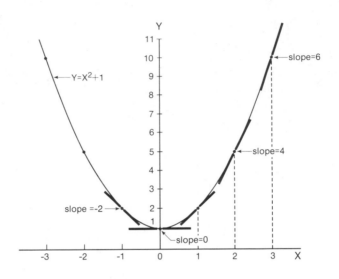

and expanding the square gives

$$\frac{dY}{dX} = \lim_{\Delta x \to 0} \frac{(x^2 + 2x\Delta x + (\Delta x)^2) - x^2}{\Delta x} \qquad 3.85$$

Then, cancel the x^2 terms.

$$\frac{dY}{dX} = \lim_{\Delta x \to 0} \frac{2x\Delta x + (\Delta x)^2}{\Delta x} \qquad 3.86$$

The denominator disappears when it is divided out of both the numerator terms.

$$\frac{dY}{dX} = \lim_{\Delta x \to 0} (2x + \Delta x) \qquad 3.87$$

Finally, we note that as Δx approaches zero, this whole function approaches $2x$, so

$$\frac{dY}{dX} = 2x \qquad 3.88$$

Verify for yourself that for any desired value of X in Figure 3.2, the slope of the $Y = X^2 + 1$ curve at that value (the derivative) is merely twice the selected value of X. Note, in particular, that *at the minimum of the function the slope must be zero,* and that we can locate the value of X providing this minimum by finding the value of X that sets the derivative (the slope at x) to zero. Namely, at the minimum the slope (derivative) $2x$ must equal 0. Dividing both sides of the equation $2x = 0$ by 2 convinces us that an X value of 0 corresponds to the minimum Y value. We can now obtain the actual minimum Y value by inserting $X = 0$ into the original equation $Y = X^2 + 1$ to obtain $Y = 0^2 + 1 = 1$. In sum, setting the equation for the derivative to zero and solving for X provides the X value (0) that minimizes the function $Y = X^2 + 1$, which is easily verified by inspecting Figure 3.2.[3]

The same procedure[4] can be used to obtain the derivative of any other curved function. You might demonstrate that the function $Y = 2X^2$ has derivative $4X$, or that the function $Y = X^2 + X$ has derivative $2X + 1$. The values of X that provide the minimums of these functions are simply the values that set the above derivatives (slopes) to zero—namely 0 and $-\frac{1}{2}$, respectively. The actual minimum Y values can be obtained by inserting these X values into the respective equations.

The only possible problem that arises in using this procedure for determining the slope (derivative) is that it may be difficult to locate the limit required by Eq. 3.83. For the preceding functions, a few simple steps are sufficient to obtain the limiting value, but for other functions obtaining the limit is considerably more complicated. We will leave the pursuit of the mathematical techniques for obtaining limits to texts on calculus, and be content with reporting the derivatives (slopes) that have been derived for some general types of functions of X.

If c is a constant and $Y = c$,

$$\frac{dY}{dX} = \frac{dc}{dX} = 0 \qquad\qquad 3.89$$

You can visualize this as the "zero" slope of a horizontal line plotted at the value c on the Y scale.

Furthermore,

$$\frac{d(cX)}{dX} = c \qquad\qquad 3.90$$

That is, if $Y = cX$, this gives a straight line having slope (derivative) c no matter what X value is considered.

If u and v are functions of X, and Y is the sum of these functions, then

$$\frac{dY}{dX} = \frac{d(u + v)}{dX} = \frac{du}{dX} + \frac{dv}{dX} \qquad\qquad 3.91$$

For example, the slope (derivative) of $Y = 3X^2 + 24X$ at any desired X

value is

$$\frac{d(3X^2 + 24X)}{dX} = \frac{d(3X^2)}{dX} + \frac{d(24X)}{dX} \qquad 3.92$$

The rightmost term simplifies to 24 (using Eq. 3.90) and the left term simplifies to $6X$ by Eqs. 3.93 and 3.94.

The derivative of c times any function of X [$u = f(X)$] is c times the derivative of that function,

$$\frac{d(cu)}{dX} = c\frac{du}{dX} \qquad 3.93$$

and the derivative of X raised to any power is that power times X raised to a one-lower power.

$$\frac{d(X^n)}{dX} = nX^{n-1} \qquad 3.94$$

Hence,

$$\frac{d(3X^2)}{dX} = 3\frac{d(X^2)}{dX} = 3(2X) = 6X \qquad 3.95$$

and the derivative of X^2 is $2X$ (which can be proven by steps identical to those in Eqs 3.84–3.88 except that the equation corresponding to 3.84 would contain no 1's). The derivatives of numerous other functions appear in most introductory calculus texts.

3.2.2 Partial Derivatives

What happens if Y is a continuous function of two or more variables, as in multiple regression or even in bivariate regression where the amount of error is a function of the values taken on by the *two* structural coefficients a and b? Figure 3.3, for example, depicts a curvilinear relationship linking Y to two other variables [$Y = f(X_1, X_2)$] and illustrates that slopes (derivatives) in the direction of either the X_1 axis or the X_2 axis may be examined. Slopes in the X_1 direction (parallel to the X_1 axis) appear if we focus on any specific constant value of X_2; slopes in the X_2 direction appear if we focus on any particular constant value of X_1. The slope in the X_1 direction is called the **partial derivative of Y with respect to** X_1 and is denoted $\partial Y/\partial X_1$. The slope in the X_2 direction is called the **partial derivative of Y with respect to** X_2 and is denoted $\partial Y/\partial X_2$. The d used in the ordinary derivative has been replaced with ∂ to emphasize that these are *slopes in only one particular direction*. To obtain the maximum

Figure 3.3   Partial Derivatives.

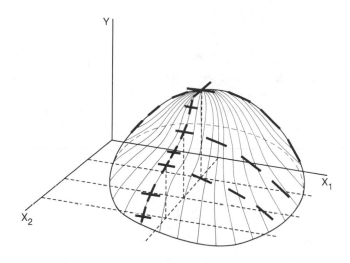

of the illustrated function Y, we would have to seek out the values of X_1 and X_2 that give zero partial derivatives (zero slopes) in both the X_1 and X_2 directions. Though several of the slopes in the X_1 direction are zero, only one of these slopes corresponds to the maximum of the illustrated function.

The procedures for obtaining partial derivatives are precisely the procedures already developed for simple derivatives, except that in calculating each partial derivative the "other variables" are treated as if they were constants. That is, if we want the slopes of the function Y in the X_1 direction, we are inquiring about the partial derivative of Y with respect to X_1, and we would treat X_2 as if it were as constant. For example, if $Y = -X_1^2 + 2X_1 - X_2^2$, the partial derivative of Y with respect to X_1 is $\partial Y/\partial X_1 = -2X_1 + 2 + 0$. The $-2X_1$ term comes from the rules illustrated in Eqs. 3.93 and 3.94, the 2 comes from Eq. 3.90, and the 0 comes from treating X_2 as a constant and using Eq. 3.89. The partial derivative of Y with respect to X_2 is $\partial Y/\partial X_2 = 0 + 0 - 2X_2$. The zeros come from rule 3.89 with X_1 treated as a constant, and the final term comes from rules 3.93 and 3.94.

The maximum value of Y for this function appears when both the partial derivatives (slopes) equal zero. Simple rearrangements of the equations implied by setting $\partial Y/\partial X_1 = -2X_1 + 2 = 0$ and $\partial Y/\partial X_2 = -2X_2 = 0$ show that the maximum occurs when $X_1 = 1$ and when $X_2 = 0$. The actual maximum Y value can be obtained by inserting these values of X_1 and X_2 into the original equation for Y.

If Y is a function of three or more variables, the same procedures apply: Y will have slopes (partial derivatives) parallel to (in the direction of) each of the constituent variables, and these partial derivatives can be obtained from the basic rules for simple derivatives in Section 3.2.1 with all the other variables treated as if they were constants. The maximum or minimum in any continuous multivariate curvilinear function of Y can be obtained by locating the set of values that simultaneously make all the slopes in all the different directions equal to zero.[3]

3.2.3 Derivatives and Matrix Algebra

The previous section allows Y to be a function of multiple variables and examines the partial derivatives of Y with respect to each of these other variables (i.e., the slope of Y in the direction of each of the variables). Since matrix algebra is well suited for keeping track of multiple entities, partial derivatives can often be conveniently expressed in matrix notation. For example, the equation $Y = b_1X_1 + b_2X_2 + b_3X_3$, can be compactly represented as either $Y = \mathbf{b}'\mathbf{x}$ or $\mathbf{x}'\mathbf{b}$ if $\mathbf{b}'$ is defined as $[b_1, b_2, b_3]$ and $\mathbf{x}'$ is defined as $[X_1, X_2, X_3]$.

The partial derivatives of Y with respect to each of the X variables are $\partial Y/\partial X_1 = b_1$, $\partial Y/\partial X_2 = b_2$, and $\partial Y/\partial X_3 = b_3$. If we further agree to represent this series of partial derivatives of Y with respect to all the X variables as a column vector, we can denote this as

$$\frac{\partial Y}{\partial \mathbf{x}} = \begin{bmatrix} \dfrac{\partial Y}{\partial X_1} \\ \dfrac{\partial Y}{\partial X_2} \\ \dfrac{\partial Y}{\partial X_3} \end{bmatrix} = \begin{bmatrix} b_1 \\ b_2 \\ b_3 \end{bmatrix} \qquad 3.96$$

If we replace Y with its alternative matrix representations, Eq. 3.96 shows that

$$\frac{\partial Y}{\partial \mathbf{x}} = \frac{\partial \mathbf{x}'\mathbf{b}}{\partial \mathbf{x}} = \frac{\partial \mathbf{b}'\mathbf{x}}{\partial \mathbf{x}} = \mathbf{b} \qquad 3.97$$

If there were several Y's, each created as a linear composite of the same set of X's, we could denote the set of Y's as the column vector $\mathbf{y}$, and the successive sets of $\mathbf{b}$ coefficients as the rows of the matrix $\mathbf{B}'$ so that

$\mathbf{y} = \mathbf{B}'\mathbf{x}$. By direct extension from Eq. 3.97 the partial derivatives of these multiple Y's will appear as successive columns of partial derivatives, so

$$\frac{\partial \mathbf{y}}{\partial \mathbf{x}} = \frac{\partial \mathbf{B}'\mathbf{x}}{\partial \mathbf{x}} = \mathbf{B} \qquad 3.98$$

The mathematics of matrix derivatives can be extended well beyond this initial formulation (cf. Lawley and Maxwell, 1971:134ff; Nel, 1980; Searle, 1982), but for our purposes it is sufficient to note that we can obtain partial derivatives of the elements in matrices and vectors, and that the matrix procedures are merely compact representations of ordinary partial derivatives (directional slopes). Matrix procedures are used in LISREL's calculations and in footnote 3 to Chapter 5, but for most of our purposes nonmatrix representations of partial derivatives (as developed in Section 3.2.2) are sufficient.

We have now completed the arsenal of mathematics required to understand LISREL, and hence we turn to LISREL itself in Chapter 4.

Notes

1.   Premultiplying both sides of **PAQ** $= \mathbf{I}$ by **Q** and postmultiplying both sides by $\mathbf{Q}^{-1}$ [which must exist because each of the elementary transformations has an inverse (cf. Searle, 1982)] gives $\mathbf{QPAQQ}^{-1} = \mathbf{QIQ}^{-1}$. Now **Q** and its inverse cancel twice to give **(QP)A** $= \mathbf{I}$. In this form it is obvious that **QP** is the inverse of **A**. Hence, we could use row and column transformations to reduce **A** to an identity matrix, with the product of the overall column and row transformations giving the inverse of **A**. Again, this procedure only works if there are row and column elementary operations that transform **A** into **I**.

2.   Two further ways of saying this are that the matrix has an *eigenvalue of zero* (Hammarling, 1970:10)—$|\mathbf{A}| = \lambda_1\lambda_2\lambda_3 \cdots \lambda_n)$—and, for symmetric matrices, that the matrix is not *positive definite* (Morrison, 1976:62). If LISREL reports that an estimated covariance matrix is not positive definite, this may mean that a variance estimate is negative (which is unacceptable), that a covariance is so large that the correlation implied by Eq. 1.54 is greater than 1 (also unacceptable), that one row of the matrix is entirely zeros (which *may* be acceptable, cf. Figure 7.19), or that one row of the matrix is a linear function of another row (which *may* be acceptable).

3.   Actually, two conditions are required before we can state that we have located a minimum: (1) The slope (partial derivative) must be zero. (2) The ends of the curve must be pointing upward. A zero slope corresponds to a maximum if the ends of the curve point downward. Taking the derivative of a derivative (i.e., a second-order derivative) can tell us if the value of X giving a zero slope corresponds to a minimum or a maximum, if this information is not directly available from visual inspection of a graph of the function (cf. Protter and Morrey, 1964, or any introductory calculus text). If the second-order derivative $\partial^2 Y / \partial X^2$ is negative at the value of X providing zero slope, that horizontal slope locates a maximum of the function. If the second-order derivative $\partial^2 Y / \partial X^2$ is positive, the horizontal

slope locates a minimum. For our example, the first-order derivative is $2x$, and the derivative of this derivative (the second-order derivative) is 2. Since this value is positive, no matter what x value provides the horizontal slope, that horizontal slope must locate a minimum.

4.   The procedure of inserting a particular function in Eq. 3.83 and then solving for the limiting value is called **differentiating** the function.

Chapter 4

In the Beginning

4.1 A New Way of Thinking:
Latent versus Measured Variables
in Causal Models

The social sciences have long recognized that the values of variables recorded during data collection do not correspond exactly to the variables of theoretical interest. The discrepancy between desired and achieved measurements is typically addressed under the topics of reliability and validity, where reliability refers to the stability of replicated measurements, and validity refers to whether the measure really measures what it is supposed to measure, as opposed to measuring some similar yet conceptually distinct variable. Indeed, the issue of measurement is so central that both the practitioners and philosophers of science often treat investigation of measurement issues as synonymous with scientific advancement (cf. Blalock, 1982).

Unfortunately, discussions of measurement typically degenerate into listing the types of reliability and validity, which amounts to listing various styles of replication (test-retest, inter-item, split-half) and strategies for checking one's measures (face validity, construct validity, and criterion or predictive validity). These discussions often stand apart from the other stages of research: theorizing, model development, model estimation, and discussion of results. LISREL integrates measurement concerns with structural equation modeling by incorporating both latent theoretical concepts and observed or measured indicator variables into a single structural equation model. Furthermore, knowledge of the methodological adequacy of the data gathering process and the quality of particular questionnaire items (measurement instruments) can be directly incorporated into LISREL models by specifying (fixing) a specific proportion of the variance in an indicator to be error variance (Section 4.5). Thus, in LISREL, measurement concerns become integrated with model development, estimation, evaluation, and interpretation (cf. Bohrnstedt, 1983).

The following section describes a general notation for keeping track of modeled concepts, the indicators of those concepts, and the postulated causal connections among the concepts. Section 4.3 illustrates how a real model can be represented within this notational framework. Section 4.4 develops the mathematical theory demonstrating how structural relations among the concepts and the indicator specifications become unified in deriving the overall model's predictions for what should be observed as covariances among the observed indicator variables. Section 4.5 advances the discussion of measurement quality by distinguishing between the reliability of measurements and measurement scales.

Throughout these discussions keep in mind that the means, variances, and covariances of the observed indicator variables are calculable and therefore make up the known data. The means, variances, and covariances of the concepts, the structural coefficients among the concepts, and the structural coefficients linking the concepts to the indicators are unknown and must be estimated. The details of estimation are postponed until Chapter 5.

4.2 LISREL is Greek to Me!

LISREL's three basic equations (containing four matrices of coefficients) and four additional covariance matrices provide a general mold into which almost any model can be fit. The matrices are denoted by Greek letters, but there is no need to memorize the Greek alphabet (Figure 4.1). Instead, try to add a new letter to your vocabulary as the use and meaning of each letter is discussed. Figure 4.2 summarizes the eight matrices that compose the general model, and this figure too should be repeatedly examined as each successive equation or matrix is discussed in the text. (Figure 4.2 is reproduced on the inside front cover for easy reference.)

The first of the three basic matrix equations demands that we locate all the concepts or latent variables in our model. Each concept is then

Figure 4.1   The Greek Alphabet.[1]

Uppercase	Lowercase	Name	Sound	English Equivalent
A	α	alpha	card	a
B	β	beta	ball	b
Γ	γ	gamma	got	g
Δ	δ	delta	dot	d
E	ϵ	epsilon	get	ê
Z	ζ	zeta	maze	z
H	η	eta	feather	e
Θ	θ	theta	think	th
I	ι	iota	beat	i
K	κ	kappa	kind	k,c
Λ	λ	lambda	like	l
M	μ	mu	man	m
N	ν	nu	net	n
Ξ	ξ	xi	axe	x
O	o	omicron	over	ô
Π	π	pi	pie	p
P	ρ	rho	red	r
Σ	σ	sigma	sell	s
T	τ	tau	town	t
Υ	υ	upsilon	youth	y
Φ	ϕ	phi	phone	ph
X	χ	chi	loch	ch
Ψ	ψ	psi	oops	ps
Ω	ω	omega	stone	o

[1] The names and pronunciations of these characters reflect current usage by statisticians/scientists, and they differ slightly from colloquial Greek.

classified as endogenous or exogenous. If a concept is directly caused or influenced[1] by any of the other concepts, it is classified as **endogenous**. If a concept always acts as a "cause" and never as an "effect," then it is **exogenous,** and fluctuations in the values of this concept are not to be explained by this model (though they may be used to explain fluctuations in the values of the endogenous concepts). Thus we locate the direct causal effects that are of interest. If a concept ever appears as the effect or dependent variable in a directed relationship of interest, that concept is endogenous. If the concept never functions as a dependent variable in our model (that is, if the fluctuations in the concept arise from causal structures that we have no intention of modeling), that concept is exogenous.

We then list the names of all the endogenous concepts in a column vector η (eta) and names of all the exogenous concepts in a column vector ξ (ksi). The number of endogenous concepts will be denoted m, so η will be an $m \times 1$ vector; the number of exogenous concepts will be denoted n, so ξ is an $n \times 1$ vector. The total number of concepts (endogenous plus exogenous) is $m + n$.

The first basic equation encapsulates all the postulated *direct effects* among the *concepts* and is[2]

$$\eta = \mathbf{B}\eta + \mathbf{\Gamma}\xi + \zeta \qquad 4.1$$

$$(m \times 1) \quad (m \times m)(m \times 1) \quad (m \times n)(n \times 1) \quad (m \times 1)$$

If the elements of $\mathbf{B}$ and $\mathbf{\Gamma}$ are denoted by the corresponding lowercase greek letters β (lowercase beta) and γ (lowercase gamma), we can rewrite Eq. 4.1 as

$$
\begin{bmatrix} \eta_1 \\ \eta_2 \\ \eta_3 \\ \cdot \\ \cdot \\ \cdot \\ \eta_m \end{bmatrix}
=
\begin{bmatrix} 0 & \beta_{12} & \beta_{13} & \cdots \\ \beta_{21} & 0 & \beta_{23} \\ \beta_{31} & \beta_{32} & 0 \\ \cdot & \cdot & & \cdot \\ \cdot & \cdot & & & \cdot \\ \cdot & \cdot & & & & 0 \end{bmatrix}
\begin{bmatrix} \eta_1 \\ \eta_2 \\ \eta_3 \\ \cdot \\ \cdot \\ \cdot \\ \eta_m \end{bmatrix}
+
\begin{bmatrix} \gamma_{11} & \gamma_{12} & \gamma_{13} & \cdots \\ \gamma_{21} & \gamma_{22} & \gamma_{23} \\ \cdot \\ \cdot \\ \cdot \end{bmatrix}
\begin{bmatrix} \xi_1 \\ \xi_2 \\ \xi_3 \\ \cdot \\ \cdot \\ \xi_n \end{bmatrix}
+
\begin{bmatrix} \zeta_1 \\ \zeta_2 \\ \zeta_3 \\ \cdot \\ \cdot \\ \zeta_m \end{bmatrix}
\qquad 4.2
$$

$$(m \times 1) \qquad (m \times m) \qquad (m \times 1) \qquad (m \times n) \qquad (n \times 1) \qquad (m \times 1)$$

In this form it is obvious that the number of endogenous and exogenous concepts, and hence the sizes of all the matrices and vectors, are determined by our substantive conceptual model.

Recalling the steps for doing matrix multiplication and addition, we see that the first endogenous concept η_1 in the vector to the left of the equal sign must equal

$$\eta_1 = 0 + \beta_{12}\eta_2 + \beta_{13}\eta_3 \ldots \beta_{1m}\eta_m + \gamma_{11}\xi_1 + \gamma_{12}\xi_2 \ldots \gamma_{1n}\xi_n + \zeta_1 \qquad 4.3$$

Since $\mathbf{B}$ and $\mathbf{\Gamma}$ are matrices containing structural coefficients (numbers we will eventually estimate), Eq. 4.3 expresses the first endogenous concepts η_1 as a linear combination of *all the other conceptual variables in our model* and an "error variable" ζ (zeta) that is comparable to e in ordinary

Figure 4.2 Summary of the General Structural Equation Model.

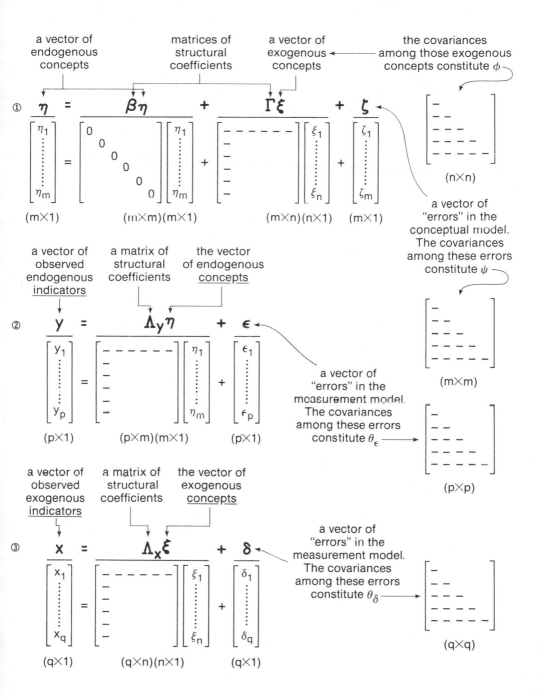

regression. That is, if we knew an individual's scores (values) on all the other concepts, and if we knew the numerical values of the β's, γ's, and ζ, we could calculate that individual's score on concept η_1. An individual's score on concept η_1 would equal β_{12} times that individual's score on concept η_2 plus β_{13} times that individual's score on concept η_3 plus . . . . Similarly, the second row of $\mathbf{B}$ and $\mathbf{\Gamma}$ gives the coefficients allowing us to calculate an individual's score on η_2 if that individual's scores on all the other conceptual variables and another error variable ζ_2 are known. Convince yourself that the equation for the second endogenous concept (η_2) will use the second row of coefficients in $\mathbf{B}$ and $\mathbf{\Gamma}$, all the other η's, all the ξ's, and the second ζ. Note that the zeros on the diagonal of $\mathbf{B}$ merely keep us from claiming that any given η has a *direct* effect on itself.

If our theory predicts that some of the endogenous variables are *not* directly influenced by specific endogenous or exogenous variables, this is accommodated by inserting zero entries in $\mathbf{B}$ or $\mathbf{\Gamma}$. For example, if concept η_1 is *not* influenced by concept η_2, β_{12} would be zero. (Write out the equation for η_2 that parallels Eq. 4.3 and convince yourself that γ_{23} would be zero if the third exogenous concept had no influence on the second endogenous concept.)

The entries in $\mathbf{B}$ and $\mathbf{\Gamma}$ are structural coefficients that express the endogenous concepts as linear combinations of all the other concepts, much as in Chapter 1 where a "new" variable was created as a linear combination of other variables. The numerical value of the coefficients in $\mathbf{B}$ and $\mathbf{\Gamma}$ are typically unknown, and one of our major tasks will be to obtain estimates of these structural coefficients. This parallels the task of estimating regression slopes.

Unlike Eq. 4.1, which links only conceptual level variables to one another, the two remaining fundamental equations *link the conceptual variables to their observed indicators*. One equation links the endogenous concepts to the endogenous indicators (Eq. 4.4), and the other equation links the exogenous concepts to the exogenous indicators[3] (Eq. 4.6). The names of the variables taken as indicators of the endogenous concepts are recorded in a column vector $\mathbf{y}$. Since we occasionally have several indicators of a single concept, it is possible for there to be more measured variables than concepts. That is, the number of endogenous indicators (denoted p) can be larger than m (the number of endogenous concepts). Similarly, the names of the q indicators of the exogenous concepts will be recorded in the column vector $\mathbf{x}$, and q may be larger than n if there are more exogenous indicators than exogenous concepts.

The values of observed indicator variables ($\mathbf{x}$'s and $\mathbf{y}$'s) are thought to arise from the underlying concepts, so we can express the observed $\mathbf{x}$ and $\mathbf{y}$ variables as linear combinations of the conceptual variables.

$$\mathbf{y} = \mathbf{\Lambda}_y\boldsymbol{\eta} + \boldsymbol{\epsilon} \qquad\qquad 4.4$$

$$
\begin{bmatrix} y_1 \\ y_2 \\ y_3 \\ \vdots \\ y_p \end{bmatrix}
=
\begin{bmatrix}
\lambda^y_{11} & \lambda^y_{12} & \lambda^y_{13} & \cdot & \cdot & \cdot \\
\lambda^y_{21} & \lambda^y_{22} & \lambda^y_{23} & \cdot & \cdot & \cdot \\
\cdot & & & \cdot & \cdot & \cdot \\
\cdot & & & \cdot & \cdot & \cdot \\
\cdot & & & \cdot & \cdot & \cdot
\end{bmatrix}
\begin{bmatrix} \eta_1 \\ \eta_2 \\ \eta_3 \\ \vdots \\ \eta_m \end{bmatrix}
+
\begin{bmatrix} \epsilon_1 \\ \epsilon_2 \\ \epsilon_3 \\ \vdots \\ \epsilon_p \end{bmatrix}
\qquad 4.5
$$

$(p \times 1)$ $(p \times m)$ $(m \times 1)$ $(p \times 1)$

$$\mathbf{x} = \mathbf{\Lambda}_x\boldsymbol{\xi} + \boldsymbol{\delta} \qquad\qquad 4.6$$

$$
\begin{bmatrix} x_1 \\ x_2 \\ x_3 \\ \vdots \\ x_q \end{bmatrix}
=
\begin{bmatrix}
\lambda^x_{11} & \lambda^x_{12} & \lambda^x_{13} & \cdot & \cdot & \cdot \\
\lambda^x_{21} & \lambda^x_{22} & \lambda^x_{23} & \cdot & \cdot & \cdot \\
\cdot & & & \cdot & \cdot & \cdot \\
\cdot & & & \cdot & \cdot & \cdot \\
\cdot & & & \cdot & \cdot & \cdot
\end{bmatrix}
\begin{bmatrix} \xi_1 \\ \xi_2 \\ \xi_3 \\ \vdots \\ \xi_n \end{bmatrix}
+
\begin{bmatrix} \delta_1 \\ \delta_2 \\ \delta_3 \\ \vdots \\ \delta_q \end{bmatrix}
\qquad 4.7
$$

$(q \times 1)$ $(q \times n)$ $(n \times 1)$ $(q \times 1)$

The entries in the $\mathbf{\Lambda}$ (lambda) matrices are lowercase lambdas (λ's) and represent structural coefficients for which we will again obtain numerical estimates. The symbols ϵ (epsilon) and δ (delta) are error variables specifying the cumulative effects of excluded variables and purely random measurement errors on the observed $\mathbf{x}$ and $\mathbf{y}$ variables, again paralleling the error variable e in regression. As in regression, we will be routinely assuming that all three error variables (ϵ, δ, and ζ) have mean zero.

So far we have introduced four matrices whose entries are structural coefficients that we will eventually estimate ($\mathbf{B}$, $\mathbf{\Gamma}$, $\mathbf{\Lambda}_y$, $\mathbf{\Lambda}_x$). Four other matrices of coefficients to be estimated are required to complete the specification of the general model. The remaining matrices $\mathbf{\Phi}$ (phi), $\mathbf{\Psi}$ (psi), $\mathbf{\Theta}_\epsilon$ (theta sub epsilon), and $\mathbf{\Theta}_\delta$ (theta sub delta), are all variance/covariance matrices, and all but Φ are variance/covariance matrices of error variables.

Matrix Φ (phi) contains the covariances among the exogenous concepts and parallels the correlations among the predictors in multiple regression. Matrix Ψ (psi) contains the covariances among the ζ's, errors, or "structural disturbance terms" influencing the endogenous concepts (Eq. 4.2). That is, if some of the concepts excluded from our model (and hence whose effects contribute to the structural disturbance terms ζ) happen to influence two of the endogenous concepts, this common omitted cause would imply a correlation between the ζ's for those concepts (recall Section 1.4.1).

Similarly, if some omitted common cause contributes to two of the measurement error variables in the equations for the observed y variables (ϵ's in Eq. 4.5), this would imply that the measurement error variables (ϵ's) would be correlated and hence would have some covariance to be recorded in Θ_ϵ. By the same argument, omission of a common cause from the determinants of the measured x variables would result in covariance among the δ variables in Eq. 4.7 and hence in a covariance term to be recorded in Θ_δ.

That these last four matrices are covariance matrices lets us make several useful observations. First, they are square matrices. Second, the diagonal elements should always be positive since they are variances and variances must be positive. Third, the matrices are symmetric, so we can eliminate the writing of redundant terms by recording only the lower triangular elements of these matrices. Fourth, if the variables are standardized, these matrices become correlation matrices (Section 1.3.5).

The foregoing general description is summarized in Figure 4.2. It has been my experience, however, that this description remains opaque until a real model is placed in the context of the three general equations (4.1, 4.4, 4.6) and the four additional covariance matrices. To this end, we develop an illustrative example in the next section.

4.3 A Real Model: Smoking Behavior and Antismoking Acts

Figure 4.3 depicts a model investigating potential causal links between smoking behavior and both antismoking (AS) views and AS actions while controlling for the exogenous concepts sex, age, education, and strength of religious convictions. This model and its numerous modifications ground the discussions in the next several chapters, but you probably will not feel you fully understand this model until you look back after reaching

Figure 4.3   A Smoking Model.

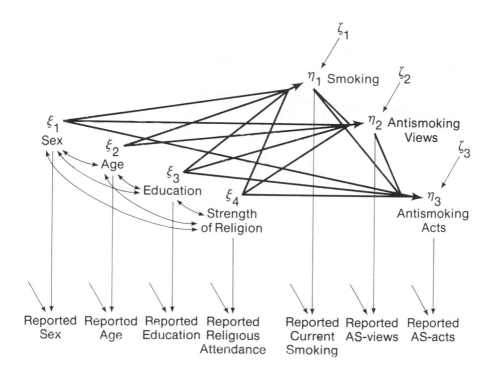

$$\eta \;=\; \beta\eta \;+\; \Gamma\xi \;+\; \zeta$$

$$\begin{bmatrix} \eta_1 \\ \eta_2 \\ \eta_3 \end{bmatrix} = \begin{bmatrix} 0 & 0 & 0 \\ ? & 0 & 0 \\ ? & ? & 0 \end{bmatrix} \begin{bmatrix} \eta_1 \\ \eta_2 \\ \eta_3 \end{bmatrix} + \begin{bmatrix} ? & ? & ? & ? \\ ? & ? & ? & ? \\ ? & 0 & ? & ? \end{bmatrix} \begin{bmatrix} \xi_1 \\ \xi_2 \\ \xi_3 \\ \xi_4 \end{bmatrix} + \begin{bmatrix} \zeta_1 \\ \zeta_2 \\ \zeta_3 \end{bmatrix}$$

Chapter 7. The data for estimating and testing this model are the responses of 432 randomly selected adults from the city of Edmonton, who in 1984 provided information to the questions presented in Figure 4.4. This data set is drawn upon repeatedly in the following chapters, so you might familiarize yourself with both the questionnaire items and the summary statistics in Figure 4.5.

The model has seven concepts—four exogenous (sex, age, education, and religion) and three endogenous (smoking, AS views, and AS acts). Each concept has a single indicator, which is provided the "same" name but is preceded by the word "Reported" to remind us that the values of the indicator variables are the actual responses provided by the respondents. The values of the concepts, which we do not know, are the corresponding true scores for these individuals—that is, the scores these individuals would have had if no errors had been introduced by coding mistakes, by the misunderstanding of questions, or by purposeful deception or exaggeration.

The conceptual variable smoking (η_1) will be given a value 0 if the person is a nonsmoker, and 1 if the person smokes. The four arrows leading to η_1 from the exogenous variables represent our suspicion that each of the four ξ's causally influences smoking behavior, although the strength of the influences of these variables will be determined only by model estimation. If our suspicions prove unfounded and none of these exogenous variables influence smoking behavior, the estimates of the structural coefficients corresponding to these arrows would be zero, and all the variance in smoking behavior would be attributed to as yet unidentified variables collectively represented by the error variable ζ_1.

The mechanisms underlying these suspected causal links are as follows: for sex (ξ_1)—males have traditionally smoked more than females; for age (ξ_2)—we hope that recent government campaigns have produced less smoking among the young; for education (ξ_3)—the better educated tend to know better because they are more likely to read about the effects of smoking and to be persuaded by technical evidence and logical argumentation; and for strength of religion (ξ_4)—religious persons are postulated to smoke less because most religions disapprove of defilement of the body and worldly dependencies.

The second endogenous concept, η_2, or AS views, was measured as the number of provisions of a city AS bylaw with which the respondent agreed. Agreement with all eight bylaw provisions indicates strong AS views, whereas agreement with fewer provisions indicates weaker AS views.

The suspected causal mechanisms justifying the arrows leading to AS views (and the corresponding structural coefficients in the equation for AS views) are as follows. Smokers should display weaker AS views (agree with fewer items) because the bylaw restrictions impede them the most and because agreement with AS statements would create dissonance for

Figure 4.4   Indicators for the Smoking Model.

Complete data on the following variables were obtained for 432 of the 452 adults (18 years +) responding to a random sample of households in Edmonton, Alberta, Canada. These questions were a part of the Edmonton Area Study (1984) conducted by the Population Research Laboratory of the Department of Sociology, University of Alberta. Details of the survey methodolgy are available in Kennedy and Kinzel (1984), and access to the data can be arranged by writing Dr. Harvey Krahn, Director of the Population Research Laboratory, Department of Sociology, University of Alberta, Edmonton, Alberta, Canada T6G 2H4.

Concept	Indicator(s)	Interview Item and/or Coding
Sex	Sex Reported	1=male, 2=female
Age	Age Reported	in years, must be 18 or over
Education	Educ. Reported	1=no schooling, 2=elementary incomplete, 3=elementary complete, 4=jr. high incomplete, 5=jr. high complete, 6=high school incomplete, 7=high school complete, 8=nonuniversity (vocational/technical) incomplete, 9=nonuniversity complete, 10=university incomplete, 11=university diploma/certificate, 12=bachelors degree, 13=medical degree (vets, doctors, dentists), 14=masters, 15=doctorate
Religion	Attendance	How often do you attend religious services? 0=never, 1=less than once a year, 2=about once a year, 3=several times a year, 4=about once a month, 5=2 or 3 times a month, 6=nearly every week, 7=several times a week.
	Strength	Would you call yourself a strong or not very strong _____ (stated religious preference inserted)? 1=strong, 2=not very strong, 3=somewhat strong (3 used only if volunteered).
	Importance	How much do you agree or disagree with this statement: My religion is important to me now. 1=strongly disagree to 7=strongly agree.
	God Exists	Respondents selected the answers coming closest to their own personal opinion about several listed religious matters. The first item listed was: There is *no* definite proof that God exists. 1=certainly true, 2=probably true, 3=uncertain if true or false, 4=probably false, 5=certainly false.

Figure 4.4   Continued.

Concept	Indicator(s)	Interview Item and/or Coding
Smoking	Current Smoker	0=no, 1=yes
	Amount Smoked	Created from number of cigarettes, cigarello's, cigars, and pipes smoked per day. 0=none, 1=light, 2=moderate, 3=heavy
	Smoking History	0=never smoked, 1=former smoker, 2=current smoker
	Physiological Strain	0=none, 1=residual from being a former smoker, 2=current light, 3=current moderate, 4=current heavy.

Respondents were informed of eight smoking activities that are illegal according to a city bylaw: smoking in public areas of stores, smoking in hospitals (except private rooms), smoking in service lines (mall food counters, bank service counters, etc.), smoking in elevators or on escalators, smoking in office reception areas (except in specially designated areas), carrying a lighted cigarette in a nonsmoking area even if one does not puff on it, stores failing to post no-smoking signs, restaurants failing to mark some seating area as nonsmoking.

For each of these activities respondents were asked if (1) they *knew* it was illegal, (2) if they *thought it should be* illegal, and (3) if they *had ever asked someone to stop* this type of violation.

AS Views	Reported AS views	This is the number (out of eight possible items) that the respondent thought should be illegal. The few "don't know" responses were counted as .5.
AS Acts	Reported AS acts	This is the number (out of eight possible items) that the respondent reported as having "ever asked someone to stop." The few "don't know" responses were again counted as .5.

Figure 4.5   Statistics for the Indicators in the Smoking Model.[1]

Indicator	Mean	S.D.	Assessed Proportion of Error Variance
Sex Reported	1.49	0.500	0.01
Age Reported	38.45	15.7	0.05
Education Reported	8.18	2.63	0.10
Religion			
Attendance	2.74	2.16	0.10
Strength	1.64	0.454	0.05
Importance	4.39	2.11	0.05
God Exists	3.25	1.47	0.20
Smoking			
Current Smoker	0.444	0.497	0.10
Amount Smoked	0.970	1.19	0.10
Smoke History	1.23	0.782	0.10
Physiological Strain	1.75	1.42	0.10
Antismoking Views	6.62	1.82	0.05
Antismoking Acts	0.605	1.14	0.05

Correlation/Covariance

	Sex	Age	Educ.	Religion Attend	Religion Strength	Religion Import	Religion God-E	Smoking Current	Smoking Amount	Smoking History	Smoking P-strain	AS-Views	AS-acts
Sex	1.000												
	.250												
Age	.021	1.000											
	.167	246.898											
Educ.	.012	-.176	1.000										
	.016	-7.251	6.909										
R-attend	.129	.213	.041	1.000									
	.139	7.239	.233	4.657									
R-strength	-.043	-.173	-.021	-.473	1.000								
	-.010	-1.232	-.025	-.464	.206								
R-import	.142	.268	-.042	.641	-.532	1.000							
	.150	8.907	-.233	2.922	-.511	4.466							
R-God-E	.120	.047	-.152	.381	-.175	.396	1.000						
	.088	1.088	-.584	1.204	-.116	1.225	2.146						
S-current	.039	-.116	-.163	-.176	.145	-.072	-.035	1.000					
	-.010	-.907	-.210	.188	.033	-.076	-.026	.247					
S-amount	-.026	-.107	-.199	-.100	.110	.078	-.000	.912	1.000				
	-.015	-2.003	-.623	-.476	.065	-.190	-.053	.540	1.417				
S-history	-.089	-.119	-.120	-.192	.178	-.094	-.039	.885	.807	1.000			
	-.035	-1.467	-.248	-.324	.063	-.156	-.045	.344	.752	.612			
S-p-strain	-.059	-.116	-.179	-.198	.146	-.090	-.032	.900	.965	.916	1.000		
	-.042	-2.599	-.667	-.606	.094	-.269	-.067	.636	1.631	1.017	2.016		
AS-views	.168	.170	.078	.141	-.086	.134	.150	-.259	-.288	-.247	-.288	1.000	
	.153	4.850	.375	.553	-.071	.516	.401	-.234	-.625	-.352	-.745	3.314	
AS-acts	.078	-.076	.118	-.062	.003	-.029	.084	-.148	-.150	-.096	-.126	.140	1.000
	.044	-1.366	.357	-.154	.001	-.069	.141	-.084	-.203	-.085	-.203	.291	1.302

[1] Listwise deletion of missing cases reduced the case base from 452 to 432. Examination of each variable's distribution and comparison of the pairwise and listwise covariance matrices uncovered no substantial changes due to case deletions.

smokers. Females, the young, the more educated, and strongly religious persons should hold stronger AS views for the same reasons these persons are thought to avoid the behavior of smoking. The justifications these groups provide for why they are nonsmokers parallel the arguments they could advance in support of an AS bylaw.

The third endogenous concept AS acts (η_3), was measured as the number of provisions in the AS bylaw for which the respondent had taken direct personal action to ensure compliance—for example, by asking someone to stop smoking in a hospital or elevator (both illegal under the bylaw), or asking a restaurant proprietor to designate a nonsmoking seating section as required by law. Clearly, taking such AS actions (η_3) should depend upon one's current smoking behavior (η_1) and one's views (η_2). Over and above these causal mechanisms, however, males, the better educated, and strongly religious persons should be most prepared to engage in the implicit confrontations. We suspect age has no effect on η_3, since the vitality of youth should counterbalance the sophistication of age when it comes to taking affirmative action.

Note that in these justifications the causal effects of the exogenous concepts on AS views (η_2) are postulated as being effects *over and above the effects transmitted through the intervening variable smoking.* Similarly, the effects of the exogenous variables on AS acts (η_3) are considered to be effects over and above (in addition to) any effects transmitted via either η_1 or η_2.

The structural relations among the concepts in this model can be expressed in the form of the basic equation (4.1):

$$\eta = \mathbf{B}\eta + \mathbf{\Gamma}\xi + \zeta$$

$$\begin{bmatrix} \eta_1 \\ \eta_2 \\ \eta_3 \end{bmatrix} = \begin{bmatrix} 0 & 0 & 0 \\ ? & 0 & 0 \\ ? & ? & 0 \end{bmatrix} \begin{bmatrix} \eta_1 \\ \eta_2 \\ \eta_3 \end{bmatrix} + \begin{bmatrix} ? & ? & ? & ? \\ ? & ? & ? & ? \\ ? & 0 & ? & ? \end{bmatrix} \begin{bmatrix} \xi_1 \\ \xi_2 \\ \xi_3 \\ \xi_4 \end{bmatrix} + \begin{bmatrix} \zeta_1 \\ \zeta_2 \\ \zeta_3 \end{bmatrix} \qquad 4.8$$

A "?" has been entered in place of each structural coefficient to be estimated. If subscripted **B**'s and **Γ**'s are used to represent the unknown coefficients, this appears as

$$\begin{bmatrix} \eta_1 \\ \eta_2 \\ \eta_3 \end{bmatrix} = \begin{bmatrix} 0 & 0 & 0 \\ \beta_{21} & 0 & 0 \\ \beta_{31} & \beta_{32} & 0 \end{bmatrix} \begin{bmatrix} \eta_1 \\ \eta_2 \\ \eta_3 \end{bmatrix} + \begin{bmatrix} \gamma_{11} & \gamma_{12} & \gamma_{13} & \gamma_{14} \\ \gamma_{21} & \gamma_{22} & \gamma_{23} & \gamma_{24} \\ \gamma_{31} & 0 & \gamma_{33} & \gamma_{34} \end{bmatrix} \begin{bmatrix} \xi_1 \\ \xi_2 \\ \xi_3 \\ \xi_4 \end{bmatrix} + \begin{bmatrix} \zeta_1 \\ \zeta_2 \\ \zeta_3 \end{bmatrix} \qquad 4.9$$

Using the row-column subscript notation from matrix algebra results in the first subscript indexing the dependent endogenous η concept, and the second subscript indexes the causal concept. If the causal concept is exogenous (a ξ), the effect coefficient is a Γ; if the causal concept is another endogenous concept (η), the coefficient is a $\mathbf{B}$. Compare Figure 4.6, where the subscripted coefficients appear in the figure, with Eq. 4.9.

Figure 4.6   Structural Coefficients in the Model.

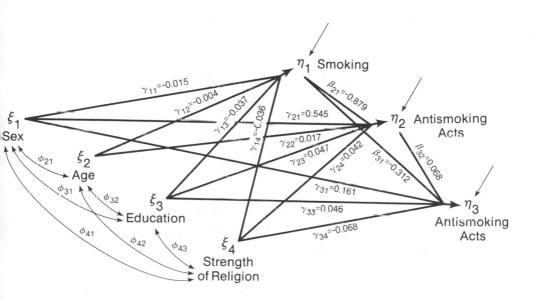

The **B** structural coefficients are "directed" in that each is associated with an arrow originating at an endogenous concept (η) and pointing toward another endogenous concept. The Γ coefficients are also directed in that they correspond to arrows originating at an exogenous concept (ξ) and pointing toward an endogenous concept (η).

The estimates of these coefficients (obtained via maximum likelihood estimation as discussed in Chapter 5) can be entered into the equation as follows.

$$
\begin{bmatrix} \eta_1 \\ \eta_2 \\ \eta_3 \end{bmatrix} = \begin{bmatrix} 0 & 0 & 0 \\ -.879 & 0 & 0 \\ -.312 & .068 & 0 \end{bmatrix} \begin{bmatrix} \eta_1 \\ \eta_2 \\ \eta_3 \end{bmatrix} + \begin{bmatrix} -.015 & -.004 & -.037 & .036 \\ .545 & .017 & .047 & .042 \\ .161 & 0 & .046 & -.068 \end{bmatrix} \begin{bmatrix} \xi_1 \\ \xi_2 \\ \xi_3 \\ \xi_4 \end{bmatrix} + \begin{bmatrix} \zeta_1 \\ \zeta_2 \\ \zeta_3 \end{bmatrix} \qquad 4.10
$$

Again, compare this to Figure 4.6. The full computer run creating these estimates is included as Appendix B, and the portion of that run providing the maximum likelihood estimates appears in Figure 4.7.

Figure 4.7   LISREL Estimates (Maximum Likelihood).

BETA

	smoking1	asviews	asacts
smoking1	0.0	0.0	0.0
asviews	-0.879	0.0	0.0
asacts	-0.312	0.068	0.0

GAMMA

	sex	age	educatio	R-attend
smoking1	-0.015	-0.004	-0.037	-0.036
asviews	0.545	0.017	0.047	0.042
asacts	0.161	0.0	0.046	-0.068

PHI

	sex	age	educatio	R-attend
sex	0.248			
age	0.166	234.553		
educatio	0.016	-7.282	6.218	
R-attend	0.139	7.269	0.233	4.191

PSI

smoking1	asviews	asacts
0.204	2.736	1.156

THETA EPS

smoking1	asviews	asacts
0.025	0.166	0.065

THETA DELTA

sex	age	educatio	R-attend
0.002	12.345	0.691	0.466

Each of the β and γ coefficients retains the usual interpretation that a unit change in the independent (causal) variable is expected to be accompanied by β or γ units of change in the dependent variable. Hence, being a smoker reduces by almost 1 the number of AS bylaw provisions with which the respondent agrees ($\beta_{21} = -.879$), and it reduces the number of AS acts by .312 ($\beta_{31} = -.312$). This reduction in acts is actually a substantial reduction because on average the respondents engaged in only about .61 of 8 possible AS actions. For each additional bylaw provision with which the respondents agreed, AS acts increased by only .068 ($\beta_{32} = .068$), so AS views display practically no transference into AS actions.

Since sex is coded 1=male, 2=female, females on average agree with about one-half more antismoking bylaw provision than males ($\gamma_{21} = .545$). Contrary to expectations, older persons are more likely to endorse the AS provisions ($\gamma_{22} = .017$), with about a 60-year age difference typically resulting in agreement with one additional AS provision. (Chapter 8 on indirect effects and Section 6.5 on standardized solutions provide further interpretations for these coefficients. Chapter 9 presents further evidence on the counterintuitive age effect.)

4.3.1 The Measurement Structure and Covariance Matrices for the Smoking Model

Turning to the measurement portion of the model, the matrix equations corresponding to Eqs. 4.4 and 4.6 for the smoking model are

$$\mathbf{y} = \Lambda_y \boldsymbol{\eta} + \boldsymbol{\epsilon}$$

$$\begin{bmatrix} y_1 \\ y_2 \\ y_3 \end{bmatrix} = \begin{bmatrix} 1 & 0 & 0 \\ 0 & 1 & 0 \\ 0 & 0 & 1 \end{bmatrix} \begin{bmatrix} \eta_1 \\ \eta_2 \\ \eta_3 \end{bmatrix} + \begin{bmatrix} \epsilon_1 \\ \epsilon_2 \\ \epsilon_3 \end{bmatrix}$$

4.11

$$\mathbf{x} = \Lambda_x \boldsymbol{\xi} + \boldsymbol{\delta}$$

$$\begin{bmatrix} x_1 \\ x_2 \\ x_3 \\ x_4 \end{bmatrix} = \begin{bmatrix} 1 & 0 & 0 & 0 \\ 0 & 1 & 0 & 0 \\ 0 & 0 & 1 & 0 \\ 0 & 0 & 0 & 1 \end{bmatrix} \begin{bmatrix} \xi_1 \\ \xi_2 \\ \xi_3 \\ \xi_4 \end{bmatrix} + \begin{bmatrix} \delta_1 \\ \delta_2 \\ \delta_3 \\ \delta_4 \end{bmatrix}$$

4.12

The 0's and 1's in the Λ matrices are not estimated but are fixed at these predetermined values. The role of the 0's is clear; they force the indicator variables to respond to only the single proper underlying concept for which a nonzero value appears. When respondents are asked to indicate their sex, we hope their responses are *not* influenced by their age or education but do reflect their true biological sex.

But the 1's do more than link specific indicators to specific concepts: they *set the scale on which the values of the underlying concepts are measured.* Since the concepts are hypothetical, we can give them any scale we wish, but it is simplest to give the concepts the same scales as the corresponding observed indicators. Thus, a unit change on the concept is scaled to be the same as a unit change on the indicator. The reason the 1's accomplish this comes directly from the interpretation of these 1's as structural coefficients. Note that each 1 is a structural coefficient indicating how many units an observed indicator (x or y) changes if the corresponding underlying concept (η or ξ) is changed 1 unit. The 1's guarantee that each unit of change in the concept corresponds to a unit of change in the indicator, and hence guarantees that the scales on which the concepts and indicators are measured have equally sized units.

Having discussed the four matrices of structural coefficients for this model ($\mathbf{B}$, $\mathbf{\Gamma}$, $\mathbf{\Lambda}_y$, $\mathbf{\Lambda}_x$), we can turn to the remaining four covariance matrices that complete the specification of the model. The $\mathbf{\Phi}$ (phi) matrix contains the variance/covariances among the four exogenous concepts (ξ's) and hence is a 4 × 4 matrix.

$$\mathbf{\Phi} = \begin{bmatrix} ? & 0 & 0 & 0 \\ ? & ? & 0 & 0 \\ ? & ? & ? & 0 \\ ? & ? & ? & ? \end{bmatrix} = \begin{bmatrix} \Phi_{11} & & & \\ \Phi_{21} & \Phi_{22} & & \\ \Phi_{31} & \Phi_{32} & \Phi_{33} & \\ \Phi_{41} & \Phi_{42} & \Phi_{43} & \Phi_{44} \end{bmatrix} \tag{4.13}$$

Locate these coefficients in Figures 4.6 and 4.3. For this model $\mathbf{\Phi}$ is estimated as

$$\mathbf{\Phi} = \begin{bmatrix} .248 & & & \\ .166 & 234.553 & & \\ .016 & -7.282 & 6.218 & \\ .139 & 7.269 & .233 & 4.191 \end{bmatrix} \tag{4.14}$$

The estimated covariances between the concepts are almost identical to the covariances among the corresponding indicators, but the variances of the concepts are slightly less than the variances of the corresponding indicators (cf. Figure 4.5). The reason for the reduced variances will become clear when we discuss measurement reliability in Section 4.5.

The remaining three matrices are variance/covariance matrices among the three types of error terms that appear in the model: errors in prediction of the endogenous concepts (η's) by the conceptual level variables (contained in Ψ); errors in measurement of the endogenous concepts (contained in Θ_ϵ); and errors in measurement of the exogenous concepts (contained in Θ_δ).

The model specifies that although some error is expected in the prediction of the endogenous concepts, no covariance is expected among these error variables. Thus Ψ (psi) is a diagonal matrix of the form

$$\Psi = \begin{bmatrix} ? & 0 & 0 \\ 0 & ? & 0 \\ 0 & 0 & ? \end{bmatrix} = \begin{bmatrix} \Psi_{11} & 0 & 0 \\ 0 & \Psi_{22} & 0 \\ 0 & 0 & \Psi_{33} \end{bmatrix} \qquad 4.15$$

and the actual estimates for this model are

$$\Psi = \begin{bmatrix} .204 & 0 & 0 \\ 0 & 2.736 & 0 \\ 0 & 0 & 1.156 \end{bmatrix} \qquad 4.16$$

The error variance corresponding to the prediction of η_1 (smoking), for example, is .204, which is about 92% of the variance in this concept. (This is $1 - R^2$ for the first equation, as reported on output page 10, or this value can be obtained directly from the standardized solution; see Section 6.5.) Hence most of the variance in this concept comes from sources other than the diagrammed causal variables, which is another way of saying that the exogenous concepts included in this model are weak predictors of smoking behavior. The prediction of antismoking views and antismoking acts is equally poor. The error variances in Eq. 4.16 make up 87% and 94% of the variances of these concepts, respectively, and hence merely 13% and 6% of the variance in these concepts are explained by the model.[4] (The stacked models in Chapter 9 improve these explained variances.)

The Θ_ϵ (theta epsilon) and Θ_δ (theta delta) matrices contain the error variances and covariances for measuring the endogenous and exogenous variables, respectively. Matrix Θ_ϵ is of the form

$$\Theta_\epsilon = \begin{bmatrix} ? & 0 & 0 \\ 0 & ? & 0 \\ 0 & 0 & ? \end{bmatrix} \qquad 4.17$$

and has actual values

$$\Theta_\epsilon = \begin{bmatrix} .025 & 0 & 0 \\ 0 & .166 & 0 \\ 0 & 0 & .065 \end{bmatrix} \qquad 4.18$$

Matrix Θ_δ is of the form

$$\Theta_\delta = \begin{bmatrix} ? & 0 & 0 & 0 \\ 0 & ? & 0 & 0 \\ 0 & 0 & ? & 0 \\ 0 & 0 & 0 & ? \end{bmatrix} \qquad 4.19$$

and has actual values

$$\Theta_\delta = \begin{bmatrix} .002 & 0 & 0 & 0 \\ 0 & 12.345 & 0 & 0 \\ 0 & 0 & .691 & 0 \\ 0 & 0 & 0 & .466 \end{bmatrix} \qquad 4.20$$

These values were not estimated but were fixed, as will be discussed in Section 4.5.

The most mysterious aspect of the foregoing is undoubtedly the transformation of the error variables ζ, ϵ, and δ from Eq. 4.1, 4.4, and 4.6 into the variance/covariance matrices Ψ, Θ_ϵ, and Θ_δ. Since an understanding of this transformation is central to how measurement reliability can be built into LISREL, we must consider this in detail. Fortunately, the moderately lengthy discussion required to clarify the disappearance of the error variables and the emergence of error covariance matrices overlaps another central aspect of structural equation modeling, namely, how and why specific structural models imply specific covariances among the observed variables. Section 4.4 addresses the larger issue of how a model implies a covariance matrix, but we interrupt this discussion at the appropriate spots to clarify precisely how the error terms (ζ, ϵ, δ) get converted into error variances and covariances (Ψ, Θ_ϵ, Θ_δ).

4.4 A Model Implies a Sigma (Σ)

The three basic equations, 4.1, 4.4, and 4.6, are structural equations indicating how the scores on some variable (η's, x's, or y's) can be obtained as the weighted sum of other variables (ξ's, η's, and error variables). From Sections 1.3 and 1.4 recall that if one or more variables can be created as weighted sums of some other "original" variables, we can express the variances and covariances of the created variables as functions of the structural equations linking the variables and the variances and covariances among the original variables. This leads us to ask: What is implied about the variances and covariances among the observed x and y indicators if these variables function in accordance with the structural equations 4.1, 4.4, and 4.6 (and the four extra matrices)?

Though it is not obvious, specific variances and covariances among the observed variables are predicted if we know (or have estimates of) the numerical values of the elements in the matrices $\mathbf{B}$, $\mathbf{\Gamma}$, $\mathbf{\Lambda}_y$, $\mathbf{\Lambda}_x$, $\mathbf{\Phi}$, $\mathbf{\Psi}$, $\mathbf{\Theta}_\epsilon$, and $\mathbf{\Theta}_\delta$. The first four are matrices of structural coefficients that come directly from the three basic equations. The remaining four are the extra matrices discussed earlier: $\mathbf{\Phi}$ is the variance/covariance matrix for the exogenous concepts; and $\mathbf{\Psi}$, $\mathbf{\Theta}_\epsilon$, and $\mathbf{\Theta}_\delta$ are matrices for the covariance among error terms in Eqs. 4.1, 4.4, and 4.6, respectively.

The *model-implied variances and covariances of the observed indicators* (which are recorded in a covariance matrix $\mathbf{\Sigma}$ (sigma) can be compared to the variances and covariances calculated from the data on the observed indicators. Comparison of the model's predictions ($\mathbf{\Sigma}$) with observed reality (the actual observed covariances) provides the fundamental basis for testing a model's adequacy (as discussed in Chapter 6) and for obtaining reasonable estimates of the model's coefficients (as discussed in Chapter 5).

To see how the prediction of specific variances/covariances arises, consider Eq. 4.1. We know $\mathbf{B}$ and $\mathbf{\Gamma}$ are matrices whose numerical entries are the amounts (weightings) by which the values of the variables on the right of the equation would be multiplied to obtain the values of the η variables on the left of the equation. The column vectors on the right (η, ξ, and ζ) are vectors of variable names indicating where an individual's true values on the various variables (concepts) would be inserted if we were to use this equation to calculate that individual's scores on the dependent η -variables (cf. the discussion of Eq. 2.2). Thus, if we imagine an individual having specific numerical values on all but the first of the η's, on all the ξ's, and on the first ζ error variable, the first of the equations implicit in 4.1 (namely, Eq. 4.3) gives that individual's score on variable η_1. Scores on the other η's could be obtained by using the second through last rows of $\mathbf{B}$ and $\mathbf{\Gamma}$ with the same set of ξ's, the other η's, and the appropriate ζ.

Similarly, in Eqs. 4.4 and 4.6, $\mathbf{\Lambda}_y$ and $\mathbf{\Lambda}_x$ are matrices of numbers (weightings), and the remaining vectors are the names of variables in whose place an individual's true scores (values) could be entered if we wanted to determine that individual's scores on the y or x variables on the left of these equations.

Thus the three basic equations describe the social forces governing the values of the various variables exhibited *by any individual*. If we assume that the same social forces (the structural coefficients in $\mathbf{B}$, $\mathbf{\Gamma}$, $\mathbf{\Lambda}_y$, $\mathbf{\Lambda}_x$, and error processes) govern the behavior of all N individuals on whom we have collected information, we can represent the data on all N individuals at once by replacing all the vectors of variable names with matrices containing the N sets of individuals' values as N successive column vectors. This change from vectors for a single individual to matrices for N

individuals does not change the overall form of the three basic equations. Convince yourself of this by examining the following enlarged versions of the basic equations. The columns containing slashes, dots, and stripes indicate where columns of values would be entered for the first, second, and last individuals (cases), respectively. For the first individual (slashes) the equations governing this individual's scores are identical to the equations that would appear if that individual's scores had been entered in Eqs. 4.1, 4.4, and 4.6. Equation 4.1 with N individuals is

$$\eta = \mathbf{B}\eta + \mathbf{\Gamma}\xi + \zeta$$

4.21

$(m \times N)$ $(m \times m)$ $(m \times N)$

$(m \times n)$ $(n \times N)$ $(m \times N)$

Equation 4.4 with N individuals is

$$\mathbf{y} = \mathbf{\Lambda}_y\eta + \epsilon$$

4.22

$(p \times N)$ $(p \times m)$ $(m \times N)$ $(p \times N)$

And Eq. 4.6 with N individuals is

$$\mathbf{x} = \Lambda_x \xi + \delta$$

4.23

$$\begin{bmatrix} \end{bmatrix} = \begin{bmatrix} \Lambda_x \end{bmatrix} \begin{bmatrix} \end{bmatrix} + \begin{bmatrix} \end{bmatrix}$$

$(q \times N)$ $\qquad$ $(q \times n)$ $\qquad$ $(n \times N)$ $\qquad$ $(q \times N)$

We are now ready to express the covariances among the *observed* variables as functions of the eight fundamental matrices. Three sets of covariances will interest us: the covariances among the $\mathbf{x}$'s, the covariances among the $\mathbf{y}$'s, and the covariances between the $\mathbf{x}$'s and $\mathbf{y}$'s.

4.4.1 Covariances among the x's

Consider first the covariances among the $\mathbf{x}$'s. From Section 3.1.1.1 recall that if the $\mathbf{x}$ scores for a set of individuals are arranged in columns, and if those scores are recorded as deviations from the means of the respective $\mathbf{x}$ variables, the covariances among the $\mathbf{x}$ variables specifying the rows of the data matrix can be obtained by

$$\begin{bmatrix} \text{Covariance} \\ \text{matrix for} \\ \text{the } \mathbf{x}\text{'s} \end{bmatrix} = \mathbf{E}(\mathbf{XX}') = (1/N)\mathbf{XX}'$$

4.24

But Eq. 4.23 of our model informs us that the observed $\mathbf{x}$ variables arise due to specific structural relations among the ξ and ζ variables, so we can replace the values in the $\mathbf{X}$ matrices with the structural equations specifying how those $\mathbf{X}$ values originated—that is, as a function of the underlying concepts and measurement errors in Eq. 4.23 (i.e., Eq. 4.6 for N persons).

$$\begin{bmatrix} \text{Covariances} \\ \text{among the } \mathbf{x}\text{'s} \end{bmatrix} = \mathbf{E}[(\Lambda_x \xi + \delta)(\Lambda_x \xi + \delta)']$$

4.25

By this simple sleight of hand we have already attained our goal of expressing the covariances among the observed $\mathbf{x}$'s (the left of Eq. 4.25) in terms of the underlying ξ concepts and the δ errors (the right side of the equation), but we will want to rearrange the right side into a more understandable form. Moving the transpose within the parentheses, multiplying out the various matrices, and moving the scalar (the multiplication by $1/N$ that provides the expected or average value) to a convenient portion of each of the resultant products give, respectively,

$$\begin{bmatrix} \text{Covariances} \\ \text{among the x's} \end{bmatrix} = E[\,(\Lambda_x\xi + \delta)(\xi'\Lambda_x' + \delta')\,] \qquad 4.26$$

$$\begin{bmatrix} \text{Covariances} \\ \text{among the x's} \end{bmatrix} = E[\,\Lambda_x\xi\xi'\Lambda_x' + \delta\xi'\Lambda_x' + \Lambda_x\xi\delta' + \delta\delta'\,] \qquad 4.27$$

$$\begin{bmatrix} \text{Covariances} \\ \text{among the x's} \end{bmatrix} = \Lambda_x[E(\xi\xi')]\Lambda_x' + [E(\delta\xi')]\Lambda_x' + \Lambda_x[E(\xi\delta')] + E(\delta\delta') \quad 4.28$$

Expressing some of these matrices in words helps make sense of the result.

$$\begin{bmatrix} \text{Covariances} \\ \text{among the x's} \end{bmatrix} = \Lambda_x \begin{bmatrix} \text{Covariances} \\ \text{among exogenous} \\ \text{concepts or } \Phi \end{bmatrix} \Lambda_x' + \begin{bmatrix} \text{Covariances} \\ \text{between } \delta\text{'s and} \\ \text{exogenous concepts} \end{bmatrix} \Lambda_x'$$

$$+ \Lambda_x \begin{bmatrix} \text{Covariances between} \\ \text{exogenous concepts} \\ \text{and } \delta\text{'s} \end{bmatrix} + \begin{bmatrix} \text{Covariances} \\ \text{among the } \delta \\ \text{errors or } \Theta_\delta \end{bmatrix} \qquad 4.29$$

If we assume the δ error terms are independent of the ξ variables (similar to assuming that e is independent of X in ordinary regression), the covariances between the ξ's and δ's become zero (Section 1.3.3), and this equation simplifies to

$$\begin{bmatrix} \text{Covariances} \\ \text{among the x's} \end{bmatrix} = \Lambda_x\Phi\Lambda_x' + 0 + 0 + \Theta_\delta$$

$$\qquad 4.30$$

$$\begin{bmatrix} \text{Covariances} \\ \text{among the x's} \end{bmatrix} = \Lambda_x\Phi\Lambda_x' + \Theta_\delta$$

which is the basic equation in factor analysis.

Surveying what we have accomplished, we note two basic phenomena. First, the matrix of error terms represented by δ has been replaced by Θ_δ (the fact we promised to draw to your attention). Thus, to discuss the implications of the structural equation model for the variances and covariances among the x's, *we do not need to know each of the individuals' error terms for each of the variables; all we need is the variance/covariance matrix for the error terms.* The averaging, or taking of expectations, implicit in the calculation of the covariances among the x's carries over to the δ in moving from Eq. 4.27 to Eq. 4.28 and subsequently implies that we need only Θ_δ but not δ to determine the specific covariances among the x's implied by the model.

Second, *any specific set of numerical values for the elements of Φ, Θ_δ, and Λ_x is sufficient to provide a specific prediction for the covariances among the indicators of the exogenous variables.* If these matrices are given specific numerical entries (the right side of 4.30), one and only one variance/covariance matrix for the x's results (the left side). Note the direct parallel to Section 1.4.1.1 and the equation expressing the variance of a newly created variable as a function of the covariances among the original variables and the structural coefficients (Eq. 1.26).

4.4.2 Covariances among the y's

What does the model imply about the covariances among the y's? As for the x's, we can begin by noting that the covariance among the y's must be of the form

$$
\begin{bmatrix} \text{Covariance} \\ \text{matrix for} \\ \text{the } y\text{'s} \end{bmatrix} = E(YY') = (1/N)YY' \tag{4.31}
$$

(cf. Eqs. 3.23 and 4.24). We now follow the same steps we used with the x's: first insert what the model claims about the formation of the y variables (namely, Eq. 4.22 which is the expanded version of 4.4), and then simplify as before.

$$
\begin{bmatrix} \text{Covariances} \\ \text{among the } y\text{'s} \end{bmatrix} = E[(\Lambda_y \eta + \epsilon)(\Lambda_y \eta + \epsilon)'] \tag{4.32}
$$

$$
= E[(\Lambda_y \eta + \epsilon)(\eta' \Lambda_y' + \epsilon')] \tag{4.33}
$$

$$
= E[\Lambda_y \eta \eta' \Lambda_y' + \epsilon \eta' \Lambda_y' + \Lambda_y \eta \epsilon' + \epsilon \epsilon'] \tag{4.34}
$$

$$
= \Lambda_y [E(\eta \eta')]\Lambda_y' + [E(\epsilon \eta')]\Lambda_y' + \Lambda_y [E(\eta \epsilon')] + E(\epsilon \epsilon') \tag{4.35}
$$

$$= \Lambda_y \begin{bmatrix} \text{Covariances} \\ \text{among the} \\ \text{endogenous concepts} \end{bmatrix} \Lambda_y' + \begin{bmatrix} \text{Covariances} \\ \text{between } \epsilon\text{'s and} \\ \text{endogenous concepts} \end{bmatrix} \Lambda_y'$$

$$+ \Lambda_y \begin{bmatrix} \text{Covariances between} \\ \text{endogenous concepts} \\ \text{and } \epsilon\text{'s} \end{bmatrix} + \begin{bmatrix} \text{Covariances} \\ \text{among the } \epsilon \\ \text{errors or } \Theta_\epsilon \end{bmatrix} \qquad 4.36$$

With the assumption that the measurement errors (ϵ's) are independent of the endogenous concepts (η's), the covariances between these entities become zero. If we define Θ_ϵ to be the covariance matrix among the ϵ's, Eq. 4.36 becomes

$$\begin{bmatrix} \text{Covariances} \\ \text{among the y's} \end{bmatrix} = \Lambda_y \begin{bmatrix} \text{Covariances} \\ \text{among the} \\ \text{endogenous concepts} \end{bmatrix} \Lambda_y' + \Theta_\epsilon \qquad 4.37$$

This parallels Eq. 4.30 and is again in the usual form for factor analysis. Did you notice the transition from the error variables ϵ's to error covariances in moving between Eqs. 4.34 and 4.35 and, hence, the replacement of the ϵ's by Θ_ϵ's?

It is tempting to provide the covariance matrix among the η's with its usual LISREL name (C) and stop at this short and understandable equation. We cannot stop here, however, since we would be left wondering about the origins of the covariances among the η's. Our model tells us specifically where these η covariances come from: they arise as a result of the structural equation 4.21 (expanded 4.1). That is, the covariances among the endogenous concepts are determined by the conceptual level model. Hence, if we really want to know what our whole model (the numerical entries in the eight basic matrices) states about the origins of the covariances between the y's, we must express the covariances among the η's as functions of the basic structural coefficients. Equation 4.37 reflects what the measurement structure portion of our model (Eqs. 4.4 or 4.22) says about the origin of the covariance between the y's, taking the covariances among the η's as "given." We must express these η covariances as functions of the basic structural coefficients before we can claim to have expressed the covariances among the y's as a consequence of the basic model.

The covariance matrix for the η's can be represented in the form

$$\begin{bmatrix} \text{Covariance} \\ \text{matrix for} \\ \text{the } \eta\text{'s} \end{bmatrix} = (1/N)\eta\eta' = E(\eta\eta') \qquad 4.38$$

Equation 4.21 (the expanded 4.1) provides for the calculation of η, but it has η's on both sides of the equation. We can rectify this by some simple rearranging:

$$\eta = B\eta + \Gamma\xi + \zeta \qquad 4.39$$

$$\eta - B\eta = \Gamma\xi + \zeta \qquad 4.40$$

$$(I - B)\eta = \Gamma\xi + \zeta \qquad 4.41$$

$$\eta = (I - B)^{-1}(\Gamma\xi + \zeta) \qquad 4.42$$

Assuming the inverse exists, we can insert this expression for η into the covariance Eq. 4.38,

$$\begin{bmatrix} \text{Covariances} \\ \text{among the } \eta\text{'s} \end{bmatrix} = E([(I - B)^{-1}(\Gamma\xi + \zeta][(I - B)^{-1}(\Gamma\xi + \zeta)]') \qquad 4.43$$

and simplify to another factor form reminiscent of our previous results.

$$\begin{bmatrix} \text{Covariances} \\ \text{among the } \eta\text{'s} \end{bmatrix} = E([\text{ same }][(\Gamma\xi + \zeta)'(I - B)^{-1'}]) \qquad 4.44$$

$$= E([\text{ same }][(\xi'\Gamma' + \zeta')(I - B)^{-1'}]) \qquad 4.45$$

$$= E[(I - B)^{-1}(\Gamma\xi + \zeta)(\xi'\Gamma' + \zeta')(I - B)^{-1'}] \qquad 4.46$$

$$= (I - B)^{-1}E[\Gamma\xi\xi'\Gamma' + \zeta\xi'\Gamma + \Gamma\xi\zeta' + \zeta\zeta'](I - B)^{-1'} \qquad 4.47$$

$$= (I - B)^{-1}[\Gamma(E(\xi\xi'))\Gamma' + (E(\zeta\xi'))\Gamma + \Gamma(E(\xi\zeta')) + E(\zeta\zeta')](I - B)^{-1'} \qquad 4.48$$

With the assumption that the ζ error terms are independent of the ξ's, with relabeling the covariance matrix of the ξ's as Φ, and with relabeling the covariance matrix for the ζ error terms Ψ, this becomes

$$\begin{bmatrix} \text{Covariances} \\ \text{among the } \eta\text{'s} \end{bmatrix} = (I - B)^{-1}(\Gamma\Phi\Gamma' + \Psi)(I - B)^{-1'} \qquad 4.49$$

Did you notice the "disappearance" of the individual error scores ζ and the emergence of an error covariance matrix between 4.47 and 4.48 and, hence, the replacement the error variables in ζ by the error covariance matrix Ψ?

Note that the central parentheses again contain the usual form of a factor analysis, with Γ functioning as the "factor loadings," Φ as the covariances among the ξ "factors," and Ψ as covariances among the error terms. That is, the ξ concepts are functioning like the factors, the η's like the items being factored, and the ζ's as the error variables. The postulation of direct effects among the η's (the B coefficients) that forced the rearrangement in Eq. 4.42 necessitates pre- and postmultiplying by an inverse involving B.

We see later that the matrix containing the covariance among the η's (called C) is useful in its own right, but our current interest in this covariance matrix is for replacement of the central piece of Eq. 4.37 (the covariances among the y variables). This replacement gives

$$\begin{bmatrix} \text{Covariances} \\ \text{among the y's} \end{bmatrix} = \Lambda_y[(I - B)^{-1}(\Gamma\Phi\Gamma' + \Psi)(I - B)^{-1'}]\Lambda_y' + \Theta_\epsilon \quad 4.50$$

which shows that a particular covariance matrix among the y variables is implied if the elements in the eight matrices constituting the model (the terms on the right of the equation) are provided numerical values.

4.4.3 Covariances between the x's and y's

We are now ready to investigate what the model implies about the covariances between the x and y variables. The general form for calculating covariances in matrix algebra (Section 3.1.1.1) is

$$\begin{bmatrix} \text{Covariances} \\ \text{between} \\ \text{the x's and y's} \end{bmatrix} = (1/N)XY' = E(XY') \quad 4.51$$

We insert into this form the model's calculations of x and y,

$$\begin{bmatrix} \text{Covariances} \\ \text{between} \\ \text{the x's and y's} \end{bmatrix} = E[(\Lambda_x\xi + \delta)(\Lambda_y\eta + \epsilon)'] \quad 4.52$$

and simplify, using the assumption that the various error variables are assumed to be independent of the concepts and of one another.

$$\begin{bmatrix} \text{Covariances} \\ \text{between} \\ \text{the x's and y's} \end{bmatrix} = E[(\Lambda_x\xi + \delta)(\eta'\Lambda_y' + \epsilon')] \qquad 4.53$$

$$= E[\Lambda_x\xi\eta'\Lambda_y' + \delta\eta'\Lambda_y' + \Lambda_x\xi\epsilon' + \delta\epsilon'] \qquad 4.54$$

$$= \Lambda_x\, E(\xi\eta')\Lambda_y' + E(\delta\eta')\Lambda_y' + \Lambda_x E(\xi\epsilon') + E(\delta\epsilon') \qquad 4.55$$

$$= \Lambda_x E(\xi\eta')\Lambda_y' \qquad 4.56$$

Inserting the model's equation for η (Eq. 4.42) and further simplifying give

$$\begin{bmatrix} \text{Covariances} \\ \text{between} \\ \text{the x's and y's} \end{bmatrix} = \Lambda_x\, E[\xi((I - B)^{-1}(\Gamma\xi + \zeta))']\Lambda_y' \qquad 4.57$$

$$= \Lambda_x E[\xi(\xi'\Gamma'(I - B)^{-1\prime} + \zeta'(I - B)^{-1\prime})]\Lambda_y' \qquad 4.58$$

$$= \Lambda_x[E(\xi\xi')\Gamma'(I - B)^{-1\prime} + E(\xi\zeta')(I - B)^{-1\prime}]\Lambda_y' \qquad 4.59$$

Noting that the independence of the exogenous concepts from the ζ error variables implies covariances of zero and that the covariances among the ξ's are called Φ gives

$$\begin{bmatrix} \text{Covariances} \\ \text{between} \\ \text{the x's and y's} \end{bmatrix} = \Lambda_x\Phi\Gamma'(I - B)^{-1\prime}\Lambda_y' \qquad 4.60$$

In this form we again see that specific numerical quantities placed in the eight model matrices (indeed in even five of the eight matrices) are sufficient to fully specify the covariances between the observed x and y indicators. None of the error variance/covariance matrices is involved (a by-product of the assumption about independence between different types of error terms), and the terms that do appear have a clear logical structure. Examining the series of matrices from left to right shows that all of the types of terms that contribute to causal chains linking the x and y variables are considered, namely, the connections between the x's and the ξ's, the connections among the ξ's (or Φ), the connections between the ξ's and η's (or Γ), the connections among the η's (B), and, finally, the connections from the η's to the y's.

Equations 4.30, 4.50 and 4.60 show that if numerical elements are placed in the eight matrices constituting a model, this implies specific covariances among the x's, among the y's, and between the x's and y's. We will frequently wish to speak of the covariances among *all* the observed indicators without distinguishing whether they are indicators of the exogenous concepts (x's) or of the endogenous concepts (y's). We denote the covariance matrix among all the observed indicators that is implied by our estimated model (i.e., by the numbers in the eight matrices) by Σ (sigma). If we list the y variables first, and then the x variables, the implied variance/covariance matrix can be written as

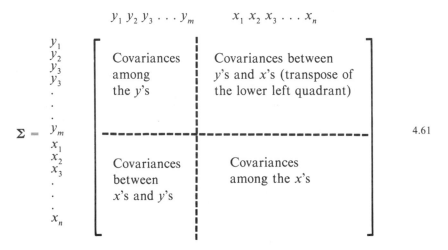

4.61

Inserting the mathematical forms of the various matrices derived earlier, we can represent/calculate Σ as

$$\Sigma = \begin{bmatrix} \Lambda_y[(\mathbf{I} - \mathbf{B})^{-1}(\mathbf{\Gamma\Phi\Gamma'} + \mathbf{\Psi})(\mathbf{I} - \mathbf{B})^{-1\prime}]\Lambda_y' + \mathbf{\Theta}_\epsilon & \begin{matrix}\text{(transpose of}\\ \text{lower left quadrant)}\\ \Lambda_y(\mathbf{I} - \mathbf{B})^{-1}\mathbf{\Gamma\Phi}\Lambda_x'\end{matrix} \\ \Lambda_x\mathbf{\Phi\Gamma'}(\mathbf{I} - \mathbf{B})^{-1\prime}\Lambda_y' & \Lambda_x\mathbf{\Phi}\Lambda_x' + \mathbf{\Theta}_\delta \end{bmatrix}$$

4.62

The symmetry of covariance matrices implies that the upper right quadrant of Σ is merely the transpose of the lower left quadrant.

It must be stressed that Σ, in Eq. 4.62, is a *model-implied variance/covariance matrix among the observed indicators x and y*. For any set of values inserted in the eight matrices on the right of this equation, one and only one covariance matrix among the observed indicators is implied by the equation. Indeed, a sophisticated theory specifying both *where* particular nonzero coefficients should appear and the *precise values* of these nonzero coefficients would imply or predict a specific set of covariances among the observed indicators.

Naturally, if we have the "wrong" model, the covariance matrix Σ implied by the model would not correspond to the actual observed covariances among the indicators (which is called S). You might expect that models implying covariance matrices (Σ's) that are very similar to what is observed (S)—namely, models whose predictions are in agreement with observed facts—would be preferred. Indeed, in Chapter 5 we will see that the closeness of the match between Σ and S gives not only a criterion for deciding which of several alternative models is better but also a criterion for determining the best estimates for the free coefficients in any given model.

You are now prepared to begin reading the LISREL manual. Numerous aspects of the manual will undoubtedly still be opaque, but you should at least skim the whole manual. Pay careful attention to the basic model specification and to the details of setting up a LISREL program, including the specification of fixed and free coefficients. The program setup for the smoking model, which appears as the first page of the LISREL output in Appendix B, can serve as a guide, and you should work through this in detail. The LISREL manual discusses how to set up LISREL programs, so there is no need to repeat the steps here. We turn instead to a discussion of more fundamental aspects of the model contained in the eight matrices (or Eqs. 4.1, 4.4, and 4.6 plus Φ, Ψ, Θ_ϵ, Θ_δ).

In particular, we focus on an aspect of modeling with LISREL that differs from traditional regression or path analysis—namely, the idea of inserting *fixed* or researcher-specified values into models. LISREL allows us to fix any structural coefficient appearing in any of the eight basic matrices to any value. This is possible because by fixing specific elements of any of the eight basic matrices, we are merely contributing to the model-implied covariance matrix Σ via the right side of Eq. 4.62. Indeed, if *all* the elements in the matrices are fixed by the researcher, a specific Σ is still implied, and the fit of the Σ to the observed covariance matrix S would provide evidence regarding the adequacy of this "fully fixed" model. It is more usual to find that prior research, well-articulated theory, or knowledge of data collection procedures determine only a few model coefficients, whereas most of the structural coefficients remain to be estimated. We will see several interesting uses of fixed coefficients in the following chapters, but we concentrate in the next section on their use in the context of measurement.

4.5 Scaling and Reliability

4.5.1 Fixed Measurement Coefficients

Our previous discussion highlighted the desirability of attending to measurement quality during model development, estimation, and interpretation. Although the measurement equations 4.4 and 4.6 assist in this effort, we have already seen that *fixing* one Λ value to 1.0 for each concept is not a statement of measurement quality, but it ensures that the concepts are measured on the same measurement scales as the corresponding observed indicators. The interpretation of these coefficients as structural coefficients linking unit changes in the concepts to unit changes in observed variables provides for scale equivalence, not for measurement quality.

Besides using fixed Λ values to set measurement scales for the concepts, we can use other fixed coefficients to specify measurement quality. Specifically, fixing the diagonal elements of Θ_ϵ and Θ_δ fixes measurement error variances for the endogenous and exogenous concepts, respectively. Consider what happens if we have a single indicator of each concept, if each indicator has its Λ value fixed at 1.0 (as in our smoking model), and if we further specify that the error variances for the indicators (the diagonal elements of Θ_ϵ and Θ_δ) all equal 0.0. This implies that the concepts and indicators are identical. To see why this is so, consider what Eq. 4.12 (4.6 for the smoking model) looks like under these conditions. Recall that the mean of the δ's is zero, so a zero variance for these variables implies that all the values of these variables must be precisely 0.0. Hence, Eq. 4.12 (4.6) becomes

$$\begin{bmatrix} x_1 \\ x_2 \\ x_3 \\ x_4 \end{bmatrix} = \begin{bmatrix} 1 & 0 & 0 & 0 \\ 0 & 1 & 0 & 0 \\ 0 & 0 & 1 & 0 \\ 0 & 0 & 0 & 1 \end{bmatrix} \begin{bmatrix} \xi_1 \\ \xi_2 \\ \xi_3 \\ \xi_4 \end{bmatrix} + \begin{bmatrix} 0 \\ 0 \\ 0 \\ 0 \end{bmatrix} \qquad 4.63$$

This claims $x_1 = \xi_1$, $x_2 = \xi_2$, $x_3 = \xi_3$, and $x_4 = \xi_4$, which is precisely a statement that the indicators and corresponding concepts are identical. (Convince yourself by examining Eq. 4.11 that $y_1 = \eta_1$ if the variance of $\epsilon = 0.0$.)

Now consider fixing an error variance (Θ_δ) at a *nonzero* value. This implies that entities other than the underlying concept can influence the indicator and, hence, acknowledges some unreliability in the measurement of the concept. Consider the form of Eq. 4.30 under these conditions, assuming we have allowed error variance but not covariances.

$$
\begin{bmatrix} \sigma^2_{x_1} & & & \mathrm{Cov} \\ & \sigma^2_{x_2} & & \\ & & \sigma^2_{x_3} & \\ \mathrm{Cov} & & & \sigma^2_{x_4} \end{bmatrix} = \begin{bmatrix} 1 & 0 & 0 & 0 \\ 0 & 1 & 0 & 0 \\ 0 & 0 & 1 & 0 \\ 0 & 0 & 0 & 1 \end{bmatrix} \begin{bmatrix} \mathrm{Var}(\xi_1) & & & \mathrm{Cov} \\ & \mathrm{Var}(\xi_2) & & \\ & & \mathrm{Var}(\xi_3) & \\ \mathrm{Cov} & & & \mathrm{Var}(\xi_4) \end{bmatrix} \begin{bmatrix} 1 & 0 & 0 & 0 \\ 0 & 1 & 0 & 0 \\ 0 & 0 & 1 & 0 \\ 0 & 0 & 0 & 1 \end{bmatrix}'
$$

$$
+ \begin{bmatrix} \Theta_{\delta_1} & 0 & 0 & 0 \\ 0 & \Theta_{\delta_2} & 0 & 0 \\ 0 & 0 & \Theta_{\delta_3} & 0 \\ 0 & 0 & 0 & \Theta_{\delta_4} \end{bmatrix} \qquad 4.64
$$

Since Λ_x is an identity matrix, neither it nor its transpose alters the matrix it multiplies, and therefore they can be ignored. The diagonal elements of the matrix to the left of the equals sign are then the sum of the corresponding diagonal elements of the matrices on the right. For example, the row-1 column-1 elements inform us that the implied variance of x_1 equals $\mathrm{Var}(\xi_1) + \Theta_{\delta_1}$. This is another way of saying that the variance of an indicator can be partitioned into variance originating in the underlying concept and variance due to error, precisely the partitioning underlying traditional discussions of reliability. Hence, if we fix a Θ variance, we are fixing the portion of variance in an indicator that is thought to arise from sources other than the concept the indicator is supposed to indicate.

4.5.2 The Philosophy and Specification of Measurement Quality

It is our view that measurement reliabilities should routinely be *fixed* rather than *free*. The researcher's familiarity with the data collection procedures provides information about measurement quality that is lost unless the researcher takes the initiative and incorporates this information by specifying particular measurement reliabilities.[5]

For the smoking model, all the Θ_ϵ and Θ_δ values were fixed to include a specific proportion of the variance in each indicator as measurement error variance, with the remaining variance being left to the corresponding underlying concept.[6] The logic used in determining the reliabilities is presented in the next few pages. The fixed values were determined by first determining the proportion of the variance in any given indicator that is error variance and then determining the values of the fixed Θ coefficients by multiplying this proportion by the variance of the indicator. The assessed proportion of error variance for each indicator is given in Figure 4.5, and we justify these reliability assessments as follows.

Reported sex should correspond almost perfectly with actual sex. The 1% error allows for an occasional keypunch mistake or slip of the interviewer's pencil, but even a 1% error rate is probably an overestimate, given the care with which the annual Edmonton Area Survey is conducted.

Age probably contains a bit more error, essentially from social pressure to underestimate age, tendencies to report ages ending in 5 or 0, and some honest "losing track" of the years. Here 5% error variance was considered an absolute upper bound on errors.

Education was thought to be less well measured and was assigned a 10% error variance. In addition to the usual coding problems, there are potential problems with "credentials" that do not fit the coding schema (due to migration from other countries), and, more importantly, the problematic ranking of some categories that could lead to some peculiar credential comparisons. Completed vocational/technical nursing, for example, is scored below even the shortest attendance at university. And consider the relative placement of categories 13–15 (in Figure 4.4). Furthermore, a person with only elementary schooling may be coded as 8 because he or she took a three-month course in welding.

The 10% error variance decision, however, was not based solely on this critique of categoric specifications. We decided to conceptualize education as "mere relative credential level" rather than as "knowledge" or "learning." Had we decided to use a knowledge/learning conceptualization of education, this variable would have required much more error variance because the current categories do not reflect amount learned or differences in informal learning. That is, *how we decided to conceptualize education played a role in determining our assessment of the adequacy of the measuring instrument employed.* This specification forces us to consistently interpret education in all the following discussions of this model as referring to relative credential level rather than to knowledge. It also implies that the model could be reestimated with a different meaning applied to the concept education (a different fixed error variance for this concept), and this might imply a better or worse fitting model.

But most importantly, *we can alter the meaning of concepts by specific decisions about measurement quality.* The strategy of fixing measurement error variance at specified values gives the researcher some direct control over the meanings of the concepts. It seems preferable to have the meanings of concepts under the researcher's conscious control than to relegate them to the vagaries of some estimation procedure that might even confound several potential meanings for a concept by "estimating" some intermediate amount of error variance.

Having recognized that an indicator may be a good or poor indicator of a concept, we should recognize the corollary that *an indicator may be a good indicator of one concept but a poor indicator of another concept, so we can help specify of the identity of a concept by forcing an indicator to have the amount of error variance that is appropriate for the concept of interest.* Specifically, we might intentionally fix the error variance for education to be substantially larger than a mere test-retest reliability would indicate, if we wanted to force the concept education to be knowledge rather than credential level. Measurement error variances should be freed only if the indicator is known to correspond to a single concept so that the only issue decided during estimation of the error variance is the quality of measurement, not the conceptual meaning.

The acceptability of the specific fixed error variances can and should be continually reevaluated at all stages of model development (e.g., by investigating the partial derivatives and modification indices discussed in Chapter 6). The multiple indicator models in Chapters 7 and 9 allow further opportunities for reconsidering measurement specifications in the context of *estimating* measurement errors, as opposed to *adjusting* our estimates of the other coefficients for researcher-specified measurement quality (Figures 4.6 and 4.7). Useful discussions of how concepts derive their meanings from indicators are provided by Burt (1973, 1976). Alwin and Jackson (1980) give a clear demonstration of how traditional meanings of reliability are challenged by various alternative specifications of multiple-indicator models.

The concept of religion has several indicators, only the first of which (frequency of religious service attendance) is used in the current model. We address all the indicators here because they are all ultimately needed and because we assess the amount of error variance as if each indicator reflects a different aspect of religion. We later investigate the possibility of using these as multiple indicators of a single concept called religion.

The 10% error variance for attendance reflects minor response category problems, the suspicion that there may be pressures to both over- or underreport attendance, and potential disagreement over what constitutes a religious service. Does this include funerals, church weddings, and youth meetings? We hope not, but the respondents might have assumed that it does.

The strength of one's religious beliefs is given a bit more error variance than justified by coding errors because of the "volunteered" category, even though it was infrequently used. The importance of one's religion is also provided slightly more error variance than could be attributed to coding errors alone. This primarily reflects the potentially ambiguous reference to "now" in the question and possible upward or downward respondent biases due to social desirability. The question regarding belief in God is provided the most error variance (20%) because we want to use it as an indicator of belief in God and not as an indicator of the certainty or uncertainty of the proof of God's existence, as the question was worded. Some persons feel that they must believe precisely because there is no solid proof: if there were proof, they would feel that the element of an individual decision grounding belief would be eliminated. Since we wish to use this as an indicator of belief in God's existence rather than of belief in the provableness of that existence, as the question is worded, we assign the higher error variance.

The four indicators of smoking are provided similar modest (10%) error variances because many of the same considerations underlie all four measures. The question sequence from which these variables were created was a bit unclear regarding how recent quitters should respond, and the indicators containing degrees of smoking were composites created from somewhat arbitrary classifications about the numbers of various kinds of materials smoked that constitute light, moderate, or heavy smoking. These items will be given more attention later.

Antismoking views and acts are conceptualized as views and acts centered precisely on the city bylaw and are not intended to capture any wider domain of smoking views or acts. Hence, the only errors are recording/coding errors and the possibility that the meaning of the bylaw to the individuals was not preserved by the strategy of questioning on which these items are based. Since neither of these is particularly likely, a low (5%) judged error is assigned.

Now that the proportion of error variance in each indicator has been specified, you should retrace the steps involved in fixing the Θ values in the program in Appendix B. Specifically, the variance for any indicator (Figure 4.5) is multiplied by the proportion of error variance (Figure 4.5) to obtain the corresponding Θ_ϵ or Θ_δ value that is fixed in Appendix B.[7]

4.5.2.1 Some general concerns about measurement

Most of the concepts and indicators in the smoking model concern events and entities routinely encountered by respondents as they engage in their daily affairs. If the respondents had been asked questions encountered only under artifical interview conditions, or if the items demanded that the respondents think about topics they do not routinely consider, this would have elevated the assessed proportion of error variance. Measurements of personality traits and opinion items are likely to encounter these types of problems.

Models containing single indicators with more than 40%–50% error variance are prone to estimation problems. With multiple indicators, estimation is likely to proceed smoothly even with 60%–70% error variance, but the meanings of the concepts using such low reliabilities deserve special attention.

Another way of seeing that decisions about measurement error variances (Θ_ϵ and Θ_δ) involve concerns for validity and not merely reliability arises from the fixed measurement error variances (Θ_ϵ and Θ_δ) indirectly fixing the variances of the concepts (recall Eq. 4.64). The concepts determined by our judgments about error variances are placed within a specific theoretical structure (Eq. 4.1). If we discover that our model ultimately provides a good fit to the observed data, we have moved beyond confirming our reliability judgments and closer to confirming the validity of our concepts, since the concepts as specified through our decisions regarding error variances will have been shown to operate in precisely the manner required for construct validity or the assessment of validity through confirmation of predicted relationships among a multitude of theoretically relevant variables.

Blalock (1979a, 1982, 1986) provides detailed discussions of why auxiliary measurement theories (the links between concepts and observed indicators) deserve increased attention in the social sciences. Several chapters in Bohrnstedt and Borgatta (1981) illustrate the investigation of measurement models using LISREL. Hoppe (1980) provides a provocative philosophical discussion that highlights why issues of measurement are inextricably intertwined with the development of conceptual models. The controversial, yet fundamental, nature of scaling issues is highlighted in the recent exchanges among Bielby (1986a,b), Williams and Thomson (1986a,b), Sobel and Arminger (1986), and Henry (1986), but we are not yet prepared to enter the fray because these works assume familiarity with the material in our Chapters 7 and 9.

4.5.3 Constrained Coefficients

Besides the ability to free coefficients for estimation, or to fix them, as in the case of Λ's equal to 1 or specific measurement error variances, LISREL provides a third option regarding the estimation of coefficients. We can constrain one coefficient to have the same estimated value as another coefficient. This allows us to model equivalent reciprocal effects, for example, or to enter the same estimated coefficient in the models for two separate groups (cf. Chapter 9). It also plays a large part in the "creative" modeling discussed in Chapter 7.

Although we can use constrained coefficients in situations where we have several "equally good" indicators of a concept, constraining Λ values to be equal should always be approached with caution. The problem is that constraining Λ values to be equal will constrain the concept to contribute an equal amount of variance into each indicator (namely, the square of the Λ's value times the variance of the concept), but this will give equivalent measurement reliability only if the variances of the indicators are also identical. If the indicator variances are not identical, different error variances will be required to accommodate the differing total variances and hence different measurement quality will be built into the model despite the seeming equality of indicators due to equivalent Λ values.[8]

4.6 Restrictions on Model Specification

Even though Eqs. 4.1, 4.4, and 4.6 are very general, there are certain types of "paths" or direct effects that cannot be entered into LISREL models. One type of restriction is unavoidable and common to all structural equation modeling strategies, not just LISREL. The restriction arises out of the distinction between endogenous and exogenous concepts. Specifically, we cannot model variables that are partially exogenous and partially endogenous. If a variable is exogenous, it is not allowed to receive any direct effects from the other model variables. If such direct effects are desired, the variable can be entered as an endogenous variable where effects are allowed. But note that if a variable is dependent in even one directed causal relation of interest, it must become fully endogenous in that it is not allowed to display mere correlations with any other variables.

For example, if we made education in our model endogenous by letting it depend on sex and age, we would have to decide upon the directed causal relations between strength of religion and education. We could allow religious strength to influence the now endogenous education, but the very reasonable possibility that education influences religious strength could not be included (religion remains exogenous). Nor could we avoid the problem of being permitted only a one-way effect (when in fact there are probably reciprocal effects) because we would no longer be allowed to model a mere correlation between these two variables. Once a variable is designated as endogenous, all the connections between that variable and other variables must be directed in that they either lead directly from, or directly to, other variables. In short, if a variable is endogenous with respect to even one variable, it must be made fully endogenous.

Other restrictions are notational and arise from the splitting of the indicators into two sets, as indicated by Eqs. 4.4 and 4.6. This implies that the endogenous concepts (η's) are not allowed to influence any of the exogenous x indicators, and that the exogenous concepts (ξ's) are not allowed to influence any of the endogenous y indicators. Similarly, although the measurement error terms are allowed to covary (correlate) within each set of indicators (Θ_ϵ and Θ_δ), there is no provision for covariance (correlation) between two indicators if one indicator is exogenous (an **x**) and the other is endogenous (a **y**). We will see in Chapter 7 that creative model specifications can circumvent the restrictions mentioned in this paragraph because these restrictions are merely notational, in contrast to the inherent analytical requirement that variables be designated as completely exogenous or endogenous.

Notes

1.   By "cause" we mean no more than that we are imagining a system in which the values of a dependent concept are derivable from, or determined by, the values of the other concepts, just as a new Y is derived from, or determined by, the X's in Figures 1.8 and 1.9. Thus, causation is part of our abstract conceptual system and is not necessarily a property of the real world. Causes and effects constitute a convenient way of thinking about the world whether or not they constitute the real world. Those interested in the philosophical foundations of causal statements can review the articles in *Synthese* on "Causality in the social sciences" (68: No. 1, 1986) and "Causation and scientific inference and related matters" (67: No. 2, 1986).

2.   This equation is identical to that used from the fifth version of LISREL onward, but differs slightly from the notation used in earlier versions of LISREL, as explained in Section 10.1.

3.   The phrase "exogenous indicators" is for "indicators of the exogenous concepts." The term "exogenous" refers to the status of the concept to which the indicator belongs, rather than claiming that the indicator itself is never a dependent variable (which would contradict Eqs. 4.4 and 4.6).

4.   Early versions of LISREL VI contained a program error that resulted in the explained variances (both as standardized Ψ values and as the "squared multiple correlation for the structural equations") being reported as 17%, 17%, and 11% (instead of the correct 8%, 13%, and 6%) for this model. Scientific Software was notified about this error early in 1985.

5.   Reviews of appropriate reliability assessments will ultimately begin to appear in the literature, but for the moment we are left to justifying reliability assessments variable by variable (cf. Bohrnstedt, Mohler, and Muller, 1987); Entwisle and Hayduk, 1982: Appendix A; Bielby and Hauser, 1977; Wolfle and Ethington, 1986).

6.   The variances of the ξ concepts could have been fixed directly by fixing the diagonal elements of Φ. There is no parallel way to fix the variances of the η concepts because these variances arise indirectly through the operation of B and Γ, and so on, as seen in Eq. 4.49.

7.   The sensitivity of this model to measurement error specification was investigated in a series of 14 runs in which each of the Θ_ϵ and Θ_δ values was individually fixed, first at half and then at double the value appearing in Appendix B. Only a handful of coefficients displayed more than a 10% change in their estimated value. The vast majority of the estimates were identical to those reported in Appendix B. The effects of these alternative error specifications were generally, but not exclusively, confined to the

corresponding portions of Φ or Ψ. The probabilities for the models' χ^2's ranged from .124 to .151, and six models had the same probability as the reported model (.131). Four models provided lower probabilities, and four provided higher probabilities. In sum, this model seems neither overly sensitive nor totally insensitive to the measurement specifications.

8.   These ideas will become clearer when we discuss multiple indicators in Section 7.3. The footnote on tau-equivalent and parallel measures is especially pertinent.

Chapter 5

Estimating Structural Coefficients with Maximum Likelihood Estimation

Chapter 4 introduced a general notation for representing structural equation models: the eight fundamental matrices ($\mathbf{B}$, $\boldsymbol{\Gamma}$, $\boldsymbol{\Lambda}_y$, $\boldsymbol{\Lambda}_x$, $\boldsymbol{\Phi}$, $\boldsymbol{\Psi}$, $\boldsymbol{\Theta}_\epsilon$, $\boldsymbol{\Theta}_\delta$); or the basic equations 4.1, 4.4, and 4.6 (covering the four matrices of effect coefficients) plus the covariances among the exogenous concepts $\boldsymbol{\Phi}$ and three error covariance matrices $\boldsymbol{\Psi}$, $\boldsymbol{\Theta}_\epsilon$, $\boldsymbol{\Theta}_\delta$. Chapter 4 also detailed why placing numerical entries in the eight matrices implies (predicts) a specific covariance matrix among the observed indicators $\boldsymbol{\Sigma}$.

This chapter demonstrates how estimates of the model's coefficients can be obtained by comparing the actual covariances among the indicators ($\mathbf{S}$) with the covariances implied by the model ($\boldsymbol{\Sigma}$). We begin by developing the logic grounding maximum likelihood estimation (MLE). Section 5.2 applies this logic to estimating model coefficients and presents the mathematical foundations for comparing the fit between the model-implied covariances $\boldsymbol{\Sigma}$ and the observed covariances $\mathbf{S}$. The chapter concludes with a discussion of potential estimation problems. Chapter 6 extends the discussion of the fit between the model-implied and observed covariances by providing a test of the goodness of fit.

5.1 Maximum Likelihood Estimation

Consider the sampling distribution for the mean as shown in Figure 5.1. If the population distribution of variable X has mean μ and

Figure 5.1   The Sampling Distribution for the Mean.

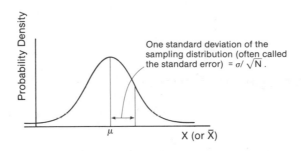

The <u>area</u> over any region on the scale of X provides the probability of finding an X̄ value falling in the specified region if a single sample is drawn at random from the population.

standard deviation σ and if a random sample of N persons is selected from this population, the mean $(\overline{X})$ and standard deviation (s) of the sample will probably differ from the corresponding population values. Random sampling fluctuations may accidentally include too many persons with high or low scores, so the sample mean and variance may be slightly above or below the corresponding population values.

The sampling distribution for the mean describes the values of the sample mean that are likely to appear if we take repeated random samples from the population, and hence it tells us how much above or below the true population mean sampling fluctuations are likely to carry the sample mean. The sampling distribution for the mean is normal, with its center falling exactly on the population mean, and with standard deviation $\sigma/\sqrt{N}$—the central limit theorem in statistics. Since 95% of the area under a normal curve falls within plus or minus two standard deviations of the center of the distribution, we can test hypotheses about the population mean or create confidence intervals reflecting a range of reasonable guesses for the population mean, given the observed sample information.

But suppose we ask, What is the single best estimate of the population mean μ if the sample mean $\overline{X}$ is observed? The trick to answering this question involves operationalizing the word "best." Consider a hypothetical example where the observed sample mean $\overline{X} = 100$ and where we are considering estimating the population mean as 200 (as shown in the top

Figure 5.2   The Sample Mean as a Maximum Likelihood Estimator.[1]

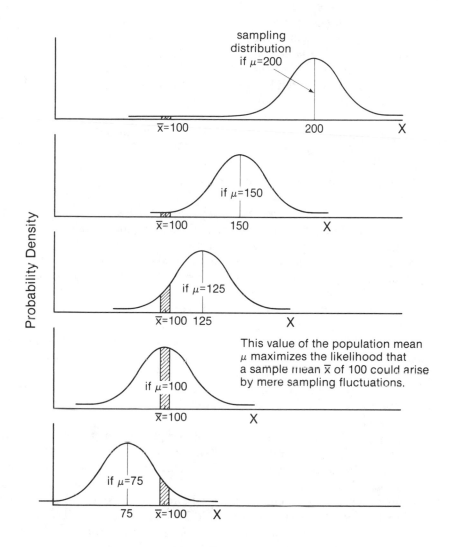

portion of Figure 5.2). To defend 200 as being the population mean, we must attribute the difference between the observed mean 100 and the supposed population mean of 200 to chance sampling fluctuations. But how likely is it that sampling fluctuations alone could provide a sample mean of 100 if the population mean was 200? It is very unlikely, because this implies an excessive number of sampled cases have low values and *it is unlikely that a truly random sampling process would lead to a systematic overrepresentation of low-valued cases.* Sampling fluctuations, by definition and dint of careful research, are random and hence are unlikely to provide a reasonable explanation for systematically larger or smaller scores.

If 200 is a poor estimate of the population mean, how about 150 or 125, as in the second and third portions of Figure 5.2? These estimates are clearly better because there is an increasing probability (increasing likelihood) that sampling fluctuations could account for the remaining difference between the observed mean and the "estimates" of the population mean. We can progressively improve our guess (estimate)—that is, increase the likelihood that the remaining discrepancy could be a mere sampling fluctuation—by making guesses progressively closer to the observed value of 100. Moving our guess to lower values (say to 75) would again decrease the likelihood that the remaining discrepancy could be a mere sampling fluctuation (random fluctuations alone would be unlikely to contribute an excess of cases scoring above the population mean), so our best guess is that $\mu = 100$.

All of this shows, nontechnically, that the sample mean is the maximum likelihood estimator for the population mean. We can choose from among several guesses at (estimates of) the population mean *by selecting as best the guess maximizing the likelihood that the discrepancy between our guess and the observed sample information could have arisen as a mere sampling fluctuation.*

The logic of choosing our estimate of the population parameter as the value that maximizes the likelihood of the observed sample data arising as a mere sampling fluctuation can be generalized to population slopes and correlations. The top portion of Figure 5.3 shows several observed cases (x's) for which we are to choose one of the four b's as our estimate of the population slope. The worst choice would be b_1, and b_4 would be the best. To defend b_1 as the true population slope, *large and systematic differences would have to be attributed to the random sampling process that selected the cases.* Slope b_4 minimizes the departures and avoids systematic differences, thereby *maximizing the likelihood* that the remaining differences could be attributed to sampling fluctuations.

Line b_4 can also be proven to be the line that minimizes the square of the distances between the data points and the line (the ordinary least

Figure 5.3   Visualizing MLE Estimates of Slopes and Correlations.

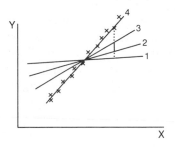

Observed Correlation Matrix			Possible Population Correlation Matrices								
			Set 1			Set 2			Set 3		
1.0			1.0			1.0			1.0		
0.0	1.0		0.8	1.0		0.2	1.0		0.1	1.0	
0.9	0.9	1.0	0.0	0.0	1.0	0.8	0.8	1.0	0.85	0.85	1.0

squares criterion), but *if we want to understand MLE, we should avoid thinking of minimizing squared errors and attempt to visualize the deviations that would have to be attributed to sampling fluctuations if b_1, b_2, b_3, or b_4 were the true population slope.*

The lower portion of Figure 5.3 applies the same strategy to a set of correlations. If the correlation matrix on the left is observed and we are to choose which of the three sets of correlations on the right is the population correlation matrix, the first set is clearly untenable. To defend this as the population correlation matrix, we must claim that the observed 0.0 correlation arose as a sampling fluctuation around a true population correlation of .8, whereas the .9 correlations arose as sampling fluctuations around two true null correlations. It is much more likely that the observed correlations could appear as sampling fluctuations around the second set, and still more likely that they could appear as sampling fluctuations around the third set, since smaller and less systematic departures must be attributed to sampling fluctuations.

The trick to MLE is the quantification of the magnitude and the seriousness of the departures between what is observed and the various estimates of the population coefficients. For example, would we prefer a matrix containing a single large correlation discrepancy or a matrix containing several small discrepancies? If the *probabilities* of both the single large departure and multiple small departures *appearing by sampling fluctuations* can be calculated, the choice will be clear. We choose the matrix that maximizes the probability or likelihood that the remaining differences are chance sampling fluctuations.

To obtain maximum likelihood estimates of the coefficients of interest, we ask, "How likely is it that the observed sample information could appear if the population parameters took on any particular set of values?" If the sample values deviate so radically from the "tentative" population values that there is only a low probability of the differences being attributable to mere sampling fluctuations, this guess is discarded, and another guess is made. We ultimately select as the best estimates those values that maximize the likelihood of any remaining differences being attributable to mere sampling fluctuations. *Maximizing the likelihood minimizes what must be attributed to sampling fluctuations.*

The following section applies this logic to the estimation of LISREL model coefficients and then develops the mathematics behind the maximum likelihood process.

5.2 Making Σ Approximate S

Consider the smoking model in Figure 4.1. The eight matrices specifying this model contain numerous fixed coefficients (mostly 0's and 1's) and several free coefficients whose values are to be estimated (primarily the coefficients in **B** and **Γ**). Imagine inserting arbitrary estimates in place of each of the free coefficients. The eight matrices would now contain only numerical entries and hence could be inserted into Eq. 4.62 to calculate a model-implied covariance matrix among the observed indicators—a Σ.

If we replace the free coefficients with a second arbitrary set of estimates, this implies a second Σ. We can decide which set of arbitrary coefficient estimates is better by comparing the observed covariance matrix **S** to the model-and-estimate–implied Σ's. The better set of estimates is the set whose Σ provides a higher likelihood (probability) that **S** could appear by random sampling fluctuations if Σ were the population covariance matrix.

Pitting the better set of estimates (and its implied Σ) against a third set of arbitrary estimates might locate an even better fitting Σ and hence a better set of estimates. Repeating this process thousands of times should eventually chance upon a set of estimates that imply a Σ as similar as possible to S, given the constraints of the model (i.e., given that free parameters are allowed in only specific locations in the eight matrices). These estimates would be the maximum likelihood estimates. The process LISREL uses to obtain maximum likelihood estimates of the free structural coefficients improves upon this multiple-guess procedure in only one important respect. LISREL makes "intelligent" guesses about the changes in the estimates that might improve the fit between S and Σ, but we postpone discussion of how this is accomplished until later in this section.

Clearly, the key to MLE is the numerical quantity summarizing how similar any particular Σ is to S. Since the likelihood of S arising as a sampling fluctuation around Σ is our criterion of similarity, we must examine the mathematical foundations of the calculation of likelihood.

5.2.1 Mathematical Foundations

Quantifying the likelihood of a covariance matrix S appearing for a sample of cases randomly selected from a population having covariance matrix Σ requires an assumption that the variables in the population are distributed according to a multivariate normal distribution. The familiar univariate (one-variable) normal distribution

$$\begin{bmatrix} \text{Univariate} \\ \text{Normal} \\ \text{Density} \end{bmatrix} = \frac{1}{(\sigma^2)^{1/2}(2\pi)^{1/2}} \; e^{-(1/2)(x - \mu)(1/\sigma^2)(x - \mu)} \qquad 5.1$$

describes a curve (probability density) enclosing a total area of 1 square unit. Area corresponds to probability in that if a case is selected at random from a normal distribution, the probability of that case having a value falling in a specific range is provided by the size of the area bounded by the horizontal axis on the bottom, the lowest value of the range on the left, the highest on the right and the normal probabilty density function (likelihood) on the top. (Recall locating the upper 5% of the *area* under a normal curve to obtain a range of scores having *probability* .05.)

The multivariate normal distribution similarly specifies the probability density of multiple variables.

$$\begin{bmatrix} \text{Multivariate} \\ \text{Normal} \\ \text{Density} \end{bmatrix} = \frac{1}{|\Sigma|^{1/2}(2\pi)^{p/2}} \; e^{-(1/2)(\mathbf{x} - \mu)\Sigma^{-1}(\mathbf{x} - \mu)} \qquad 5.2$$

Though we use the properties of the multivariate normal distribution only indirectly (namely, through what it implies about the sampling fluctuations in covariances) a clear understanding of these properties is necessary if we are to grasp precisely what is involved in making the assumption of a multivariate population distribution.

Unlike the univariate normal distribution, the multivariate normal distribution must allow for covariances between variables. Figure 5.4 depicts a bivariate (two-variable) normal distribution. In this figure vertical height represents the bivariate normal probability density. The bivariate distribution appears as a "hill" with altitude contours that are concentric ellipses if the standardized variables covary and that are concentric circles if the variables do not covary. Any vertical slice through either a "round" (no-relationship) or an "elongated" (some-relationship) bivariate normal distribution produces a curve that is normal [Mood, Graybill, and Boes (1974:163)]. In particular, all vertical planes parallel to the X_1 or X_2 axes (i.e., conditioning on particular values of X_2 or X_1, respectively) will be normal, even though the area of the normal curves on such vertical slices will be less than 1 square unit.

Multivariate normal distributions for more than two variables cannot be diagrammed and can be described only by equations depicting specific marginal or conditional distributions. If three variables are trivariate normal, for example, each pair of bivariate relationships will be bivariate normal, as shown in Figure 5.4. In general, if a set of variables is multivariate normal, any subset of those variables will also be multivariate normal, and, in particular, each variable individually must be normal, each pair bivariate normal, and each triplet trivariate normal.

The reasonableness of the assumption of a multivariate normal distribution becomes apparent if we are challenged to provide an alternative description for a population distribution that could be more widely applicable and hence routinely assumed. Clearly we need a distribution that allows the occasional extreme case while the majority of cases cluster near the means. The distribution should be symmetric (as opposed to routinely expecting all distributions to be skewed in one particular direction), and it should lack any sharp corners (as opposed to routinely expecting sharp transitions in probabilities at any specific point in the distribution). Furthermore, the variables in the distribution should be allowed to display stronger or weaker relationships with one another without disrupting the basic nature of the assumed distribution. These criteria are satisfied by a multivariate normal distribution, and hence assuming multivariate normality is often a reasonable approximation. The multivariate normal distribution may not apply to any particular data set, but we would be hard pressed to specify a more widely applicable type of distribution. (The only real alternative is to avoid making assumptions about the population distribution. DeLeeuw (1983), DeLeeuw, Keller, and

Figure 5.4   A Bivariate Normal Relationship.

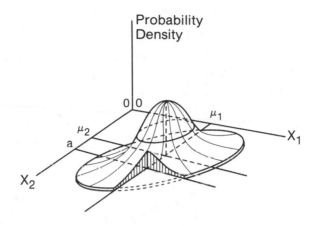

Wansbeek (1983), and Bentler (1983c) discuss recent developments in asymptotically distribution free estimates, and LISREL VII will include two estimation procedures that relax the multivariate normality assumption.)

Paralleling the univariate normal distribution, we can inquire about the probability of a specific set of values appearing for a single case selected at random from a multivariate normal distribution. The probability is obtained by calculating the proportion of the total "volume" (of 1 unit) that is covered by the ranges of scores of interest. In Figure 5.4 the volume of the carved-out section gives the probability that a randomly selected case has an X_1 score greater than μ_1 and an X_2 score greater than a (obtained by integrating the bivariate normal density for X_1 from μ_1 to ∞ and X_2 from a to ∞).

We can also ask about the probabilities of particular types of values appearing for samples of several individuals randomly selected from a multivariate normal distribution, and hence we can examine the sampling fluctuations in the vector of means and matrix of covariances for all the variables in the multivariate distribution. We will not pursue the multivariate analog to the central limit theorem for the mean, but we will direct our attention instead to *the sampling distribution of* **S**, the covariance matrix. This probability density, first calculated by Wishart (1928) and, hence, called the Wishart distribution, is

$$W(\mathbf{S};\mathbf{\Sigma},n) = \frac{e^{-\frac{1}{2}n\text{tr}(\mathbf{S}\mathbf{\Sigma}^{-1})} \; |n\mathbf{S}|^{\frac{1}{2}(n-p-1)}}{|\mathbf{\Sigma}|^{\frac{1}{2}n} 2^{\frac{1}{2}np} \; \pi^{\frac{1}{4}p(p-1)} \; \prod_{i=1}^{p} \Gamma(\frac{1}{2}(n+1-i))} \qquad 5.3$$

The notation $W(\mathbf{S};\mathbf{\Sigma},n)$ denotes that the right side of Eq. 5.3 is the probability density (derived by *W*ishart) that a sample with covariance matrix **S** appears if a random sample of N individuals ($n = N - 1$) is selected from a multivariate normal population having covariance matrix $\mathbf{\Sigma}$ and mean μ. The number of variables is p, and Γ is the gamma function discussed by Mood, Graybill, and Boes (1974:534). (See also Anderson, 1958:154.)

Fortunately, this formula can be simplified, given that we are interested in maximizing the function (maximizing the fit between **S** and $\mathbf{\Sigma}$) rather than in calculating its precise value. The matrix $\mathbf{\Sigma}$ is the only aspect of Eq. 5.3 that we can vary (by selecting different sets of coefficient estimates) in attempting to maximize the likelihood. Most of the terms in the formula are constants once the data have been gathered (notably the sample size and the sample covariance matrix **S**). If we combine all the constants in Eq. 5.3 (all the terms except those involving $\mathbf{\Sigma}$) and call this combined value C, Eq. 5.3 can be simplified to

$$W(\mathbf{S};\mathbf{\Sigma},n) = \frac{e^{-\frac{1}{2}n\text{tr}(\mathbf{S}\mathbf{\Sigma}^{-1})}}{|\mathbf{\Sigma}|^{\frac{1}{2}n}} \cdot C \qquad 5.4$$

$$= e^{-\frac{1}{2}n\text{tr}(\mathbf{S}\mathbf{\Sigma}^{-1})} \; |\mathbf{\Sigma}|^{-\frac{1}{2}n} C$$

Here we insert a logical step whose full implications will not be apparent until Chapter 6, but it has the side benefit of eliminating the constant C. Suppose we chanced upon just the right model (the one locating the true causal forces) and the perfect set of coefficient estimates for this model. The $\mathbf{\Sigma}$ implied by this perfect model would be identical to **S** and would provide the absolute maximum value for Eq. 5.4 because such a fit could never be improved. The following equation divides Eq. 5.4 by the absolute maximum likelihood appearing for the best possible fit. In other words, Eq. 5.5 expresses the likelihood for our particular model and tentative parameter estimates, as a proportion of (or in ratio to) the maximum likelihood that could be achieved with a perfect model.

$$\text{likelihood ratio} = \frac{\text{likelihood for any given model}}{\text{likelihood with a perfectly fitting model}} = \frac{e^{-\frac{1}{2}n\text{tr}(S\Sigma^{-1})}|\Sigma|^{-\frac{1}{2}n}C}{e^{-\frac{1}{2}n\text{tr}(SS^{-1})}|S|^{-\frac{1}{2}n}C} \qquad 5.5$$

The denominator is the same as Eq. 5.4 except that Σ has been replaced by S (for a perfect model $\Sigma = S$). The constants C now cancel, and reversing the signs of the exponents of the denominator terms allows us to move these terms to the numerator $(1/10^{-1} = 10^1)$:

$$\text{likelihood ratio} = e^{-\frac{1}{2}n\text{tr}(S\Sigma^{-1})}|\Sigma|^{-\frac{1}{2}n}e^{\frac{1}{2}n\text{tr}(SS^{-1})}|S|^{\frac{1}{2}n} \qquad 5.6$$

We can further simplify this equation by taking the natural logarithm of both sides of the equation:[1]

$$\text{log likelihood ratio} = -\frac{1}{2}n\,\text{tr}(S\Sigma^{-1}) - \frac{1}{2}n\log|\Sigma| + \frac{1}{2}n\,\text{tr}(SS^{-1}) + \frac{1}{2}n\log|S|$$

$$5.7$$

$$= -\frac{1}{2}n[\text{tr}(S\Sigma^{-1}) + \log|\Sigma| - \log|S| - \text{tr}(SS^{-1})]$$

Maximizing this new function, called the log of the likelihood ratio, also maximizes the likelihood because the largest value the logarithm can take on must correspond to (be the logarithm of) the largest value of the likelihood ratio.

In LISREL notation there are $p + q$ observed indicators (p indicators of the endogenous concepts, q indicators of the exogenous concepts), so S and S^{-1} are square matrices containing $p + q$ rows and columns. Hence, SS^{-1} is a $p + q$ identity matrix whose trace is the sum of the $p + q$ 1's on the diagonal.

$$\text{log likelihood ratio} = -\frac{1}{2}n[\text{tr}(S\Sigma^{-1}) + \log|\Sigma| - \log|S| - (p + q)] \qquad 5.8$$

Since a minus sign precedes this quantity, maximizing the likelihood is equivalent to *minimizing* the function in brackets, called F, which appears in the LISREL IV manual on p. 13 and the LISREL V and VI manuals on p. I-28.

$$F = \log|\Sigma| + \text{tr}(S\Sigma^{-1}) - \log|S| - (p + q) \qquad 5.9$$

The two negative terms originated with the "perfect" model against which any real model's predictions (Σ) are evaluated, and they are constants once the sample covariance matrix has been obtained. If our model also happened to provide a perfect fit, the first and third terms would be identical, as would the second and fourth (since Σ would equal S), so the function F would become zero, which is another way of saying that the likelihood ratio (Eq. 5.5) would equal 1.

$$F = 0 = \log|S| + \text{tr}(SS^{-1}) - \log|S| - (p + q) \qquad 5.10$$

The further Σ is from S, the farther $|\Sigma|$ will depart from $|S|$ and the further $(S\Sigma^{-1})$ will depart from an identity matrix (and hence have a trace differing from $p + q$). The important point is not that we have developed a function that reflects the departures between our model-implied Σ and S, but that Eqs. 5.8 and 5.9 *represent the specific function that characterizes these departures as the likelihood of S appearing as a sampling fluctuation around our model-implied Σ* (relative to the likelihood that even a perfectly fit model would result in the observed covariance matrix). This is precisely the function that, if maximized, provides reasonable coefficient estimates, as argued in Section 5.1.

5.2.2 The Smoking Model

Returning to the smoking example, consider the values of the fitting function (see Appendix B) for the iterations (sets of guesses) LISREL produced in converging to the final estimates reported in Figure 4.7. Two features of this progression are notable: relatively few sets of guessed values are used (there are few iterations), and the fit function is progressively reduced so that LISREL never seems to encounter a set of "worse" estimates than the set considered immediately prior to the current set. These are signs of the intelligent searching for estimates mentioned previously.

The first set of estimates provided by LISREL VI is created using two-stage least squares estimates of the coefficients (which often approximate the maximum likelihood estimates), so LISREL begins with an initial set of fairly reasonable estimates.[2] These "good" first guesses are then progressively improved upon by using first steepest-descent iterations and then Fletcher-Powell iterations to fine tune the initial estimates. Steepest descent iterations operate by obtaining the partial derivative of F with respect to each of the free coefficients—that is, the slope or rate of change of the fit F at the current guessed value of each of the coefficients. The coefficients having the largest partial derivatives (slopes) represent the places where changes in coefficient values can provide the greatest improvement in fit F.[3]

Such steepest-descent iterations continue until all the partial derivatives approach zero, indicating the maximum is near. LISREL then switches to a more finely tuned method of locating the precise maximum. This procedure was developed by Davidon, Fletcher, and Powell (Davidon, 1959; Fletcher and Powell, 1963) and uses both the first- and second-order derivatives to provide a more precise location of the maximum likelihood (minimum F). Detailed discussions of the iterative estimation procedures appear in Joreskog (1967, 1973a) and Mulaik (1972:Chap. 7).

5.2.3 Model and Data Constraints

The preceding discussion implies that maximum likelihood estimates *result from a combination of model and data constraints.* The theoretical model constrains the estimates by specifying the variables included in the model and the location of the free coefficients linking the variables. The data constrains the estimates since comparison of the model-and-estimate-implied Σ with the *data* S ultimately selects the best set of coefficient estimates.

This implies that errors in either theoretical specification or data collection can invalidate the resultant estimates. LISREL has no way of knowing if you have adequately converted Professor X's theory into structural equations, or if you accidentally included a wrong variable during creation of the covariance matrix S. All LISREL can do is provide the numerical values making the free coefficients in your "possibly nonsensical" model as consistent as possible with your "possibly erroneous" data. Unbelievable coefficient estimates may signify an erroneous model or incorrect data. Unfortunately, the absence of wild estimates does not guarantee that we have located either the proper model or entered the proper data. Even a close fit between Σ and S does not *prove* that the correct model has been estimated with sound data, a point we return to later.

Another disconcerting fact is that even combining the best available theory and data may be insufficient to provide useful estimates. The following section investigates this phenomenon in more detail.

5.3 Identification and Colinearity

The general issues of identification and colinearity may be illustrated by a simple non-LISREL example. Consider a theory claiming that the sum of two coefficients, *a* and *b*, must equal some specific number, while the available data indicate that number is 10. Hence $a + b = 10$. What then are the values of a and b? Mimicking LISREL's maximum likelihood procedure, we try multiple estimates of a and b in search of the pair of estimates providing a sum closest to 10. We try the pairs 4–20, 6–7, 1–9, 2–8, and 9–1. At first, things progress well, 6 and 7 are preferred over 4 and 20 since their sum is closer to 10 than is the sum of 4 and 20. The pair 1 and 9 does even better, and so it eliminates the 6–7 pair.

Then comes the snag. Several pairs of estimates fit equally well. The pairs 1–9, 2–8, and 9–1, among others, all imply sums of 10 and hence maximize the likelihood of observing 10 as data. Thus the model constraints (requiring the sum of a and b) and the data constraints (the sum equaling 10) eliminate some sets of estimates of a and b (4–20 and 6–7), but the combined force of both types of constraints is insufficient to determine unique values for a and b. *It is the failure of the combined model and data constraints to identify (locate or determine) unique estimates that results in the name "the identification problem."* What can be done? The only option is to seek *more model or data constraints* in hopes of eliminating more of the remaining estimate pairs.[4]

The identification problem, in general, arises because *the combined force of the theory and data constraints are insufficient to determine unique estimates of the structural coefficients.* This is not a problem of sample size or quality, because the problem may occur even if we have population data. The estimates obtained when there are insufficient constraints (when the coefficients are not identified, underidentified, or unidentified) may be better than no estimates at all (recall the elimination of certain pairs of estimates at the beginning of the preceding example) but little confidence can be placed in the estimates because the coefficients remain free to take on a range of values. *Any selection from the remaining set of acceptable estimates is purely arbitrary because it is dictated neither by theory nor data,* whether the selection is made by a researcher or by a computer. The solution to the identification problem is to place further theoretical or data constraints on the coefficients. We will consider detailed suggestions for doing this.

One way to visualize identified and unidentified coefficients in LISREL arises in the context of the fit function. Recall that LISREL creates maximum likelihood estimates by selecting as estimates those coefficient values implying a Σ maximizing the likelihood that the observed data S could arise by chance sampling fluctuations. If more than one set of coefficient estimates imply the same best fitting Σ, we have encountered the identification problem. Figure 5.5 depicts what the likelihood function might look like for a hypothetical identified coefficient β_{12} and a nonidentified coefficient β_{13}. The straight ridge along the maximum of the likelihood function (the top of the hill) illustrates that only a single value

Figure 5.5   Identification and the Log Likelihood Ratio.

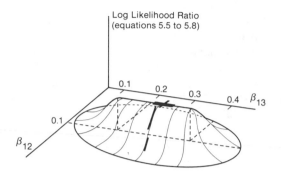

of β_{12} can provide the maximum likelihood, whereas a range of values of β_{13} could provide the maximum likelihood.[5] Had two nonidentified parameters been plotted, this figure would appear as a hill with a flat top.

Since LISREL reports only a single set of estimates in its output, how can we determine if any particular estimate is unique (identified) or is merely one of several equally acceptable estimates (unidentified)? Equivalently, *what are the signs of identification problems*? If a coefficient is unidentified, the "acceptable" values of that coefficient can range considerably (Figure 5.5). This implies the standard error of the coefficient would be large, and thus large standard errors may signify unidentified coefficients.

The range of "acceptable" values for nonidentified coefficients has a more mathematically fundamental implication, which also can be seen in the context of Figure 5.5. Since the zero slope of the fit function parallel to the β_{13} axis (the partial derivative of the fit with respect to β_{13}) does not change over a range of values along the maximum likelihood ridge, the rate of change of that slope is zero (i.e., the second-order partial derivative of the fit function with respect to β_{13} is zero). The second-order partial derivatives of the fit function are fundamental to both the calculation of the standard errors of the estimates, and to the iterative steps used to obtain the maximum likelihood estimates (the Fletcher-Powell iterations). Both these calculations require that LISREL calculate what is called the **information matrix** (the probability limit for the matrix of second-order partial derivatives of the fit function with respect to the estimated coefficients). If the second-order partial derivative for some coefficient is zero, this indicates the fit function has a flat top and the coefficient is nonidentified, as in Figure 5.5.

If the information matrix (containing the second-order partials of all the coefficients) has an inverse (a nonzero determinant), all the estimated coefficients are identified and none of the coefficients have infinitely large standard errors. If this matrix has no inverse (i.e., is singular or has a zero determinant), one or more of the coefficients may not be identified and LISREL searches for the identity of the offending coefficient and prints a warning message. Thus LISREL *itself seeks out and reports identification problems* using a procedure that is fundamentally connected to both the maximization of the fit function and the calculation of standard errors.[6]

Wildly unreasonable estimates provide another clue to identification problems (recall the 9010 and -9000 pair in footnote 4 continuing the hypothetical example) as do *impossible estimates*: negative error variances (Ψ, Θ_ϵ, or Θ_δ), negative Φ variances, or standardized coefficients exceeding 1.0 (see Section 6.6). Unfortunately, these symptoms are not unique to identification problems. Substantive disagreements between the data and theory and mistakes in setting up the LISREL program can produce similar diagnostics. Whenever these symptoms arise, they are serious and demand attention, including the investigation of potential identification problems.[7]

A further check for identification problems arises from the matrix of the correlations among the estimates (Section 6.4.1). Consider the a and b example with which we began. If we plot the multiple sets of unidentified pairs of estimates for a and b on a scatter plot, these values fall on a negatively sloped line. As one estimate increases, the other must decrease. That is, the possible estimates of a and b are perfectly negatively correlated or **colinear**. In general, correlations between coefficient estimates exceeding about $\pm.9$ may signify identification problems.

5.3.1 Traditional Approaches to Identification

Diagnosing identification problems in the context of actual models, as discussed earlier, reflects a recent trend in structural equation modeling. Much of the traditional literature on structural equation modeling treated the question of identification as "logically prior" to the question of estimation and therefore held that the identification of all the model coefficients had to be demonstrated before attempting estimation. This section first characterizes the essentials of the traditional discussions of identification and then summarizes why this tradition is on the decline.

The traditional procedures for checking identification begin by writing the equation for each covariance/variance as a function of the structural coefficients (cf. Eq. 4.62). This gives a series of equations containing some unknowns (the coefficients to be estimated) and some knowns (if the observed covariances are inserted in place of the model-implied covariances). Our task, then, is to rearrange these equations so that each unknown coefficient is expressed in terms of the known covariances. That is, we try to "solve" the equations so that each coefficient appears alone on the left side while only known covariances appear on the right side, thereby demonstrating that the coefficient's value can be calculated from the known information.

If there exists one and only one function of the covariances by which a coefficient can be calculated, that coefficient is said to be **just identified**. If there exists no combination of covariances providing the value for a coefficient, the coefficient is **not identified** (underidentified or unidentified). If there is more than one way to calculate a particular coefficient, the coefficient is **overidentified**.[8] If *all* the coefficients in a model are identified (or overidentified), the entire model is identified. If even a single coefficient in a model is not identified, the entire model is not identified.

Because determinants of matrices are involved in solving sets of equations (Section 3.1.3), the presence or absence of solutions to the equations can be framed as questions about the rank (nonzero determinant, invertibility, or nonsingularity) of particular matrices. Indeed, the rank condition amounts to saying that if some matrix has a nonzero determinant (and hence is invertible), certain coefficients can be solved for by the procedures in Section 3.1.3 and, hence, are identified. Demonstrating that the relevant matrix has the desired rank is sufficient for demonstrating that certain coefficients are identified.[9]

We will not discuss the details of the rank and order conditions; refer to Berry (1984) or any of the references at the end of this section. We turn instead to the reasons why we find statements such as the following beginning to appear in the literature:

> Even the well-known rank and order conditions and their generalization (Monfort, 1978) do not provide a simple, practicable method for evaluating identification in the various special cases that might be entertained under the general model. (Bentler and Weeks, 1980:295)

First, no general conditions have yet been enumerated that guarantee the identifiability of the diverse types of models that can be accommodated within LISREL's general equations (cf. Bollen and Joreskog, 1985; Dupacova and Wold, 1982; Joreskog, 1981b; Long, 1983a,b; McDonald, 1982). Although the identification of several specific models has been investigated (e.g., Burt, Fischer, and Christman, 1979; Greenberg and

Kessler, 1982; Werts, Joreskog, and Linn, 1973; and see the references in Joreskog, 1981b), the general conditions for identifiability remain elusive. Indeed, when we encounter models in later chapters that illustrate interactions and nonlinearities among the concepts, models placing equality, proportionality, and squaring constraints between coefficients, models for multiple groups with constraints between the groups, and models containing estimates of means and intercepts, you will appreciate why "it can be virtually impossible to determine from the specification of the hypothesis whether the parameters in the model are identified, and hence uniquely estimable" (McDonald and Krane, 1977).

Second, the procedures for checking identification are not standardized. One suggestion is to first try to identify all the variances and covariances among the concepts using the variances/covariances of the indicators as knowns and then to use the identified concept covariances as knowns to solve for the structural coefficients among the concepts (cf. Long, 1983b). Unfortunately, the procedures are not standardized beyond this. If a coefficient resists expression as a function of the known covariances, there is no unambiguous check on whether we have exhausted all the possibilities for calculating this coefficient. To make matters worse, some of the widely disseminated rules for sufficient conditions for identification are wrong, as Bollen and Joreskog (1985) show.

Third, the nonstandardized steps require lengthy hand calculations, so we are tempted to avoid doing them. The burden is compounded because the addition of even one coefficient to an identified model requires complete rechecking of the identification status of the whole model. Deleting coefficients from an identified model usually results in an identified model, because this reduces the number of unknowns being sought from a given pool of covariances.

A further type of problem with proving identification occurs for models that McDonald (1982) describes as "locally identified." Such models may have all their coefficients identified in the neighborhood of some set of values of interest (e.g., in the neighborhood of the top of the maximum likelihood function, so the maximum likelihood estimates are clearly obtainable), even though the coefficients in the model are not identified if coefficient values far from the maximum likelihood values are considered. Thus, a model might not be globally identified even though the coefficient values are well identified in the neighborhood of (locally to) the maximum likelihood values. McDonald (1982:103) concludes that although the problem of

> identifiability can be said to be logically prior to the problem of estimation, it does not at all follow that investigating it at the MLE is theoretically inappropriate.

Quite to the contrary, in the present state of knowledge it seems eminently reasonable to recommend just such investigations.

Finally, there is the problem of empirical underidentification (Kenny, 1979). Even if all the coefficients in a model can be proven to be identified by showing that some combination of observed covariances gives the model parameters, the data may provide quirks that defy the seeming proof. For example, if the proof of a coefficient's identification involves division of covariances, both the proof and the calculation of the coefficient's value will fail if the denominator covariance happens to be zero. Similar problems can occur if a variable supposedly functioning as an "instrumental variable" happens to have "zero effect" (Duncan, 1975:87) or if perfect correlations appear between variables (Burt, Fischer, and Christman, 1979:116). Consequently, we must be vigilant for signs of identification problems, whether or not we have a proof of identification. This also implies that identification is not an all-or-nothing proposition. For example, a model may be weakly identified if a key covariance or effect approaches, but is not quite, zero. Rindskopf (1984a) clearly discusses the problem of empirical underidentification.

For further reading on the debate over data based versus a priori evaluations of identification, see McDonald and Krane (1977, 1979), Bentler and Weeks (1980), and McDonald (1982). For more general discussions of identification, see Bielby and Hauser (1977), Bentler and Weeks (1980), Berry (1984), Bollen and Joreskog (1985), Dupacova and Wold (1982), Fisher (1966), Geraci (1977), and Werts, Joreskog, and Linn (1973).

5.3.2 Sources of, and Solutions to, Identification Problems

Where should we expect identification problems? Models with large numbers of coefficients relative to the number of input covariances are candidates for problems. It is never possible to estimate more coefficients (unknowns) than there are data covariances/variances (knowns).[10] We should *avoid inserting coefficients (effects) into a model merely because effects are possible.* There should be some reasonable and considered grounds for an effect before it warrants estimation, and *some piece of substantive theory should fail each time an included effect is found to be insignificant.*

Reciprocal effects and causal loops are another common source of identification problems. To estimate reciprocal effects, we must break the symmetry of the reciprocal relationship by including variables thought to cause one or the other, but not both, of the reciprocally related variables

Figure 5.6   Underidentified and Identified Models.[1]

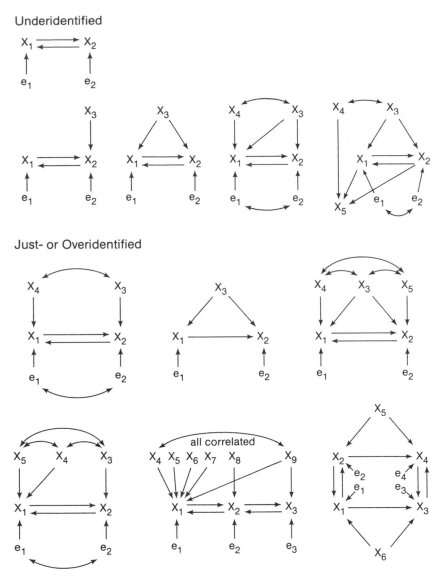

Underidentified

Just- or Overidentified

[1]The three right underidentified models and the upper and lower left identified models are discussed in detail by Duncan (1975). Berry (1984) discusses the three lower identified models.

(cf. Duncan, 1975). Figure 5.6 illustrates some underidentified, identified, and overidentified models containing reciprocal effects.

Do not eliminate expected reciprocal causal coefficients merely to attain identification. It is better to be known for having developed the correct model (even if you could not get estimates of its coefficients) than for having identified and estimated the wrong model.

Another common identification problem concerns the variance of a conceptual level variable and the λ's linking that concept to some indicators. If the variance of the concept (e.g., a Φ) is free and if all the λ's linking that concept to observable indicators are free, the λ's and the concept variance are not identified. The variances and covariances of the observed indicators can be modeled by a whole set of incrementally increasing concept variances if the λ's are correspondingly decreased (cf. Long, 1983a). The solution to this problem is to always *fix at least one λ for each concept*. We saw in Chapter 4 that this provides a measurement scale for the concept, so we are not only avoiding identification problems by fixing a λ, we are increasing the interpretability of the ultimate model.

Models containing many similar concepts or many error covariances are also potentially troublesome. If two concepts are nearly identical, exceptionally precise and error-free data may be required to attribute unique effects to these concepts. The elimination of error covariances (e.g., by including a common, even if uninteresting, cause within the model, or by creating clearly differentiating wordings at the time of questionnaire construction) is well worth the effort because the estimation of error covariances taps the same limited supply of indicator covariances required for estimating more interesting structural coefficients.

5.3.2.1 Prevention as the best cure

The best way to solve the identification problem is to avoid having it! But how? Occasionally we can add extra data constraints by adding more indicators of concepts or by replacing poor indicators with stronger indicators, but the primary prevention strategy is to emphasize theoretical constraints. Be sure you have included all suspected absences of effects (fixed zero coefficients), effects of specifiable magnitude (fixed nonzero coefficients), and equality or proportionality constraints (equal or proportional effects, equal or proportional error variances) (see Chapter 7). Greenberg and Kessler (1982), for example, use equality constraints "between time periods" to identify coefficients while modeling a process thought to be stable over time (cf. Thornberry and Christenson, 1984). Researchers accustomed to using ordinary regression are often not attuned to fixing particular slopes, or constraining slopes to be equal, and they subsequently overlook this useful source of identifying restrictions.

One of the nicest things a theory can do is to indicate where we *should not expect effects*, especially if it avoids a reciprocal effect. As a corollary, be sure to include variables known to influence only a few of the variables entwined in dense causal webs. This produces a net increase in usable information. We pay the price of estimating a few more structural coefficients but, in return, gain the numerous covariances of the variables with all the other modeled variables. We should also *initially build our model with the minimum necessary coefficients* and later add coefficients of dubious use or of interest to others. The necessity of such emendations can be tested with the procedures discussed in Section 6.1.2.

Another form of constraint, which may be considered either a theoretical or a data constraint, is to fix measurement error variances on the basis of known data collection procedures (e.g., the known strengths and weaknesses of particular questionnaire items) or published reliabilities. We argued in Chapter 4 that measurement error variances should routinely be fixed, but this is not yet standard operating practice. Often the relevant information exists but is overlooked.

A routinely useful strategy is to determine the time order in which the values of the relevant variables appear. If a variable's values are determined before the values of some other variable, an unambiguous causal specification occurs. The later appearing variable cannot possibly cause the earlier appearing variable.

We might also comb the literature to see if there are any studies documenting the size of any of the effects in the model. If an experimental study has demonstrated that some variable influences another with a specific magnitude of effect, inserting this size of effect as a fixed coefficient may help identify the model. As a last resort, we might simplify the model by eliminating troublesome variables. The problem with simplifying models is that this is likely to lead to misspecification of the model and biased estimates. Model simplification is discussed in Section 5.4.

Occasionally we encounter models where the combined force of the foregoing strategies is insufficient to identify all the coefficients. If a particular pair of coefficients (e.g., a reciprocal causal relation) is problematic, we can estimate a series of models where first one and then the other of the problematic coefficients is fixed at each of a series of values covering the full range of reasonable values for that coefficient. Recording the amount of variation in the estimates of all the other model coefficients gives the *sensitivity* of these estimates to any future determination of the true value of the incrementally fixed coefficients. This strategy has been dubbed *sensitivity analysis* since it depicts the sensitivity of the remaining model estimates to the range of reasonable values for some troublesome coefficients. In effect, we create an additional "artificial" theoretical constraint (the fixing of one coefficient's value) and then investigate the sensitivity of the other estimates to the choice of value for the artificially fixed coefficient (cf. Bielby and Hauser, 1977; Kessler and Greenberg, 1981:125ff; Land and Felson, 1978; Reilly, 1981). The local sensitivity of the model to a coefficient's current fixed value is provided by the partial derivative, as discussed in Section 6.5 (see also Kim, 1984).

5.3.3 Procedures for Testing Identification

If we are concerned about identification, either with or without any "symptoms," three check procedures are available. The first involves reestimating the model coefficients several times, starting each estimation from a different set of start values for the coefficients.[11] If LISREL's maximum likelihood procedure repeatedly converges to the same set of final estimates from diverse initial estimates, we can be reasonably confident that the estimated coefficients are distinguishable from their neighboring values and, hence, are identified.

Incidentally, this procedure provides strong evidence that the maximum located by the iterative search is not a local maximum. If the likelihood function has several humps or hills, we might wonder whether the iterative procedure has located the highest of the peaks for the true maximum or whether we have reached the summit of some lower mount. The convergence of multiple start points to the same maximum likelihood estimates does not guarantee that the maximum is not local, but it does radically reduce the possibilities of it. For further discussions of local maxima see Rubin and Thayer (1983), Fink and Mabee (1978), and Joreskog (1966) on the dynamics of LISREL's search procedures.

The second procedure involves recovering a set of known coefficient values. We first choose a set of reasonable values for all the model coefficients and obtain the Σ implied by these values. (LISREL can do this calculation if all the coefficients are specified as fixed.) This created Σ is then entered as "data" into LISREL (i.e., as S) in a run using the same model but with the coefficients *freed and started from a different set of initial estimates*. LISREL is thus challenged to recover the "known" values of the coefficients using only the model (whose identification status is in doubt) and the artificially created covariance matrix S. If LISREL is able to recover the coefficient values used in creating the artificial data, the model coefficients are identified.

A third type of testing is appropriate if we are concerned about the identification status of some particular coefficient. The model is estimated twice: once with the coefficient in question free, and once with the coefficient fixed at a value thought to be minimally yet substantially different than the estimated value. If fixing the coefficient at the different value results in a substantial decrement in the fit of the model (which can be judged by a procedure discussed in Section 6.1.2), we again have evidence that this coefficient's value is uniquely determined by the combined model and data constraints.[12]

5.4 On Simplifying Models

It is not uncommon for novices to create huge models that ultimately require simplification. Eliminating unnecessary concepts and eliminating all but the best one or two indicators for the remaining concepts usually reduces the model to manageable size, but further reductions are occasionally required. Model simplification can also be required to eliminate reciprocal effects or to eliminate concepts for which no indicators could be found.[13]

Unfortunately, simplifying a model can be difficult because simplification can result in misspecification of the model and, subsequently, biased estimates of the effect coefficients. *A model is* **misspecified** *if the contribution of a true common cause to the covariance between two variables is modeled as a direct or indirect effect between those variables, or if an incorrect causal sequencing is used.*[14] Misspecification implies biased estimates of effects because spuriously arising covariance between variables (Section 1.4) is attributed to the effects of the variables, or because covariance arising from one true causal sequencing is attributed to a different causal sequence. If X_1 causes X_2 and we model this as X_2 causing X_1, the model would be misspecified, not because of a confounding of effect and spurious covariance, but because the direction of effect is incorrectly specified. Any occurrence of an incorrect causal direction is serious and likely to lead to biased estimates.

Since the need for the proper causal ordering is intuitively understandable (any estimate of effects in the wrong direction are obviously wrong), we focus our comments on the confounding of effect covariance with spurious covariance. Figure 5.7 illustrates several acceptable and unacceptable types of simplifications. Figure 5.7A shows that solitary causes or effects can be eliminated, and indirect effects can be incorporated into direct effects. Eliminating a solitary cause transforms an intervening variable into an exogenous variable, and eliminating a solitary effect merely eliminates a variable from the model. Eliminating an intervening variable eliminates the specification of the mechanism by which a "direct" effect is thought to operate.

Figure 5.7B illustrates the basic error to be avoided—replacing a spurious cause with a direct or indirect effect. Clearly, there is no effect of X_2 on X_3 or of X_3 on X_2, so if the covariance between X_2 and X_3 is modeled as an effect in either direction, this is in error, and any estimate overestimates the true zero direct effect between these variables. Figure 5.7C illustrates the same point. Eliminating the common cause X_1 leaves all the covariance between X_2 and X_3 to be attributed to the effect of X_2 on X_3, and this is unjustified if the common cause X_1 really accounts for part of the covariance between these variables. Eliminating X_2 from this model is acceptable because this makes the estimated effect of X_1 on X_3 the sum of the direct effect of X_1 on X_3 plus the indirect effect of X_1 through X_2, but the total effect of X_1 on X_3 would remain unchanged (see Chapter 8), and all of the covariance between X_2 and X_3 continues to be modeled as resulting from the effects of X_1.

Figure 5.7D illustrates that although all the covariance between X_3 and X_4 is spurious, eliminating the common cause is acceptable as long as that elimination results in the continued attribution of the covariance between X_2 and X_3 to a spurious cause. The effect of X_1 on X_3 in the simplified model will equal the product of the effects of X_1 on X_2 and of X_2 on X_3 in the original model (see Chapter 8), but the spurious nature of the covariance between X_3 and X_4 is maintained.

The first simplification in Figure 5.7E is acceptable because it replaces the more completely specified direct and indirect effects of X_1 with the total effects of X_1 on X_3 and X_4. The second simplification in E is unacceptable because the effect of X_2 on X_3 is not properly modeled. X_1 previously contributed a spurious component to the covariance between these variables, whereas all the covariance between these variables is now modeled as an effect.

Figure 5.7F provides simplifications that are acceptable even though the coefficients linking X_1 to X_2 would differ between the original model and the second simplified model, which will become clear when we discuss the implications of loop enhancements in Chapter 8. Note that identification problems would be aggravated, not improved, if the third simplified model (the reciprocal effect between X_2 and X_3) were used.

The first two simplifications of Figure 5.7G will become self-explanatory as "total enhanced effects" after you have read Chapter 8. The third "simplification," besides being wrong (due to omission of the spurious contribution of X_1), does not improve the identification status of the model. Figure 5.7H should be self-explanatory.

Figure 5.7I illustrates the difference between *deleting paths* versus *deleting variables*. Omission of any real path is serious and leads to specification errors (biased estimates of effects) unless that path is a solitary link whose elimination also results in the elimination of a variable

Figure 5.7   Model Simplifications.[1]

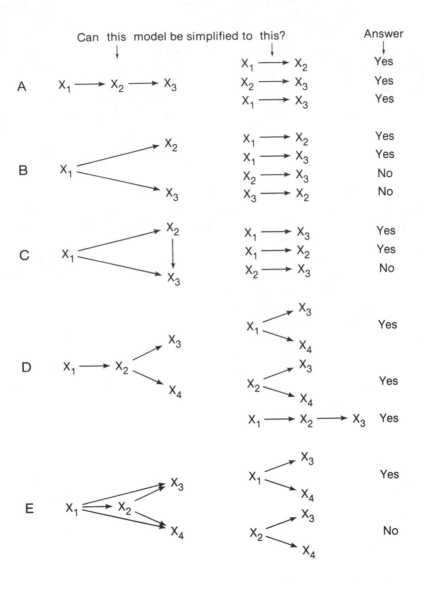

Figure 5.7   Continued.

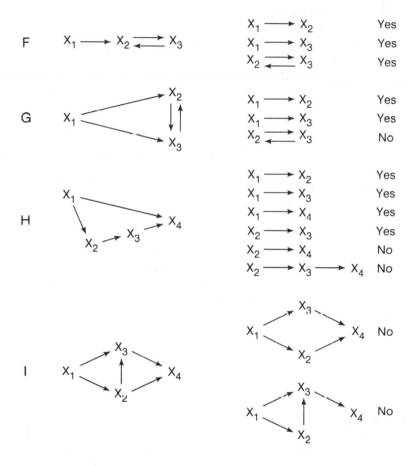

[1] The independent error term assumed to be present for each of the endogenous variables is omitted for diagrammatic simplicity.

(recall Figure 5.7A). The first simplification of Figure 5.7I is unacceptable because the spurious contribution of X_2 to the X_3-X_4 relationship is missing, and because all of the X_2-X_3 covariance is now modeled as spurious even though only part of this covariance was originally modeled as spurious (cf. Duncan, 1975:109). The second simplification of Figure 5.7I is unacceptable because X_2's contribution of spurious covariance to the X_3-X_4 relationship is no longer recognized by this simplification.

Of all the simplifications depicted in Figure 5.7, only the first two in 5.7F and G provide any improvement in the identification status of the model. All the models on the left (except F and G) are recursive models (they contain no loops or reciprocal causation) and are therefore identified. The improvements in identification in Figure 5.7F and G are accomplished by replacing a basic direct effect or a basic indirect effect with loop-enhanced effects, and the effects estimated in the acceptably simplified models are *not* the same effects as the correspondingly labeled arrows in the figures on the left. These points will become clear after reading Chapter 8. Clearly, *model simplification is of limited use in improving identification. The addition of variables having a few clear effects will, in general, be much more beneficial.*

Our comments about acceptable and unacceptable model simplifications are as relevant to model creation as they are to model simplification. If we read the left side of Figure 5.7 as being the real world, and the right side as being the model we are estimating, the question of misspecification now asks whether the estimated model misrepresents the magnitudes of effects in the real world.[15] The key issues remain: the misrepresentation of spuriously produced covariance as covariance attributed to effects, proper causal sequencing, and the combination of effects into total and enhanced effects (which necessitated the frequent references to Chapter 8).

5.5 In Closing

The best solution to the identification problem is never to have it. We should aim at strongly overidentified models—models where numerous variables are linked with a few structural coefficients. Researchers inclined to let everything influence everything else are not only headed for identification problems, but they have abandoned the parsimony canon of science. The fewer the structural coefficients required to achieve an acceptable data fit, the more parsimonious is our explanation of the observed data. Ways of judging the acceptability of the fit between S's and Σ's are discussed in the next chapter. Chapter 8 continues the discussion of estimation by emphasizing that successful estimation is not the end of the modeling process; it merely sets the stage for a detailed investigation of the estimates and their implications. Coefficient estimates are not results. The *results are the substantive implications of the estimates* obtained through the combined force of the best available theory and data.

Notes

1.   A natural logarithm is a nonlinear transformation in which the original values are transformed into the corresponding powers of the number e (which is 2.72, the base of the natural logarithms). For example, the number 2.72 would be transformed to (has a natural logarithm of) 1, because $2.72 = 2.72^1$. The number 7.39 has natural logarithm 2 because $7.39 = 2.72^2$. Similarly, 20.12 is transformed to 3, 54.7 to 4, and 148.8 to 5. The value 1 would be transformed to 0 since $2.72^0 = 1$; $.368 = 1/2.72 = 2.72^{-1}$, so .368 is transformed to -1. Since probabilities and the likelihood ratio (Eqs. 5.5 and 5.6) are fractions, they will have negative logarithms. Draw a figure with logarithms plotted on the horizontal axis and with the corresponding actual numbers on the vertical axis so that you see the curvature implied by transforming numbers into natural logarithms. Values between the whole powers of 2.72 are obtained with the aid of calculus but you should be able to see that the numbers 4, 5, 6, and 7 could be transformed to the values (natural logarithms) 1.39, 1.61, 1.79, and 1.95, respectively.

The joy of logarithms is that multiplication of the original numbers corresponds to adding their logarithms, whereas division of the original numbers corresponds to subtracting their logarithms. For example, $(2.72)(7.39) = 20.12$, for which it is easy to see that the sum of the logs for the multiplied numbers (i.e., $1 + 2$) equals the log of the product (3). Similarly, $20.12/2.72 = 7.39$, so the difference of the logs of the divided numbers $(3 - 1)$ is the log of the result (2).

2.   Boomsma (1985) reports there is little difference between the ultimate coefficient estimates obtained if one starts with LISREL's automatic start values, very good initial estimates, or poor but reasonable initial estimates. Better initial estimates imply that fewer iterations are required to reach the maximum of the likelihood function; hence they result in cheaper runs. Wild and improper estimates (e.g., negative variances) can lead to nonconvergence of the estimation procedure and/or costly runs.

3.   The partial derivatives of the fit function with respect to all the potentially free coefficients in the general LISREL model are given by Joreskog (1977a). Unfortunately, they are presented using the old $\mathbf{B}$, as explained in Chapter 9, so wherever $\mathbf{B}$ appears in Joreskog's (1977a) formulas, we insert $(\mathbf{I} - \mathbf{B})$. (Old $\mathbf{B}$ has 1's on the diagonal, instead of 0's, and off-diagonal elements that are the negatives of the current $\mathbf{B}$ values.) We have also slightly altered the formulas by adding three transposes (two in the partial derivative for $\mathbf{\Phi}$ and one in the derivative for $\mathbf{\Psi}$) to make the matrix multiplications conformable and to provide consistency with earlier unpublished versions of the derivatives.

The residuals $\mathbf{S} - \mathbf{\Sigma}$ represent the core of the partial derivatives and are frequently invoked. We can simplify the presentation by defining $\mathbf{\Omega}$ as the matrix

$$\Omega = \Sigma^{-1}(\Sigma - S)\Sigma^{-1} = \begin{bmatrix} \Omega_{yy} & \Omega_{yx} \\ \Omega_{xy} & \Omega_{xx} \end{bmatrix} \qquad 5.11$$

Ω is partitioned by the number of y and x indicators included in any given model, just as Σ or S are partitioned.

The covariance matrix among the endogenous concepts (Eq. 4.49, or p. III-66 of the LISREL VI manual) also appears relatively frequently, and we can avoid repeated reference to the matrix calculations producing this covariance matrix by naming it **C**. Thus

$$\mathbf{C} = (\mathbf{I} - \mathbf{B})^{-1}(\Gamma\Phi\Gamma' + \Psi)(\mathbf{I} - \mathbf{B})^{-1\prime} \qquad 5.12$$

With these definitions, the partial derivatives of the fit function F with respect to the potentially free model coefficients are as follows:

$$\partial F / \partial \Lambda_y = \Omega_{yy}\Lambda_y \mathbf{C} + \Omega'_{xy}\Lambda_x \Phi\Gamma'(\mathbf{I} - \mathbf{B})^{-1\prime} \qquad 5.13$$

$$\partial F / \partial \Lambda_x = \Omega_{xy}\Lambda_y(\mathbf{I} - \mathbf{B})^{-1}\Gamma\Phi + \Omega_{xx}\Lambda_x \Phi \qquad 5.14$$

$$\partial F / \partial \mathbf{B} = -(\mathbf{I} - \mathbf{B})^{-1}\Lambda'_y(\Omega_{yy}\Lambda_y \mathbf{C} + \Omega'_{xy}\Lambda_x \Phi\Gamma'(\mathbf{I} - \mathbf{B})^{-1\prime}) \qquad 5.15$$

$$\partial F / \partial \Gamma = (\mathbf{I} - \mathbf{B})^{-1\prime}\Lambda'_y(\Omega_{yy}\Lambda_y(\mathbf{I} - \mathbf{B})^{-1}\Gamma + \Omega'_{xy}\Lambda_x)\Phi \qquad 5.16$$

$$\partial F / \partial \Phi = \Gamma'(\mathbf{I} - \mathbf{B})^{-1\prime}\Lambda'_y\Omega_{yy}\Lambda_y(\mathbf{I} - \mathbf{B})^{-1}\Gamma + \Lambda'_x\Omega_{xy}\Lambda_y(\mathbf{I} - \mathbf{B})^{-1}\Gamma$$
$$+ \Gamma'(\mathbf{I} - \mathbf{B})^{-1\prime}\Lambda'_y\Omega'_{xy}\Lambda_x + \Lambda'_x\Omega_{xx}\Lambda_x \qquad 5.17$$

$$\partial F / \Psi = (\mathbf{I} - \mathbf{B})^{-1\prime}\Lambda'_y\Omega_{yy}\Lambda_y(\mathbf{I} - \mathbf{B})^{-1} \qquad 5.18$$

$$\partial F / \partial \Theta_\epsilon = \Omega_{yy} \qquad 5.19$$

$$\partial F / \partial \Theta_\delta = \Omega_{xx} \qquad 5.20$$

The presentation of a closely related set of partial derivatives and two appendixes on derivatives is given in Joreskog (1973a). Discussions of the second-order partial derivatives (the information matrix) appear in Joreskog (1973a, 1977a). See also Joreskog (1966, 1967, 1971a, 1977c, 1978a), Schonemann (1985), and Lawley and Maxwell (1971).

4.   For example, reconsidering the $a + b = 10$ theory might uncover that a must also be greater than b. This eliminates 1-9 and 2-8 but still leaves pairs such as 6-4, 10-0 and 9010-(−9000). Pressing the theory further might reveal that the product ab must also be some special number, which further data observations determine is 24. Now the *only* acceptable and well-identified estimate for a is 6, and b is 4, because they are the only values satisfying the theoretical constraints (that $a + b =$ an observable sum, that $a > b$, and that $a \times b =$ an observable product) and the data constraints (the sum is 10, the product is 24).

5.   You should be able to visualize the partial derivatives of the likelihood function with respect to the coefficients on the axes of Figure 5.5 as the slopes of the hill parallel to the axes. Along the maximum likelihood ridge, the slopes in both directions (both partial derivatives) are zero, but several values of β_{13} give the same zero slope (thus several values of β_{13} can maximize the fit between Σ and S), but only one value of β_{12} gives zero slope.

6.   This test of the identifiability of model coefficients is limited in two respects, one of which is potentially under the researcher's control and one which is merely a question of numerical accuracy. The accuracy question amounts to determining whether the top of the fit function in Figure 5.5 is "really" flat or "very nearly" flat. Clearly, there will be some limit to the numerical accuracy with which the β_{13} slope (partial derivative) can be calculated, and hence there will be some limit to the accuracy with which the rate of change in this slope (the second-order partial derivative) can be calculated. It is possible that a function having a minimal curvature might be called flat merely because current computers are not accurate enough to detect the differences in slope. This problem appears in LISREL as limitations on the accuracy with which the elements and inverse of the information matrix can be calculated. Such numerical accuracy limitations imply that LISREL's check of identification is not "perfect," but the double precision calculations of most computers seem to reduce the problem beyond the interest of most users.

The problem under researcher control begins from the fact that the calculation of partial derivatives (slopes), the second-order derivatives (changes in slopes), and hence the information matrix and its inverse are always made at particular values of the structural coefficients (above particular points on the "floor" of Figure 5.5). If the researcher specifies a totally wild start value for a model coefficient, it is possible that the fit function could be flat *in the neighborhood of this wild value* (slightly more or less wild values being equally unacceptable). Hence, the second-order derivative would be zero, the information matrix would have no inverse, the iterative search for maximum likelihood estimates would fail, and an error message would be printed. Clearly, if the iterative procedure fails, a careful review of the start values is warranted. Similarly, if a flat portion of the fit function is encountered on the way to a well-defined maximum, such a local nonidentification might be avoided either by specifying a different set of start values (taking a different route from the base to the top of the hill—the maximum of the fit function) or by specifying more precise start values (which avoid the flat spot by starting closer to the top of the hill). Either way, the researcher has some control over the problem.

7.   LISREL also checks to see if the estimated covariance matrices (Ψ, Θ_ϵ, Θ_δ, Φ) have zero determinants (i.e., if the estimates in any row can be expressed as a linear function of the estimates in the other rows, or if one row contains only zeros). If colinearity is found, LISREL prints a warning that some of the elements in the offending matrix may not be identified. This particular diagnostic message is not always informative because it can appear in perfectly reasonable models (e.g., if a diagonal Ψ has one of its values *fixed* at zero, which occasionally happens in the creative modelings discussed in Chapter 7), and LISREL does not pinpoint which of the coefficients in the offending matrix are problematic.

8.   If the multiple ways of estimating a coefficient result in different estimates, there is some inconsistency in the model's implications, and we must find some way of averaging the estimates to find the model's best estimate. Maximum likelihood estimation avoids this problem by selecting as best the estimate maximizing the likelihood of all the observed data. Overidentified coefficients provide the degrees of freedom for the χ^2 test discussed in Chapter 6. If there are no overidentified parameters (i.e., insufficient model restrictions to imply multiple ways to calculate some coefficient), χ^2 has zero degrees of freedom, and no overall model test is possible.

9.   A slightly weaker condition is that the relevant matrix be of the right size (or order). If the relevant matrix is not large enough, even a nonzero determinant cannot guarantee a solution, and hence the relevant matrix being of the proper order (size) is a necessary condition for identification. The relevant matrix referred to involves the variables excluded from any given equation, and the order condition can be summarized as saying that the coefficients in an equation for a variable are identified if the total number of causal variables in the equation is less than or equal to the total number of predetermined variables for that equation (the predetermined variables are those that occur earlier in the causal

ordering, so they are never directly or indirectly influenced by the variable whose equation is being considered) (cf. Duncan, 1975).

10.   In the language of Chapter 6, this says models with negative degrees of freedom must be nonidentified.

11.   LISREL provides a "no start values" option that lets the user specify start values (Joreskog and Sorbom, 1984:II23-II26) rather than depending on LISREL's default two-stage least squares initial estimates as discussed in Section 5.2.

12.   The partial derivative of the fixed coefficient provides much the same information. If the partial derivative for a coefficient that has been fixed at an incorrect value is zero, this strongly indicates identification problems. If a particular coefficient in the current model has a nonzero partial derivative, this implies that the coefficient is probably identified (and hence could be estimated in an expanded model that includes this coefficient), since the nonzero partial derivative indicates the value of the fitting function changes if the value of the coefficient is changed from its current value.

13.   It is possible to estimate models having concepts with no indicators (e.g., Bielby, 1981; or any article having "second-order factor analysis" in its title), but such models should not be attempted until you have read Chapter 7 and estimated several regular LISREL models.

14.   Actually, a model can be said to be misspecified if anything about the model fails to correspond to the real world. Specification errors include: omitting important paths (coefficients), including paths having incorrect causal directions, using incorrect functional forms (using an additive model when an interactive model is required or using a linear model when a nonlinear model is required), omitting spurious causes or causally effective correlates of ineffective but included exogenous variables, incorrect error specifications (unjustifiably assuming the error variables are independent from one another or from the exogenous variables), failing to achieve an interval level of measurement, and modeling data sets that have not reached equilibrium. We prefer to discuss most of these issues as separate topics even though these misspecifications share the common characteristic of potentially leading to biased estimates.

15.   The necesary equivalence between a complex model and a correctly simplified version of the model containing fewer variables is used by Herting and Costner (1985) as a diagnostic tool for testing correct model specification. If the effects independently estimated for the simplified model correspond to the total effects (see Chapter 8) implied by the original model, we have an additional reason to believe in the veracity of the overall model specification.

Chapter 6

Hitting Paydirt

Having specified a model in Chapter 4 and estimated its coefficients in Chapter 5, we turn now to an assessment of the adequacy of estimated models. Even models bearing little resemblance to reality can be specified and estimated. And, if maximum likelihood estimation has been used, we can even claim that the estimates maximize the likelihood of S arising as a sampling fluctuation around Σ. The catch is that the obtained estimates *provide the best fit between S and Σ, conditional on the theoretically dictated placement of the free coefficients in the model.* If the model omits important effects (incorrectly inserts fixed zero coefficients), even the best estimates of the free coefficients may be unable to create a Σ closely resembling S.

This chapter introduces a χ^2 test assessing the fit between Σ and S and hence provides an omnibus test of the model and the maximum likelihood estimates of its free coefficients. If even the best (maximum likelihood) estimates result in a poor match between Σ and S, the model should be revised or discarded. If the estimated model provides a Σ closely matching S, the model has survived a potential discreditation, but we cannot claim the model has been proven. Several models might give equally acceptable fits.

This chapter also introduces several other ways of assessing the adequacy of the estimated model, including examination of residuals, correlations among the estimates, the significance of the estimates, partial derivatives, and modification indices. The chapter concludes with a discussion of the standardized solution, one of whose many uses is the detection of unacceptable models.

6.1 Chi-Square

6.1.1 The Likelihood-Ratio Chi-Square

In Chapter 5 we developed the function maximized in obtaining maximum likelihood estimates of the free coefficients in a model. We saw that if a random sample is selected from a multivariate normal population having covariance matrix Σ, the likelihood of finding a sample with covariance matrix S is given by the Wishart distribution (Eqs. 5.3 and 5.4).

Rather than maximizing the Wishart distribution directly to obtain the maximum likelihood estimates, we introduced two additional steps: we developed the ratio of the likelihood for "our" model to the likelihood for a perfectly fitting model in Eq. 5.5, and we took the logarithm of this ratio in Eq. 5.7. We argued that neither of these steps altered the values of the coefficients that maximized the likelihood, because the first step amounts to dividing by a constant, and the second step is a monotonic transformation that merely changes the scale on which the maximum is located. Thus, rather than directly maximizing the likelihood of S given Σ (Eq. 5.3), we maximized the logarithm of the likelihood ratio (Eq. 5.7).

The reason we maximized the logarithm of the likelihood ratio rather than the likelihood itself was to allow us to capitalize on a statistical proof developed by Neyman and Pearson (1933) and discussed by Morrison (1976) and Wilks (1938, 1962). They showed that *minus twice the logarithm of a likelihood ratio is asymptotically distributed as a χ^2 variable with degrees of freedom equaling the difference between the number of free coefficients in the models represented in the denominator and numerator of the ratio.* That is, Eq. 5.8 times -2 yields a quantity that should be distributed as a χ^2 variable and thus can be used for a χ^2 test. Hence,

$$\chi^2 = -2(-\tfrac{1}{2}n[\text{tr}(S\Sigma^{-1}) + \log|\Sigma| - \log|S| - (p + q)]) \qquad 6.1$$

This has the following rearrangements:

$$\chi^2 = 2n(\tfrac{1}{2}[\text{tr}(S\Sigma^{-1}) + \log|\Sigma| - \log|S| - (p + q)])$$

$$= 2n(\text{value of fit function minimized within LISREL}) \qquad 6.2$$

$$= n[\text{tr}(S\Sigma^{-1}) + \log|\Sigma| - \log|S| - (p + q)]$$

$$= nF$$

LISREL calculates this likelihood ratio χ^2, its associated degrees of freedom ($d.f.$), and probability whenever maximum likelihood estimates are computed.

For the smoking model developed in Chapter 4 and estimated in Chapter 5, for example, LISREL reports a $\chi^2 = 2.28$ with 1 $d.f.$, which has probability .131. Recalling that n is the sample size minus 1, and checking the fit function on page 344 in Appendix B confirms that the χ^2 of 2.28 is $2n$(minimum of fit function) $= 2(432 - 1)(.002641)$. But what does this tell us about our model? Since the probability of .131 does not exceed the usual .05 criteria, we must accept (fail to reject) that the *differences between the model-implied covariance matrix Σ and the observed S are small enough to be sampling fluctuations.* The probability of .131 indicates that S's differing this much or more from Σ should appear in about one in every eight samples if Σ was the true population covariance matrix. Note that *smaller χ^2 values indicate better fitting models*, and that *an insignificant χ^2 is desirable*, since it says the model's predicted Σ is sufficiently close to the observed data S for the remaining differences to be mere sampling fluctuations.

Adopting the usual .05 level of significance amounts to accepting models as adequate if the observed sampling fluctuations could appear in about 1 in every 20 samples. This is not particularly strong confirmation of a model. Since accepting the null hypothesis amounts to accepting one's theory (a reversal of the usual role of the null hypothesis), it would seem preferable to use a .1 or even a .2 level of significance (cf. Fornell, 1983).

To be precise, interpretation of the χ^2 test should acknowledge the existence of an implicit alternative hypothesis. Recall that in creating the likelihood ratio (Eq. 5.5), we introduced an imaginary perfectly fitting model whose implied Σ matched S perfectly. If we think of models as consisting of a multitude of constraints (on the location of coefficients, on the values of particular coefficients, and on the functional form of the relationships between the variables), the hypothetical perfect model might be likened to an "unconstrained" model where there is nothing to keep (constrain) the imaginary model from perfectly reproducing the observed covariance matrix. From this perspective, we can inquire whether the package of constraints embodied in the estimated model gives a significantly worse fitting Σ than the Σ for the hypothetical unconstrained alternative. For the smoking model, the insignificant χ^2 tells us that though our model's constraints have produced an implied Σ differing slightly from the perfect Σ, the differences are not large enough to significantly reduce the likelihood of S arising as a mere sampling fluctuation. That is, our model is not significantly worse than the perfect model capable of predicting a Σ that is identical to S.

The degrees of freedom for the χ^2 test are calculated as the difference between the total number of unique entries in the covariance matrix (the observed variances/covariances) and the total number of coefficients estimated in our model. The formula for degrees of freedom is

$$d.f. = \frac{1}{2}[(p + q)(p + q + 1)] - t \qquad 6.3$$

where t is the total number of estimated coefficients and $p + q$ is the total number of observed indicators (p endogenous, q exogenous). The quantity $\frac{1}{2}[(p + q)(p + q + 1)]$ is the total number of nonredundant elements in a covariance matrix made with $p + q$ variables. This can be seen by rewriting this as $\frac{1}{2}[(p + q)(p + q) + (p + q)]$ and subsequently as $\frac{1}{2}(p + q)(p + q) + \frac{1}{2}(p + q)$. The first of these terms is half of the total entries in a $(p + q) \times (p + q)$ covariance matrix (say the lower triangular set of covariances and half the variances), and the second term gives the other half of the variances $[\frac{1}{2}(p + q)]$.

This equation becomes more meaningful if we stress the idea that the degrees of freedom for χ^2 are the difference between the number of free coefficients in the denominator and numerator of the likelihood ratio, as mentioned in conjunction with Eq. 6.1. A perfect model constituting the denominator of this ratio might be created by assigning one model coefficient to account for each of the observed variances and covariances. Estimating this model would guarantee a perfect fit because each model coefficient would provide a perfect prediction for one of the covariances. The total number of estimated coefficients in such a hypothetical perfect model would equal the total number of variances/covariances, that is, $\frac{1}{2}(p + q)(p + q + 1)$. Thus Eq. 6.3 is the difference between the number of coefficients that would have to be estimated for a perfectly fitting model and the total number of coefficients estimated in our real model (t).

From this perspective, the *d.f.* represents the degree of representational compactness of one's model. If a well-fitting model has many degrees of freedom, it indicates the model has come close to duplicating the observed covariance matrix much more parsimoniously than merely allocating a single model coefficient to each covariance. *One should strive for models having many d.f. (few estimated coefficients) since the larger the d.f. the more parsimonious is the prediction (implication) of an acceptably fitting Σ.*

If our model has as many estimated coefficients as input variances/covariances, there will be no degrees of freedom left for a χ^2 test. Indeed, with the smoking model if γ_{32} is freed, then there are zero degrees of freedom, χ^2 becomes zero, and the model implies a Σ that is identical to S, as does the perfect comparison model with the same number of estimated coefficients. Again, even the ability to perfectly predict/reproduce S does not guarantee that the correct model has been specified and estimated; it merely means we have located a set of estimates that make the model consistent with the observed data (cf. Stelzl, 1986).

Observing that the freeing of γ_{32} produces a perfect model seems to imply that the less than perfect χ^2 for the model with γ_{32} fixed at zero provides a χ^2 test of the necessity of just the γ_{32} coefficient (sometimes worded as a test of the overidentifying restriction implied by fixing γ_{32} at zero). Strictly speaking, this is not true. The χ^2 tests whether this model, with all its constraints (including γ_{32} equaling zero), implies an acceptable Σ, but not whether γ_{32} is necessary in and of itself. For example, consider leaving γ_{32} fixed at zero, but free one additional coefficient and fix one additional coefficient. Fix β_{21} at zero to allow no effect of smoking on AS views, and simultaneously free the covariance between the structural disturbance terms influencing AS views and AS acts (ψ_{32}). This model also has 1 *d.f.* and γ_{32} fixed at zero, but the χ^2 is a highly significant 20.09 in comparison to the previously insignificant 2.28 for the original model. Clearly, the χ^2 test depends on the particular model examined and not merely on the single coefficient γ_{32}.

You may be tempted to rebut the preceding comments by pointing to "two" fixed coefficients in this revised model (γ_{32} and β_{21}) compared to the "single" fixed coefficient in the original model (γ_{32}). If you empathize with this, you are demonstrating precisely the myopia grounding our admonition against thinking of χ^2 as a test of any single fixed coefficient (γ_{32} in this case). Are there "two" fixed coefficients in the revised model and "one" in the original model? Both models contain the *same large number of fixed coefficients*. Both models fix all the Ψ, Θ_ϵ, and Θ_δ covariances at zero (with the exception of freeing ψ_{32} and the offsetting fixing of β_{21} in the revised model). Both models fix all the Λ values to either 0.0 or 1.0, and both models fix the measurement error to the same predetermined portions of the variance in the various indicators and about half the B coefficients. The point is, all models include numerous specifications of fixed coefficients, and there is no justification for claiming the model χ^2 is more a test of any one of these coefficients than any other, whether the fixed zero falls in B, Γ, Ψ, Θ or Λ. We should acknowledge χ^2 as an omnibus test of all the model constraints.

Tests of significance for single coefficients are best made via the t-tests discussed in Section 6.4 or as a special case of the difference between χ^2's for nested models, as discussed in the following section.

Finding an insignificant χ^2 does not prove that one has located the right model. It does indicate we have located a model and set of coefficient estimates that are consistent with the observed covariances, and that the model has survived a challenge that results in the failure of many models.

6.1.2 Chi-Square and Nested Models

To illustrate another use of χ^2, we introduce an alternative model of smoking behavior suggested by a skeptic whose personal observations convince him or her that attending religious services makes no difference to smoking behavior or engaging in AS acts. Although our skeptic feels that attendance is unrelated to what people do, the skeptic concedes that attendance could possibly influence what they say—their antismoking views.

The skeptic's beliefs center on two coefficients in the smoking model: both γ_{14} and γ_{34} should be zero if the skeptic's beliefs are justified. To test the skeptic's model, we fixed these two coefficients to zero values and reestimated the model. The χ^2 for the skeptic's model is 15.96 with 3 $d.f.$ and probability .001. Hence, the model including the skeptic's views fails in that even the maximum likelihood estimates for the coefficients in this model are unable to imply a Σ that could easily account for the observed covariances S.

But the argument is unlikely to end here. A true skeptic will undoubtedly retort that the χ^2 test is an omnibus test testing *all* the model assumptions, including measurement assumptions, the "zero" impact of age on acts, the structural disturbance specifications, and, only incidentally, the skeptic's hypothesis. Any of these assumptions (model constraints) could be wrong and hence provide the unacceptably large χ^2. We might remind the skeptic that the inclusion of all the other model constraints resulted in a model that fit acceptably even if not perfectly. But the skeptic insists, and we must agree, that the omnibus nature of the χ^2 test implies that we contaminated our test of the skeptic's hypothesis with a test of, from the skeptic's perspective, irrelevant hypotheses.

Fortunately, we can purify our test of the skeptic's views by adopting the general strategy to which this section is dedicated. Consider estimating two models, one of which is nested within the other in that it can be created from the other model by imposing additional model constraints. These constraints may be the fixing of specific coefficients at zero (as in the skeptic's model), but constraining coefficients to be equal or to have specific nonzero values are also acceptable. (Models based on different sets of observed variables cannot be nested.)

Imagine further that both models have been estimated, and the χ^2's calculated such that χ_1^2 with $d.f._1$ appears for the basic model and χ_2^2 with $d.f._2$ appears for the model with the additional restrictions. Model 2 with its additional restrictions should have a larger χ^2 than model 1, and the $d.f.$ for model 2 should be larger than the $d.f.$ for model 1, because fewer coefficients are estimated for the more restricted model. That is, $\chi_2^2 > \chi_1^2$ and $d.f._2 > d.f._1$. For the basic smoking model and the more restricted skeptic's model, $\chi_2^2 = 15.96 > \chi_1^2 = 2.28$ and $d.f._2 = 3 > d.f._1 = 1$.

We now use the fact that *the difference between the two χ^2's is also distributed as a χ^2 with degrees of freedom equal to the difference between the degrees of freedom for the two models.* That is, we can create a χ_3^2 with $d.f._3$ as

$$\chi_3^2 = \chi_2^2 - \chi_1^2$$
$$13.68 = 15.96 - 2.28$$

6.4

and

$$d.f._3 = d.f._2 - d.f._1$$
$$2 = 3 - 1$$

6.5

where we recognize that the 2 $d.f.$ occur because the skeptic's model contains two fewer coefficients to estimate—the two coefficients fixed to zero.

Testing to see if $\chi_3^2 = 13.68$ is significant with $d.f._3 = 2$ indicates whether the skeptic's additional constraints have significantly reduced the model's ability to fit the data. In this case the observed χ_3^2 has $p < .01$, so the fit of the model has been significantly hindered by introducing the additional constraints. It seems intuitively reasonable that if all the other model constraints contributed a misfit of magnitude 2.28 (as measured by χ^2), only the extra 13.68 can be attributed to the "erroneous" skeptic's views, but the statistical basis for this difference between χ^2's providing a new χ^2 is worth investigating.

One way of seeing that *the difference between χ^2's is also a χ^2* arises in the context of the likelihood ratio. Recall that $-2 \log(\text{likelihood ratio})$ is distributed as a χ^2 variate (Eqs. 6.1, 5.5, and 5.7).

$$\chi^2 = -2 \log\left(\frac{\text{L model}}{\text{L perfect model}}\right)$$

6.6

Hence the difference between χ^2's for two nested models can be expressed as a difference in likelihood ratios, where the more restricted model has the larger χ_2^2.

$$\chi_2^2 - \chi_1^2 = -2\left[\log\left(\frac{\text{L model 2}}{\text{L perfect model}}\right) - \log\left(\frac{\text{L model 1}}{\text{L perfect model}}\right)\right]$$

6.7

Since the difference between logs corresponds to division, this can be rewritten as

$$\chi_2^2 - \chi_1^2 = -2 \log\left[\frac{\dfrac{\text{L model 2}}{\text{L perfect model}}}{\dfrac{\text{L model 1}}{\text{L perfect model}}}\right]$$

6.8

from which the likelihood of the perfect model cancels, leaving

$$\chi_2^2 - \chi_1^2 = \chi_3^2 = -2 \log\left(\frac{\text{L model 2}}{\text{L model 1}}\right) \qquad 6.9$$

Hence the difference between the χ^2's for the original and more restricted models is itself a χ^2, because it corresponds to minus twice a log(likelihood ratio) (cf. Eqs. 6.1, 6.2, and 5.7). Furthermore, the implicit comparison tested by the difference χ_3^2 centers on only the basic model 1 and the restricted model 2 rather than involving comparison to the hypothetical model providing a perfect fit of Σ to S. Subsequently, the specific hypothesis tested by χ_3^2 is whether *the restrictions added during the creation of the restricted model 2 significantly reduce the fit compared to the fit attainable with all the model restrictions incorporated in the basic model 1.*

For our example, the $\chi_3^2 = 13.68$ with 2 *d.f.* tests whether the two additional zero restrictions result in a significant worsening of the fit attainable with the basic smoking model and all its attendant restrictions. Since with 2 *d.f.*, a χ^2 of 13.68 appears with probability less than .01, we must conclude that the skeptic's claim results in a significant worsening of the model's ability to fit the data, and hence the skeptic's claim is untenable.[1]

Although the "difference χ^2" test of the skeptic's hypothesis of null effects no longer confounds the misfit of the original model with the misfit created by demanding null effects, the test is not completely free of the basic smoking model. The basic smoking model provides the denominator in Eq. 6.9. The skeptic's constraints produce a significant worsening of *this* model, and hence if we had a completely wrong model, the test might still be invalid. The onus, however, is clearly on the skeptic to provide another well-fitting model to which addition of the two extra constraints provides no significant decrement in fit.

The strategy of using the difference between model χ^2's as a test is applicable whenever one can create a more restricted model by placing additional constraints on some basic model. This is inefficient, however, if only a single coefficient is fixed at zero, because an equivalent test of the significance of a single coefficient is available within LISREL, as discussed in Section 6.4. If a test of fixing a single coefficient to a *nonzero* value is required, the difference χ^2 procedure is appropriate. Testing whether a significant reduction in fit results from fixing a whole set of coefficients at nonzero values is comparable to reporting on the power of the test to distinguish between the estimated and specified values. Another procedure for estimating the power of the test to distinguish between obtained estimates and researcher-specified values has recently been proposed by

Satorra and Saris (1985) (cf. Saris and Stronkhorst, 1984:202), but the statistical foundations of the procedure are too advanced for presentation here. (On power, see Matsueda and Bielby, 1986.)

The difference χ^2 procedure is *also useful for testing the significance of improvements in an initially ill-fitting model*. If one has an initially poor model with huge χ^2, an obvious next step is to free one or more coefficients (reduce the number of model constraints) in hopes of improving the model's fit. If the reduction in χ^2 (the χ^2 difference) is large relative to the difference in *d.f.* between the two models, we have achieved a significant improvement in fit. Several model revisions may be required before an acceptable model χ^2 is achieved.

We caution that though the original model χ^2 and each of the difference χ^2 tests for previously planned model modifications are legitimate, the χ^2 for the ultimate model may be compromised as an omnibus test of the ultimate model's fit, just as scanning hundreds of correlations to locate a few significant correlations compromises the selected tests of significance. That is, there is *an inherent and fundamental difference between using χ^2 for testing a model and using χ^2 as one of the tools for incrementally improving the fit of a model*. From the viewpoint of pure testing, we have compromised the ability of χ^2 to test the ultimate model the instant we change anything about the model on the basis of the observed covariances or on the basis of previous attempts to fit the model to the data (and specifically the χ^2 for any previous attempts). *Once the data has been used to fix the model, that data no longer provides a pure test of the model.*

This principle provides an answer to another frequently encountered problem. Suppose we estimate an original model and find that several of its coefficients are insignificant (Section 6.4). Should we omit the insignificant coefficients and reestimate the model (with a reduced number of estimated coefficients and hence increased degrees of freedom), or should we report the model containing the insignificant coefficients? Unless there is some reason for omitting the coefficients beyond their mere insignificance, the foregoing argues that *the model should not be reestimated with the insignificant coefficients deleted*. The insignificance of the coefficients is a function of the input data. Hence, using this information to alter the model (deleting the coefficients) compromises the model χ^2 test because it amounts to incorporating data-prompted model revisions.

6.1.3 Chi-Square and *N*

One feature of χ^2, and indeed most tests of significance, is that *with large sample sizes even minute differences tend to be detectable as being more than mere sampling fluctuations and hence significant*. Consider what would happen to the smoking model if we had taken a sample that was 100 times as large as the real sample (namely 43,200 = 100 × 432) and found exactly the same covariance matrix among the indicators. Since the covariance matrix has not changed, LISREL would provide exactly the same estimates of the free coefficients as reported in Figure 4.3, but the model χ^2 (and the t-tests for the coefficients in Section 6.4) would have changed. Since $\chi^2 = -2nF$ and since the same fit and, hence, minimum of the fitting function would appear (the same data covariance matrix and model are entered), the χ^2 for the hypothetical 100-fold larger sample would be

$$\chi^2 = -2(100n)F = 100(-2nF) = 100(\text{original } \chi^2) \qquad 6.10$$

This implies that large data sets are likely to produce significant χ^2's, not because the fit between Σ and S is any worse, but because *with larger sample sizes, smaller differences are detectable as being more than mere sampling fluctuations.*

This feature of χ^2 has received considerable attention in the literature and prompted several suggestions for corrective strategies. Joreskog (1969) proposed expressing χ^2 relative to the degrees of freedom—that is, calculating $\chi^2/d.f.$—as a more appropriate measure if N is unusually large. The issue then became one of deciding how many times larger χ^2 must be than the degrees of freedom (i.e., how big the $\chi^2/d.f.$ ratio must be) before rejecting a model as ill fitting. Wheaton et al. (1977) suggest that a χ^2 five times the degrees of freedom is reasonable, and Carmines and McIver (1981) suggest two or three times is more acceptable.[2]

Hoelter (1983a) *argues against use of the* $\chi^2/d.f.$ *procedure* and suggests instead that we *refocus our attention on the issue of the size of N rather than on d.f.* (which are a function of the number of variables and estimated coefficients, not of N). He provides a formula for what he calls the "critical-N," which is the size of the sample that would be required to make the observed difference between Σ and S just significant at a typical critical level of significance such as .05. He then reports that "examination of numerous models" convinces him that critical-N's of 200 or more are a reasonable cutpoint. Essentially the same decision criterion can be obtained, without the use of Hoelter's formula, by simply inserting $N = 200$ into the LISREL program and using the χ^2 that results. If the real N exceeded 200, this is tantamount to ignoring the extra sensitivity or precision provided by the extra cases. If the real N is less than 200, this is tantamount to asking if the observed differences are large enough to be detected by a reasonable sized sample ($N = 200$).

My preference is to use χ^2 as one of several (see Sections 6.2, 6.3, and 6.5) indicators of the quality of model fit, paying most attention to it if N is modest. If N is large, even trivial departures between Σ and S can be significant, but if N is small, χ^2 may have insufficient power to detect substantial differences. My experience suggests that χ^2 is instructive for N's ranging from about 50 to 500, but I suspect this range depends on the kinds of models estimated. Anderson and Gerbing (1984) summarize what is known about the sensitivity of various goodness of fit measures and find that χ^2 compares favorably with several other indices.[3]

The issue of minimum sample size may ultimately be decided on the basis of considerations other than χ^2. Boomsma (1985) reports that improper solutions (nonconvergence of the iterative procedure, or negative error variance estimates) become bothersome if N is less than 100. Between 1% and 35% of his LISREL runs based on $N=50$ failed to converge, and between 0% and 16% of the error variance estimates were negative. The substantial difference between the results for different types of models (for some models N's of 50 produced convergence in 99% of the runs) makes generalization difficult, but N's less than 100 definitely deserve extra attention.

6.2 Residuals

The quality of model fit should be assessed in the context of the substantive concerns motivating model construction. One way to do this is to examine the discrepancies between the observed covariances and the model-implied covariances $S - \Sigma$. To place these residual covariances in the context of the smoking model, you might locate, in Appendix B, Σ, which is labeled "fitted moments," and $S - \Sigma$, which is labeled "fitted residuals."

In the smoking model the only substantial residual is for the covariance between age and antismoking acts with an observed covariance of -1.366 (Appendix B, LISREL p. 4), a model-implied covariance of $-.195$ (Appendix B, LISREL p. 17), and hence a "residual" of -1.171. The placement of this residual is not surprising, because it corresponds to the zero γ_{32} that eliminated any direct effect of age on AS acts. Even this "largest" residual is small enough to be of little concern. The covariances in this case correspond to an observed correlation (using Eq. 1.54) of $-.076$ being fitted by an implied correlation of $-.011$. Clearly, we would be hard pressed to justify extended discussion of even this largest difference.

We could reduce this residual by including an effect of age on AS acts (by freeing γ_{32} for estimation), but we know this would not significantly improve the model. Estimating this coefficient results in a perfectly fitting model: $\Sigma = S$, $\chi^2 = 0$, and $d.f. = 0$ (since this is a recursive model containing completely fixed measurement errors), but the *improvement* in χ^2 remains insignificant because the basic model χ^2 (with 1 *d.f.*) was insignificant. Since complete elimination of the discrepancy between the implied and observed covariances between age and AS acts produces only an insignificant improvement in the model, this largest residual (the poorest fit covariance) is statistically insignificant as well as being substantively unimportant.

Residual covariances, in general, are sufficiently important that either the full set of residuals or a discussion of the pattern in the residuals should be reported if no acceptably fit model has been located. Even insignificant residuals (for models with acceptable χ^2's) occasionally display informative residual patterns. Residuals may be diagnostic of model problems (Costner and Schoenber, 1973), but the largest residuals do not always pinpoint a model's deficiencies (Sorbom, 1975). Numerous model features contribute to the modeling of most covariances, and any of these features may be the "problem" creating the large residual (e.g., the absence of a direct effect, the absence of indirect effects, an omitted reciprocal effect, an omitted common cause, covariances among the measurement errors or structural disturbances, improperly fixed effect coefficients, mismodeled covariances among the background causal variables, use of a completely wrong model, entering the wrong data matrix, or entering a data matrix containing incorrectly coded data). Though residuals may not be totally diagnostic, *they perfectly describe the fit between the data and the model, and they should always be carefully examined.*

6.2.1 Standardized Residuals

LISREL reports residual covariances in their real metrics and as standardized residuals which help locate the largest residuals and patterns that may appear in the residuals. What LISREL reports as "normalized residuals" are actually "standardized residuals," because each is "the residual covariance . . . divided by the square root of its asymptotic variance" (Joreskog, 1981a:91). The standardized residuals are estimates of the number of standard deviations the observed residuals are away from the zero residuals that would be provided by a perfectly fitting model. If only random errors remained in these residuals, we would expect most (all but about 5%) of the standardized residuals to be within two standard deviations of zero (a perfect fit) (cf. Mosteller, Rourke, and Thomas, 1961:203). Furthermore, if each residual is assumed to arise as the sum of several similarly distributed random events (a requirement that is incidental and not a formal requirement of the LISREL model), the residuals will also be approximately normally distributed (cf. Mosteller,

Rourke, and Thomas, 1961:280).

To assist the researcher in assessing if the residuals are normally distributed, LISREL provides what is called a Q–Q plot (see Appendix B or LISREL VI:III-16). Q–Q plots are discussed by Wilk and Gnanadesikan (1968) and Gnanadesikan (1977: Chap. 6). They are created as follows. For each standardized residual (plotted on the horizontal axis and labeled "normalized residuals") a corresponding value selected from a standard *normal* curve will be plotted on the vertical axis (labeled "normal quantiles"). The "corresponding" value from the standard normal curve is the value having the same relative placement within the normal distribution that the standardized residual has within the distribution of residuals. That is, if 70% of the standardized residuals have values smaller than the standardized residual located on the horizontal axis, the vertical placement of the point is determined as the value from the standard *normal* curve that has 70% of the normal distribution below it.

If the residuals are normally distributed, we get a line of points rising at 45° in the Q–Q plot. If the points in the Q–Q plot are nonlinear, then the residuals are not normally distributed. If the points in the Q–Q plot fall on a line sloped more steeply than 45°, the residuals are normally distributed and less variable than would be expected on the basis of the asymptotic variances used in standardizing the residuals. If the points fall on a line sloped less steeply than 45°, the residuals are normal and more variable than would be expected on the basis of the asymptotic variances used in standardizing the residuals. Boomsma (1982:162, 169) provides some comparative illustrations that assist in visualizing the information conveyed by Q–Q plots.

Outliers in the Q–Q plot provide a convenient way of locating the most poorly fit covariances. (The value on the horizontal axis for an outlier should equal the value in the standardized residual matrix, and the corresponding location in the covariance matrix provides the offending covariance.) Values greater than $+2$ or less than -2 on the Q Q plot (or in the standardized residual matrix) indicate substantial residuals and hence demand attention. Nonlinearities in the Q–Q plot may also signify problems, but no firm guidelines are available.[4]

6.3 Fitting Better Than the Competition

Another way to assess model fit, one that emphasizes substantive concerns, is to assess the fit of our favored model in the context of the fit attainable with competing substantive models. Bentler and Bonett (1980) and Bentler (1982), for example, suggest seeking out a "null model" to see if our favored model fits significantly better than no model at all. This may be useful with extremely small N to assure ourselves that the χ^2 test has sufficient power to discriminate between models, but there is a problem of deciding exactly what a null model is (Sobel and Bohrnstedt, 1985). In exploratory factor analysis, it is a model with only zero factor loadings, so there are no "common" factors. In models such as our smoking model, a null model might set all $\mathbf{B}$ and $\mathbf{\Gamma}$ coefficients to zero so that "no effects" are allowed, but even this is not quite a nonmodel because the measurement error specifications remain as defended during development of the model in Chapter 4. Although the failure of this model would ensure that some significant effect coefficients could be found, this same conclusion can be reached by investigating the significance of the individual coefficients (Section 6.4).

An insignificant χ^2 difference between a basic model and a nested yet substantially different model implies there is no significant difference between the quality of the fit provided by the models. It therefore clearly indicates the ambiguity in the support for the original model, whether or not the alternative model is a null model.

A preferable strategy is to seek real competing models from the literature or to create meaningful alternative models if the literature provides no competitors (just as we created the imaginary skeptic's model). Sobel and Bohrnstedt (1985) provide a practical example of how to create a substantively meaningful comparison model, and they conclude that creation of such models requires careful attention to what is already known in the relevant research domain. This stands in stark contrast to the procedure for developing alternative models as embodied in the recent TETRAD program (Glymour and Scheines, 1986), where a demand for consistency with the data drives the search for a better model.

If the competing models are not nested, we cannot use the procedure of creating and testing a difference χ^2. The likelihood ratio χ^2 procedure demands that the models compared in the numerator and denominator of the likelihood ratio be nested, or in statistical jargon, that the coefficients in one model have a parameter space that is a restricted form of the parameter space for the other model (cf. Morrison, 1976:21).

For models that are not nested (where we cannot convert one model into the other by adding restrictions), the model with the smaller χ^2 is automatically preferred, because a smaller χ^2 implies that the model has a Σ that more closely matches $\mathbf{S}$. *If many more estimated coefficients are required by the better fitting model, however, we must choose between parsimony and fit.* Here again, we will want to investigate the location and size of the residual covariances in making a decison. The adequacy of a model also depends on what we were seeking from the model in the first place, so assessments of model adequacy should also reflect the substantive concerns that prompted model development. Blalock (1986), for example, is willing to sacrifice parsimony if it is attained at the expense of generalizability.

6.4 Significance of Structural Coefficients

Having obtained estimates of the model coefficients and tested the fit of the overall model, we now inquire about the significance of each model coefficient individually. As in multiple regression, some of the estimated coefficients may imply large effects, whereas others are sufficiently close to zero that they are likely to be mere sampling fluctuations around a zero population parameter.

The key to assessing whether any estimated coefficient's value can be reliably distinguished from zero resides in determining the size of estimates likely to arise as mere sampling fluctuations. If we repeatedly took random samples from a population having zero effect between two variables, the estimates of that effect coefficient might not be identically zero but could be slightly above or below zero, depending on which particular sample of cases provided the covariance matrix S. Plotting the values of the repeated estimates would provide a sampling distribution for the estimate, and the variance (standard deviation) of this distribution informs us about how far sampling fluctuations are likely to carry the estimate above or below the true population value.

Following traditional hypothesis testing procedures, *we can reject a hypothesis that the population parameter is zero if we observe an estimate that is more than about two standard deviations (standard errors) away from zero.* Estimates closer than this are likely to be mere sampling fluctuations and not real effects. And, following traditional confidence interval procedures, *we can create an interval estimate of the population parameter by creating an interval covering about two standard deviations (standard errors) above or below the observed estimate.*

The key to testing for the significance of estimates is the standard deviation of the sampling distribution for the estimate. Fortunately, statisticians have developed procedures for obtaining an estimate of the standard deviation of the sampling distribution without repeated sampling, but unfortunately the mathematics is so complex that we can only summarize the steps here. Boomsma's (1982:150) summary of the basic proofs by Cramer (1946:500ff.) runs as follows. If we estimate the model coefficients using MLE and if the basic assumptions underlying the model are true (we have a multivariate normal distribution and have located the correct model), then

1. The estimates are consistent in that they converge in probability to the true population parameters as N increases.
2. The estimates are efficient in that with increasing N their sampling distributions have minimum variance.
3. Asymptotically, the sampling distribution of the estimates is multivariate normal with means equaling the true population parameter values and a variance/covariance matrix that is the inverse of Fisher's information matrix.

Fisher's information matrix is the probability limit of, or the asymptotic mean of, the matrix of second-order partial derivatives of the fit function with respect to the estimated coefficients. Joreskog (1973a, 1977a) discusses this in detail. Hence, the standard deviations of the sampling distributions of the estimates are the square roots of the diagonal elements (variances) of the variance/covariance matrix for the estimates. LISREL reports these as the "standard errors" of the estimates. See Appendix B for the standard errors for the smoking model.

All the estimated coefficients, not just the effect coefficients in **B** and **Γ**, are given standard errors, so we can test hypotheses or create confidence intervals for measurement error variances and covariances (if they are estimated), structural disturbance variances and covariances (if estimated), and the measurement loadings of the indicators on the concepts (if estimated).

To assist with the routine tests of the hypothesis of zero effects (or zero error variance, etc.), LISREL provides optional output labeled "T values," which are the coefficient estimates divided by their standard errors. The T values thus provide the number of sampling distribution standard deviations the estimate is away from zero and hence can be used to test the null hypothesis that the true parameter value is zero. We select a desired level of significance (alpha or type I error) and then enter a normal probability table (not a t-table, which we might expect from the labeling of LISREL's output) to obtain the corresponding critical value. If LISREL's T value is greater than the critical value, we can reject the null hypothesis of a zero parameter at the preselected level of significance.[5] If a coefficient is insignificant, even though its magnitude is large enough to be substantively important, this may reflect a lack of power in testing the implicit hypothesis of zero effect, and we should withhold judgment about the coefficient until further data are available.

Since the sampling distribution of the estimates is multivariate normal, each estimate's sampling distribution is normal (Section 5.2), and tables of normal probabilities can be used to create confidence intervals of any desired accuracy. For example, the interval "estimate ± 2.6 standard errors" provides a 99% confidence interval for the corresponding population parameter.

A caution is in order regarding the word "asymptotically" in point 3 earlier. The mathematical proof requires only that the sampling distribution of the estimates attain the stated standard deviations and multivariate normal shape when N reaches infinity. Though we might expect estimates based on reasonably sized samples to approach multivariate normality with the stated standard deviations, no general criteria are yet available to determine precisely how large N must be before a reasonable approximation is attained. (A similar comment applies to χ^2, because the log likelihood ratio is only asymptotically distributed as a χ^2 variate.) Gerbing and Anderson (1985) report a recent Monte Carlo study in which LISREL's standard errors are found to be very acceptable for N's ranging from 50 to 300, but this is based on a known multivariate normal population distribution and a factor analytic (as opposed to path analytic) type of model.

One response to the "only asymptotically" problem is to routinely require a higher level of significance than the usual .05 level (2 standard errors), say a nominal .012 level or 2.5 standard deviations. The inability to justify this or any other particular higher level, however, suggests we might do equally well to adopt a less rigid philosophy of statistics and conscientiously avoid making firm decisions on the basis of whether a test statistic falls just above or just below any arbitrary critical value.

6.4.1 Correlations among the Estimates and Colinearity

The preceding section emphasizes the variances of the estimates. The inverse of Fisher's information matrix also provides the covariances among the estimates (point 3 in the preceding section). If the entire variance/covariance matrix among the estimates is standardized (Eq. 1.54), it becomes a correlation matrix among the estimates, which LISREL provides as optional output (again see Appendix B). A low correlation in this matrix says that LISREL's estimate for one coefficient is only minimally related to the estimate it provides for the other coefficient. A correlation near 1.0 (or -1.0) indicates that the value estimated for one coefficient almost perfectly predicts the value estimated for the other, merely another way of saying there is colinearity among the estimates of the two coefficients (recall Section 5.3). Thus, scanning the matrix of correlations among the estimates is a handy way to check for colinearity problems.[6]

Here we encounter another "how big is big" problem. The most frequently cited value in the literature is that a correlation of .9 or more between two estimates might indicate colinearity problems. But I have seen models where correlations of .8 made me very uncomfortable (especially when I was not expecting even a .2 correlation among the estimates), and I have also seen models where even values of .95 did not disturb me—an example being models where both a Λ and corresponding Θ are estimated, but where the measurement structure is well established in that both the estimated Λ and Θ are suspected of falling in predictable narrow ranges; that is, where one was tempted to enter a fixed Λ or Θ.

We close this section by pointing out that the standard errors underlying the calculation of T values and correlations among the estimates depends on the calculation of the *inverse* of Fisher's information matrix. If this inverse is not calculable, as happens when severe modeling problems (such as perfect colinearity) keep the iterations discussed in Chapter 5 from converging to a maximum likelihood, none of the output discussed in Section 6.4 will be available. This has the unfortunate consequence that less diagnostic information is available for the models requiring the most help!

6.5 Partial Derivatives and Their Uses

The partial derivatives reported as "first-order derivatives" in LISREL's output are the partial derivatives of the fit function (Eq. 5.9) with respect to all of the coefficients in the model (both fixed and free). These partial derivatives give the slope of the fit function at the current values of the coefficients and hence indicate the rate at which the fit would change if the coefficients were allowed to take on values slightly larger or smaller than their current values (recall Figure 5.6, the discussion in Section 5.2, and the formulas provided in footnote 3 to Chapter 5). They provide useful diagnostic information for improving models, because large positive or negative partial derivatives indicate that increasing or decreasing a particular coefficient's value is likely to substantially improve the model fit and decrease the model χ^2.

Two features of partial derivatives should be noted. First, they reflect the real metrics of the relevant variables. Thus a unit change on a scale with many scale points (such as "dollars of income") is likely to produce a minimal change in the fit F (and hence a small partial derivative) compared to a coefficient associated with a variable measured on a scale with fewer scale points. Subsequently, judgments about what constitutes a large or small partial derivative must consider the real metrics of the variables.

Second, a partial derivative gives the slope only at the current value of the coefficient. If the coefficient is allowed to increase or decrease slightly, the slope of the fit function (the partial derivative) might change. Consequently, a coefficient with a small partial derivative could radically change the fit if that slope persisted over a substantial change in the coefficient's value, whereas a small change in fit could appear even with a large partial derivative if that slope flattened out after a minor change in the coefficient's value.

These problems have prompted Joreskog and Sorbom to supply other indicators of the amount by which the model fit is likely to change if a coefficient is freed. The precise definition of these "modification indices" has changed from time to time, but the indices have all been based on the partial derivatives and they are designed to adjust for both the scales on which the variables are measured and the rates of change of the partial derivatives (i.e., the second-order partial derivatives).[7]

We should resist making model revisions merely because the model can be improved by freeing a coefficient. That is, do not alter your model merely to take advantage of a particularly large partial derivative or modification index.[8] Whenever we introduce changes in a model on the basis of any model output, we compromise our ability to test the model on that data set. Such compromising can be minimized by making as few changes as possible in arriving at the final model, and by specifying (prior to any estimation) the changes that would be undertaken if the model happened to give a poor fit. Any strategy that reduces the number of changes made or the ability of the data to dictate the changes helps maintain the purity of the overall test of the model's fit. *Model modifications should be nine-tenths theory driven and only one-tenth data driven.* MacCallum even suggests we should be willing "to continue a search beyond the point of finding a model with a nonsignificant χ^2" (1986:118).

Although we should be familiar with the modifications suggested by the partial derivatives (modification indices) in case they suggest overlooked, yet theoretically reasonable, changes, we should strongly resist making any change merely to improve a model's fit. From a scientific vantage point, it is better to reject a model that truly represents a particular theoretical perspective than to accept a model with no discernible theoretical orientation. *Researchers should be strongly encouraged to provide a summary of the modeling history that produced their final published model.*

Splitting the data into random halves is strongly recommended if we anticipate a period of model development that is data coordinated.[9] We develop the model using one half of the data, and we reserve the untouched half for an uncompromised test of the ultimate model. Even with split-half data, we should use few modifications and not rely on data instigated revisions. If the split-half strategy is followed, the procedures in Chapter 9 for "stacking" models can test whether data prompted modifications have significantly capitalized on chance fluctuations in the correlations among the observed variables (cf. Cudeck and Browne, 1983; Entwisle and Hayduk, 1982).

All the partial derivatives for the free coefficients should be zero in LISREL's output. A free coefficient with a nonzero partial derivative implies that the *maximum* of the likelihood function (top of the hill) has not yet been reached, so LISREL's estimates are not truly maximum likelihood.

A zero partial derivative for a *fixed* coefficient indicates that the current fixed value is the best (maximum likelihood) value or that the coefficient would be unidentified if freed. The current and neighboring coefficient values give the same maximum likelihood fit. Thus all partial derivatives in a model can be zero, so no coefficients can be added to improve the model, even though we may not have located the true model. For example, if γ_{32} is freed in the smoking model, a perfect match between the model-implied Σ and S is achieved, and all the partial derivatives are zero. For free coefficients this indicates that the estimated coefficients locate a maximum where the slopes of the fitting function are zero (recall Figure 5.5). For fixed coefficients this indicates that no further coefficients are identified. Adding even one coefficient leads to an unidentified model. Thus the zero partial derivatives indicate that as many coefficients as possible have been estimated; they do not indicate that a perfect model has been located.[10]

Herting and Costner (1985) give an excellent discussion of the diagnostic usefulness of modification indices, residuals, Q–Q plots, and χ^2 in the development of models. They suggest using a two-stage diagnostic analysis: first locate misspecifications in the measurement model, and then locate misspecifications in the conceptual portions of the model. Measurement misspecifications are located by estimating a confirmatory factor analysis model providing *only correlations among the measured concepts*, and hence no effects among the concepts that might be misspecified, so that all the diagnostic power of the residuals, modification indices, and so on, are focused purely on the measurement structure of the model. With the measurement structure suitably altered, we reintroduce the postulated causal structure among the concepts and use the "new" residuals, modification indices, and so on, for diagnosing the conceptual causal structure.

As reasonable as this procedure sounds, it is clearly at odds with the strategy of forcing some concepts to take on researcher-determined meanings—for example, determining the identity of a concept by forcing it to enter into fixed relations with other concepts. That is, stage 1 of the process may lead us toward locating concepts whose meaning is at odds with the "intended" meanings of the concepts. Thoughtful reconsideration of the relevant theory and data gathering procedures remain the best guide to model revison.

6.6 Standardized Solutions

6.6.1 Standardizing the Concepts

Section 1.3.4 discussed standardizing as a rescaling of variables from some original scale to one with mean 0 and variance (standard deviation) 1.0. Such rescalings are convenient because a 1-unit increase (or change) on the new scale can be interpreted as an increase (or change) of one standard deviation. Section 2.2.2 illustrated how standardized regression slopes (structural coefficients) can be obtained from unstandardized slopes: we multiply the unstandardized slopes by the standard deviation of the independent variable and divide by the standard deviation of the dependent variable. Multiplying an unstandardized slope by the number of real metric units in the *independent* variable's standard deviation guarantees that the reported effect is the number of real metric units of change resulting from "one standard deviation's worth" of change in the independent variable. Dividing by the number of real metric units in the dependent variable's standard deviation guarantees that the resulting effect is measured as multiples or fractions of standard deviations of that variable rather than as multiples or fractions of real metric units.

We will want to standardize some of the variables (and hence coefficients) in LISREL. But which variables should be standardized, the concepts or the indicators? Standardizing the indicators is easily done by entering the correlation matrix rather than the covariance matrix among the indicators as the matrix to be analyzed (S). We recommend that correlation matrices not be entered as S, because doing so destroys the information about the real scales on which the indicators are measured and interferes with calculating χ^2 (see LISREL V and VI, p. I-39).

Standardization of the underlying concepts (ξ's and η's) while leaving the observed indicators in their observed metric is slightly more complex but has many uses (Section 6.6.2). The "standardized solution" LISREL reports is a *rescaling* of the maximum likelihood estimates such that *all the concepts are given variance 1.0, but the indicators remain in their original scales.*[11]

6.6.1.1 The Mathematics of Standardization

Obtaining the standardized solution involves rescaling six of the eight basic matrices, denoted Λ_y^s, Λ_x^s, B^s, Γ^s, Φ^s, and Ψ^s when standardized. Matrices Θ_ϵ and Θ_δ are the same in the standardized and unstandardized solutions because the X and Y variables remain in their original metrics, and these matrices continue to capture the amount of measurement error variance in these variables in their real metrics. The standardizing equations that follow were presented in LISREL III, p. 24 and LISREL IV, p. 60, but they have been omitted from later versions of the LISREL manual.

Paralleling the discussion of standardizing regression coefficients, our first step of standardizing is to obtain the standard deviations of the variables (concepts) in order to multiply/divide the unstandardized structural coefficients to obtain the standardized coefficients. The standard deviations of the ξ variables are easily obtained after we recognize that Φ contains the variance/covariance matrix for these variables. A matrix containing the standard deviations of the ξ concepts as diagonal elements (which we will denote A_ξ) can be obtained by selecting only the diagonal elements (variances) of Φ and taking their square roots. In matrix form,

$$A_\xi = (\text{diagonal } \Phi)^{1/2} \qquad \qquad 6.11$$

Thus A_ξ is a diagonal matrix composed of the standard deviations of the ξ concepts.

A diagonal matrix A_η containing the standard deviations of the η's as the diagonal elements can be similarly obtained by starting with the covariance matrix among the η's $[\text{Cov}(\eta)]$ rather than the ξ's. The covariance matrix for the η's can be calculated from the estimates in B, Γ, Φ, and Ψ, as shown by Eq. 4.49. Thus a diagonal matrix of the standard deviations of the η concepts can be obtained as

$$A_\eta = (\text{diagonal Cov}(\eta))^{1/2} \qquad \qquad 6.12$$

Since the inverse of a diagonal matrix is a diagonal matrix containing the inverses of the diagonal elements, A_ξ^{-1} and A_η^{-1} are diagonal matrices containing the reciprocals of the standard deviations of the ξ and η concepts, respectively.

We are now ready to standardize B. We begin by conceptualizing our problem as replacing the matrix B, whose elements are the number of units of change in the row η expected to follow a unit change in the column η, with a matrix $\mathbf{B}^s$, whose elements are the number of standard deviations of change in the row η expected to follow a standard deviation change in the column η. Multiplying each column of $\mathbf{B}$ by the number of units in a standard deviation of the appropriate causal (column) variable gives the amount of change (in real units) expected in each row η following a one standard deviation change in the causal η. Dividing each row of $\mathbf{B}$ by the standard deviation of the corresponding row η expresses the expected amount of change in standard deviations rather than real units. Thus the standardized effects are simply the unstandardized effects rescaled by multiplying by the standard deviation of the column (causal) η and dividing by the standard deviation of the row (effect) η. Note the parallel to Eq. 2.20 and convince yourself that this is accomplished for all the elements of $\mathbf{B}$ by premultiplying by $\mathbf{A}_\eta^{-1}$ and postmultiplying by $\mathbf{A}_\eta$. That is, the standardized effects $\mathbf{B}^s$ equal

$$\mathbf{B}^s = \mathbf{A}_\eta^{-1}\mathbf{B}\mathbf{A}_\eta \qquad 6.13$$

Standardizing $\mathbf{\Gamma}$ for the effects of the exogenous ξ concepts on the endogenous η's proceeds similarly. We multiply each column of $\mathbf{\Gamma}$ by the standard deviation of the corresponding ξ to replace the real unit change with a one standard deviation change. We then divide each row of $\mathbf{\Gamma}$ by the standard deviation of the corresponding η to replace the number of real units of expected change with the number of standard deviations change in the dependent η. Thus,

$$\mathbf{\Gamma}^s = \mathbf{A}_\eta^{-1}\mathbf{\Gamma}\mathbf{A}_\zeta \qquad 6.14$$

Next we standardize the measurement matrices $\mathbf{\Lambda}_y$ and $\mathbf{\Lambda}_x$. Consider a single concept with a single indicator where the corresponding λ is fixed at 1.0. The variance contributed by the concept to the indicator under these conditions (according to Eqs. 1.23 or 1.34) is

$$1.0^2(\text{variance of concept}) \qquad 6.15$$

The problem is to maintain exactly this same contribution to the variance of the indicator if the variance of the concept is rescaled from its original variance to 1.0 (i.e., standardized). We do that by multiplying the appropriate λ by the standard deviation of the original concept. To demonstrate why this is so, we again use Eq. 1.23 to obtain the contribution of the concept to the variance of the indicator, but we use the concept's standardized variance (1.0) and the standardized λ^s (the old λ multiplied by the original standard deviation of the concept). We get

$$(1.0 \times \text{std.dev.})^2(\text{variance of standardized concept or 1.0}) \qquad 6.16$$

which equals

$$1.0^2(\text{variance of concept})(1.0) \qquad\qquad 6.17$$

We recognize this as precisely the contribution to the variance made by the concept in its unstandardized form, as we just saw.

In general, the concepts' contribution to the variances of the indicators is preserved if each element of Λ is multiplied by the standard deviation of the corresponding concept while the variance of the concepts is simultaneously "reduced" to 1.0. Thus,

$$\Lambda_x^s = \Lambda_x A_\xi \qquad \text{and} \qquad \Lambda_y^s = \Lambda_y A_\eta \qquad\qquad 6.18$$

Next consider the standardized version of Φ or Φ^s. Since Φ is a covariance matrix among the exogenous concepts, Φ^s will merely be the correlations among the exogenous concepts, in direct correspondence to Eq. 1.54. That is, if each column of Φ is divided by the standard deviation of the concept associated with that column, and if each row is divided by the standard deviation of the concept associated with that row, the resultant elements will be of the form $\text{Cov}(\xi_i\xi_j)/(\text{std.dev.}\xi_i)(\text{std.dev.}\xi_j)$, which is a correlation coefficient according to Eq. 1.54. Hence,

$$\Phi^s = A_\xi^{-1}\Phi A_\xi^{-1} \qquad\qquad 6.19$$

The covariance matrix among the endogenous concepts can be standardized analogously so that

$$(\text{correlations among } \eta\text{'s}) = A_\eta^{-1}(\text{Cov}(\eta))A_\eta^{-1} \qquad\qquad 6.20$$

The appropriate standardization for Ψ, the variances/covariances among the structural disturbance terms, is

$$\Psi^s = A_\eta^{-1}\Psi A_\eta^{-1} \qquad\qquad 6.21$$

Although this parallels the form of standardizing a covariance matrix, it is not a correlation matrix because it is not the standard deviations of the error variables in the matrix but the standard deviations of the unstandardized associated concepts that are used. Each error variance is reduced in proportion to the reduction in variance of the corresponding η, and each covariance is reduced in proportion to the reductions in the two corresponding concepts.

The total effect matrices (discussed in Chapter 8) may also be standardized, but since we have not yet discussed total effects, we merely note that the total effects of the η's on other η's can be standardized by premultiplying the appropriate total effect matrix by A_η^{-1} and postmultiplying by A_η (i.e., the procedure that standardizes B). This matrix is not provided in LISREL's output but is easily obtained by hand calculations. The total effects of the ξ's on the η's can be standardized by premultiplying the appropriate total effect matrix by A_η^{-1} and postmultiplying by A_ξ (i.e., the procedure that standardizes Γ). This matrix is provided by LISREL.

6.6.1.2 Standardizing in context

Implicit in the preceding discussion is the fact that all elements of **B**, Γ, Λ_x, and so on, are rescaled during calculation of the standardized solution. Previously fixed coefficients (such as 1.0 λ's) will take on new values, and coefficients previously constrained to be equal may become unequal. That does not mean the corresponding coefficients have been set free or become unconstrained. *These changes appear because equality constraints specified in terms of real metrics, and values fixed in terms of real metrics, take on different and incorrect meanings if the same restrictions are applied to standardized concepts.* If this statement is puzzling, reconsider the discussion surrounding Eqs. 6.15 and 6.17.

The significance or insignificance of standardized coefficients is determined by the significance or insignificance of the corresponding unstandardized coefficients. Beware that the standard errors of the estimates discussed in Section 6.4 apply only to the unstandardized solution.

Caution: When two models are stacked (see Chapter 9) into a single run and estimated together, each model can be standardized. Rather than estimating a separate A_ξ and A_η for each group and proceeding as before, LISREL uses a single A_ξ and a single A_η matrix for the two groups. These matrices are calculated as the weighted average of the A_ξ and A_η matrices for the groups individually. That is, the standardizing matrices A_ξ^* and A_η^* are

$$A_\xi^* = \Sigma \left(\frac{N_g}{N}\right)A_{\xi g} \qquad\qquad A_\eta^* = \Sigma \left(\frac{N_g}{N}\right)A_{\eta g} \qquad\qquad 6.22$$

where N_g is the number of cases in any group and $A_{\xi g}$ is the A_ξ matrix for that group.

Acock and Fuller (1985) question the use of this weighting procedure on the grounds that it is inconsistent with traditional procedures for standardizing. If a coefficient is constrained to be equal in two groups, but the standard deviations of the relevant variables differ between the groups, the standardized coefficients should also differ between the groups (recall Eq. 2.20); the foregoing procedure, however, would provide the same standardized coefficients in the groups.

I also believe this standardizing procedure is questionable, but for different reasons. First, it implies that if two models are estimated in separate runs and the same models then estimated in a stacked run, different standardized solutions will appear even though all coefficients receive identical estimates. Second, the weighted standardizing leads to standardized Φ matrices that are not correlation matrics as they should be. Diagonal elements larger or smaller than 1 are routine, despite the labeling of the standardized Φ as a "correlation matrix." A similar comment applies to the standardized version of the covariances among the η variables.

Fortunately, these problems are not insurmountable, because within-group standardizing is always easily obtainable by hand calculations (by multiplying by one standard deviation and dividing by another). On the positive side, some versions of LISREL VI (those distributed to MTS operating systems since March 1985) have supplied two standardized outputs for stacked models: the regular LISREL output with the weighted standardizing, and a within-group standardizing where the A_ξ and A_η appropriate for each group are used in standardizing the groups.

6.6.2 Uses of the Standardized Solution

The standardized solution gives an easily grasped picture of effect sizes. Effects (B and Γ values) near 0 are small, and those near 1.0 are huge. In addition, the proportion of explained/unexplained variance in each η is readily apparent. Since the total variance in each η is 1.0, the Ψ^s variance corresponding to each η gives the proportion of error variance in the prediction of that η. Equivalently, $1 - \Psi^s$ gives the proportion of variance in the η explained by the whole model.

The standardized solution can also clearly indicate estimation problems. Elements of B^s, Γ^s, or Ψ^s exceeding 1.0 are signs of severe estimation problems because standardized effects greater than 1 and "error" variances contributing more than the total variances in concepts are "impossibilities" in good models. Observing variances in Φ^s or variances of the standardized η's other than 1.0 is equally distressing, for this claims LISREL was unable to standardize the model with the preceding calculations. Correlations within Φ^s or η's correlation matrix that exceed or even approach 1.0 can also be signs of problems. The closer these correlations are to 1.0, the closer LISREL is coming to saying that it "thinks" the two concepts are identical because they are perfectly correlated.[12]

If any of these problems arise, check to see if the model has been properly entered into LISREL (e.g., be, sure a "wrong" coefficient was not fixed at 1.0, or that some unreasonable coefficient was not freed). Also recheck the data matrix (are the proper variables included, and are they in the proper order?) Finally, reconsider the substantive theory focusing on the problem coefficient. Reconsidering the problematic standardized value in light of Eqs. 1.26 and 1.34 often helps.

6.6.3 The Standardized Smoking Model

Figure 6.1 gives the standardized solution for the smoking model. Even a quick glance at the standardized coefficients in B^s, Γ^s, and Ψ^s suffices to show that only weak effects link the concepts in the smoking model and, hence, that only a small portion of the variance in the endogenous concepts is explained by the postulated causal structures. Less than 15% of the variance in each η is attributable to modeled effects, the remaining percentage being "error" variance.

The elements of B^s and Γ^s may be interpreted as the number of standard deviations change in an η expected to follow a one standard deviation increase in another η or ξ. For example, a one standard deviation increase in education is expected to lead to an increase of .102 standard deviations in the number of antismoking acts engaged.

Although standard deviations in interpretations are often convenient, they are not always helpful. For example, with dummy variables such as sex or smoking status, the meaning of one standard deviation is obscure. The entries in the first column of Γ^s are the effects likely to follow a one standard deviation change in sex. A unit change in sex using the original 1–2 scale amounts to comparing the sexes, but a one standard deviation change corresponds to .497 (the square root of .248) units of change on the real scale, and that seems to defy description. Chapter 8 discusses how to salvage some meaning for less than unit changes on dummy variables by couching the interpretation in terms of aggregate group differences, but it is awkward to have to revert to discussions of real units in the midst of discussions of standardized effects. Effects involving dummy variables are most easily interpreted as *un*standardized effects, where the real units of the scales are focal and hence can be invoked without digression.

The remaining aspects of the standardized solution for the smoking model are unproblematic but not particularly informative. The correlations among the standardized ξ's and η's display no signs of problems, and the standardized Λ values are, as expected, the standard deviations of the unstandardized concepts. For example, the standardized λ of 15.315 for age is the square root of the variance of the concept age appearing in Φ for the unstandardized solution (234.55).

Figure 6.1   The Standardized Solution for the Smoking Model.

```
STANDARDIZED SOLUTION
        LAMBDA  Y

                smoking1    asviews    asacts
smoking1          0.472        0.0        0.0
asviews            0.0        1.774       0.0
asacts             0.0         0.0       1.112

        LAMBDA  X

                  sex         age      educatio   R-attend
sex              0.498        0.0        0.0        0.0
age                0.0      15.315       0.0        0.0
educatio           0.0        0.0       2.494       0.0
R-attend           0.0        0.0        0.0       2.047

        BETA

                smoking1    asviews    asacts
smoking1           0.0        0.0        0.0
asviews          -0.234       0.0        0.0
asacts           -0.132      0.108       0.0

        GAMMA

                  sex         age      educatio   R-attend
smoking1         -0.015      -0.126     -0.197     -0.155
asviews           0.153       0.147      0.066      0.049
asacts            0.072       0.0        0.102     -0.126

        PHI

                  sex         age      educatio   R-attend
sex              1.000
age              0.022       1.000
educatio         0.013      -0.191      1.000
R-attend         0.136       0.232      0.046      1.000

        PSI

                smoking1    asviews    asacts
                 0.918       0.869      0.935

        CORRELATION MATRIX FOR ETA

                smoking1    asviews    asacts
smoking1         1.000
asviews         -0.280       1.000
asacts          -0.159       0.147      1.000

        TOTAL  EFFECTS    ETA ON KSI (STANDARDIZED)

                  sex         age      educatio   R-attend
smoking1         -0.015      -0.126     -0.197     -0.155
asviews           0.156       0.176      0.112      0.085
asacts            0.091       0.036      0.141     -0.096
```

Notes

1.   Another way to see why the difference between χ^2's is also a χ^2 originates in the fundamental definition of χ^2 variates (cf. Hays, 1963). If the probability density equation for the normal curve is standardized and then squared, the resultant curve is a χ^2 distribution with 1 $d.f.$ This distribution describes the values likely to appear if a single case is randomly selected from a standardized normal distribution and then squared. For example, since values more than 1.96 standard deviations above or below the mean appear with probability .05 for the standard normal curve, values of χ_1^2 greater than $(+1.96)^2 = (-1.96)^2 = 3.84$ appear with probability .05.

If two cases are independently and randomly selected from a standardized normal distribution and then squared, the sum of such paired scores is distributed as a χ^2 distribution with 2 $d.f.$ In general, the sum of N squared standardized scores selected independently and randomly from a normal distribution is distributed as a χ^2 distribution with N $d.f.$ The difference between χ^2 distributions with N and $N+1$ $d.f.$ is the inclusion of just one more independent squared standard normal variate, which is another way of saying it is a χ^2 distribution with 1 $d.f.$ Thus, the fact that a χ^2 with N $d.f.$ is the sum of N independent χ^2's with 1 $d.f.$ implies that the difference between two χ^2's is itself a χ^2 with $d.f.$ equal to the difference between the $d.f.$ for the two χ^2's.

2.   The $\chi^2/d.f.$ notion is also the basis of the Tucker-Lewis (1973) coefficient for assessing the goodness of fit of *exploratory* factor analysis models.

$$TL = (\chi_0^2/d.f._0 - \chi_i^2/d.f._i)/(\chi_0^2/d.f._0 - 1) \qquad 6.23$$

Bentler and Bonett (1980) and Bentler (1982) propose a generalization of this measure as

$$TL\text{-}BB = (\chi_i^2/d.f._i - \chi_i^2/d.f._i)/(\chi_0^2/d.f._0 - 1) \qquad 6.24$$

where model j is a model nested within model i and the null (0) model is "the most restrictive, theoretically defensible model" (Bentler and Bonett, 1980:600). This measure is discussed by Ioolter (1983a) and has three basic drawbacks: first, its denominator is likely to be undefined, given the absence of any general agreement as to what constitutes a null model; second, "the scale of the fit indices is not necessarily easy to interpret" (Bentler and Bonett, 1980:600); and third, it depends on sample size (Bollen, 1986).

Bentler and Bonett (1980:599) also discuss a fit index δ, which is defined as

$$\delta = (\chi_j^2 - \chi_i^2)/\chi_0^2 \qquad 6.25$$

This index expresses the improvement in χ^2 when moving between nested models as a proportion of the χ^2 for the null or worst possible model. We suspect the proportional reduction in χ^2 would be more meaningful if the baseline model represents some reasonable substantive model (having χ_k^2) rather than a null model.

$$\delta2 = (\chi_j^2 - \chi_i^2)/\chi_k^2 \qquad 6.26$$

This expresses the improvement achieved by moving from model j to model i as a proportion of the χ^2 for some substantive model (acceptable or not) rather than as an improvement over some hypothetical and possibly trivial null model. All such proportional reduction in χ^2 measures are most useful if we are considering multiple similar yet alternative models (cf.

Marsh and Hocevar, 1985), and they all suffer the drawbacks already cited. There may be disagreement over what constitutes a null or baseline model, and the scale of such fit indices remains model specific and hence inherently nongeneralizable (despite the indices' values ranging between 0 and 1.0 due to the "proportion" nature of these measures).

The goodness of fit index (GFI) and adjusted goodness of fit (AGFI) provided in LISREL VI are sufficiently clearly discussed in the manual to obviate repetition here.

3.   Their explanation for why their data contradict Boomsma (1982), who suggests χ^2 should not be used with N's smaller than 100, is enlightening (Boomsma included both "proper and improper" solutions in his analysis), but I still hesitate to specify rules regarding N. The "simulations" underlying these recent discussions are strongly model dependent, stressing factor models as opposed to causal models. I have run models with as few as 22 cases and found no discernible problems and results replicable with larger data sets (Hayduk, 1985), but such small N's are not likely to be acceptable unless we are modeling experimental data. See also Boomsma (1983, 1985), Gallini and Mandeville (1984), Gerbing and Anderson (1985), Geweke and Singleton (1980), and Tanaka (1987).

4.   I have seen models that are completely acceptable except for modestly nonlinear Q–Q plots, and models that are unacceptable (e.g., with impossible estimates) with reasonably linear Q–Q plots.

5.   The significance of indirect effects (discussed in Chapter 8) can be obtained by side calculations outlined by Sobel (1982, 1986). See also Wolfle and Ethington (1985).

6.   The correlations among the estimates also imply that the tests of significance for the coefficients (in the previous section) are not independent of one another. In general, the higher the correlation between two estimates, the less independent are the tests of significance for those coefficients.

7.   In LISREL VI the modification index for any given coefficient is $N/2$ times the ratio of the squared first-order partial derivative to the second-order partial derivative (Joreskog and Sorbom, 1984:I-42). The dependence of these indices on N parallels the dependence of χ^2 on N. See also Sorbom (1986).

8.   Indeed, the inclusion of automatic model modification corresponding to the freeing of the coefficient with the largest modification index (scaled partial derivative) in the latest version of LISREL is retrograde for the vast majority of LISREL users. I do not doubt that this provides the easiest way of improving model fit, but I have severe reservations regarding the quality of the models ultimately located. I would rather see two or three theoretically sensible coefficients freed than one nonsensical coefficient. I have encountered situations where I would rather accept a model having a marginally significant χ^2 than free the coefficient with the largest modification index (or partial derivative). I recommend that all fixed coefficients be treated as "never to be freed" without an explicit substantive decision by the researcher (cf. MacCallum, 1986).

LISREL frequently requests the freeing of measurement error covariances in poorly fitting models, assuming Θ_ϵ and Θ_δ have been declared as symmetric so that modification indices are calculated for the off-diagonal elements. But error covariances are not the kinds of things that the data itself should persuade us to enter. Measurement error covariances arise from the data collection procedures, which the researcher presumably knows in detail and in advance of modeling. Knowledge of the data collection procedures and not an ill-fit covariance should be the first prod to the entry of covariances among measurement errors.

The reason the partial derivatives (modification indices) frequently suggest the freeing of error covariances is that these covariances provide a direct link between the variables whose covariance is ill fit. If the real model problem requires a chain of effects in the substantive model, the partial derivatives will not detect this because each of the paths constituting the necessary chain may be relatively useless by itself, and the partial derivatives reflect this "useless" freeing of any one coefficient at a time.

Some models seem improvable by the inclusion of larger than expected measurement error variances. It is possible that unexpectedly low-quality measurement can lead to this diagnostic, but an alternative explanation is that one has done a good job of measuring a different but closely related concept. Altering the causal connections for the concept to reflect the concept measured as opposed to the concept we hoped to measure might well produce a more informative model than blindly consenting to huge measurement error.

The moral is, beware of models whose acceptable fit has been gained by the post-hoc freeing of error covariances. Indeed, checking to see if such covariances are reasonable, by checking the magnitude of the implied *correlation*, frequently uncovers implied correlations greater than 1.0. The ability to request a standardized Ψ, Θ_ϵ, and Θ_δ as optional LISREL output would aid this task immeasurably.

9.   Splitting on odd/even case numbers is a fair substitute if the data set is unordered. Naturally, if several equivalent data sets are available, this reduces the need for splitting data sets.

10.   Sorbom (1975) illustrates the use of partial derivatives in model development, but note that he assumes you know that partial derivatives are based on real scale units.

11.   Even if the indicators are standardized via analysis of a correlation matrix, this does not standardize the concepts. If the total variance in an indicator is 1.0 and if this variance is partitioned into measurement error variance and true concept variance, the concept's variance must be less than 1.0 (assuming the scale of the concept is set by a λ of 1.0).

The fact that it is a rescaling implies that the fit of the model and the significance of the individual coefficients remain unchanged, the same covariance matrix is implied by the rescaled (standardized) solution, and the same theory is embodied in the model.

12.   A point of interest: The appendix to the LISREL IV manual presents two models containing standardized Γ values that exceed 1.0 (pp. A17, A27). Since LISREL V and VI continue these examples but do not provide the full LISREL output, it is unkown if these problems persist to the later manuals.

Chapter 7

Becoming a LISRELITE: Some Tricks of the Trade and Learning to Play

This chapter is designed to free the reader from a myopic dependence on the formal LISREL model developed in Chapter 4 (Eqs. 4.1, 4.4, and 4.6). Some aspects of the formal model are unavoidable; others are not. By learning to circumvent the arbitrary aspects of model specifications, we are freed from all but the truly fundamental and immutable restrictions on LISREL modeling. This is where LISREL becomes fun in the sense that we begin to "play" with LISREL by creating tricks that "fool" LISREL into doing what we want it to do, instead of being slavishly bound to what LISREL seems to want to do. The practical benefits of this freedom include being able to constrain a model coefficient to be a multiple of another (e.g., one effect to be double another effect), to constrain some effect to be equal to or greater than another, to constrain a coefficient to be the square root of another (e.g., so both a variance and the corresponding standard deviation can be used as model coefficients), to constrain some model coefficient to be negative or positive (e.g., to avoid negative variance estimates), and to allow exogenous concepts (ξ's) to influence endogenous indicators (Y's) and endogenous concepts to influence exogenous indicators (recall the "limitation" mentioned in Section 4.6).

The formal outline of this chapter reflects a series of useful tips and extensions of the basic LISREL procedures, but the chapter's hidden curriculum is cultivation of freedom and fluidity in your ability to think about models. The chapter is designed to help you see that there are many ways to do certain things and that creative use of some tricks of the trade both liberates us from slavish application of LISREL and deepens our understanding of, and respect for, the fundamental core of unavoidable essentials. We begin by considering four minor changes to the smoking model that leave the basics of the model unchanged. We conclude by using these tricks to illustrate how conceptual level nonlinear effects and interactions can be modeled in LISREL.

7.1 Four Simple Replacements

7.1.1 Replacing Measurement Errors with Concepts

7.1.1.1 Replacing ϵ's

In the basic smoking model (Figure 4.1) the proportion of measurement error variance in each of the indicators was determined a priori using knowledge of the data collection procedures and the intended meanings of the concepts (Figures 4.4 and 4.5). In this section we focus on the measurement of antismoking acts η_3, which was measured as the number of AS bylaw provisions for which the respondent reported having taken personal affirmative action (Y_3). It was decided (Figure 4.5) that 5% of the variance in this measure should be considered error variance, so .065 (5% of the total variance of 1.302 for this indicator) was fixed as measurement error variance by fixing $\Theta_{\epsilon3}$ to .065 in the LISREL program (Appendix B).

Figure 7.1 presents an equivalent LISREL model where $\Theta_{\epsilon3}$ has been fixed at zero, but the measurement error variance has been reintroduced indirectly through the additional concept η_4. The implications of adding the concept η_4 are highlighted in Figure 7.1 (cf. Figures 4.3 and 4.7). The Λ matrix has an additional column containing fixed zeros except for the fixed 1, which forces any change in the value of η_4 to be perfectly transmitted to Y_3, the indicator whose error variance is being replaced. The B matrix has one additional column, filled with fixed zero elements to guarantee that η_4 influences no other η's, and one additional row, also filled with zero elements to guarantee that η_4 is not influenced by the other η's. Similarly, the extra row of fixed zero elements in Γ permits no effects from any of the exogenous concepts (ξ's) on the newly created η_4. The $\Theta_{\epsilon3}$ has been fixed at zero, and Ψ_4, corresponding to the new η_4, has been fixed at .065 (the value $\Theta_{\epsilon3}$ previously had).

The LISREL estimates for all of the free parameters in the model

Figure 7.1   Replacing an ϵ with an η.

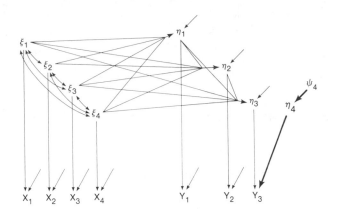

```
LISREL ESTIMATES (MAXIMUM LIKELIHOOD)

      LAMBDA Y

                ETA 1        ETA 2        ETA 3        ETA 4
smoking1        1.000        0.0          0.0          0.0
asviews         0.0          1.000        0.0          0.0
asacts          0.0          0.0          1.000        1.000

      BETA

                ETA 1        ETA 2        ETA 3        ETA 4
ETA 1           0.0          0.0          0.0          0.0
ETA 2          -0.879        0.0          0.0          0.0
ETA 3          -0.312        0.088        0.0          0.0
ETA 4           0.0          0.0          0.0          0.0

      GAMMA

                sex          age          educatio     attend
ETA 1          -0.015       -0.004       -0.037       -0.036
ETA 2           0.545        0.017        0.047        0.042
ETA 3           0.161        0.0          0.046       -0.068
ETA 4           0.0          0.0          0.0          0.0

      PHI

                sex          age          educatio     attend
sex             0.248
age             0.166        234.553
educatio        0.016       -7.282        6.218
attend          0.139        7.269        0.233        4.191

      PSI

                ETA 1        ETA 2        ETA 3        ETA 4
                0.204        2.736        1.156        0.085

      THETA EPS

                smoking1     asviews      asacts
                0.025        0.166        0.0

      THETA DELTA

                sex          age          educatio     attend
                0.002        12.345       0.691        0.466


CHI-SQUARE WITH    1 DEGREES OF FREEDOM IS        2.28 (PROB. LEVEL = 0.131)
```

containing η_4 (**B**, **Γ**, etc., and χ^2 with its *d.f.*) are identical to the values estimated in Figure 4.7. Why is this model with the extra η identical to the original model? In short, it is because we have merely replaced the error variable previously called ϵ_3, whose variance was $\Theta_{\epsilon 3}$ (.065), with a similarly behaved "error" variable called η_4, whose variance is ψ_4 (.065). The diagram at the top of Figure 7.1 emphasizes this parallel. The independence of the "error" variable is reflected in η_4 not influencing or being influenced by any variable other than Y_3. Fixing ψ_4 at .065 fixes the variance of η_4 at .065 because ψ_4 is the only contributor to the variance of η_4. Furthermore, fixing λ^y_{34} to 1.0 implies that the variance η_4 contributes to Y_3 is (according to Eq. 1.23) $1.0^2(.065) = .065$, which is precisely the variance contributed by $\Theta_{\epsilon 3}$ in the original model.

This reparameterization of the basic model informs us that *the sequestering of "measurement error" variances within the Θ_ϵ and Θ_δ terms in LISREL models is purely a matter of arbitrary convention.* Exactly the same function can be fulfilled by specifying "unnamed" concepts functioning independently of all the other variables in the model and having the proper variance (or having a variance to be estimated if the measurement error variance had been estimated).

Consider the equation for Y_3 in the original and new models. Originally, according to the last row of Λ_y in Eq. 4.11

$$AS \; acts = Y_3 = 1.0\eta_3 + \epsilon_3 \quad (\text{original}) \qquad 7.1$$

Using the last row of Λ_y in the revised model (Figure 7.1) and the fact that the variance $\Theta_{\epsilon 3}$ is fixed at zero, we get

$$AS \; acts = Y_3 = 1.0\eta_3 + 1.0\eta_4 + 0 \quad (\text{revised}) \qquad 7.2$$

In the revised model, only ζ_4 influences η_4 (no other η's or ξ's influence η_4 because the last rows of **B** and **Γ** contain zeros), so we can rewrite Eq. 7.2 as

$$AS \; acts = Y_3 = 1.0\eta_3 + 1.0\zeta_4 = 1.0\eta_3 + \zeta_4 \quad (\text{revised}) \qquad 7.3$$

Comparing Eqs. 7.1 and 7.3 demonstrates that ϵ from the original model has been replaced by ζ_4 in the revised model, and, hence, fixing the variance of ζ_4 (namely ψ_4) at .065 in the revised model is equivalent to fixing the variance of ϵ_3 (namely $\Theta_{\epsilon 3}$) in the original model to .065.

7.1.1.2 Replacing δ's

How does model specification change if we replace the measurement error in an exogenous indicator with an unnamed ξ that is independent of

Figure 7.2   Replacing a δ with a ξ.

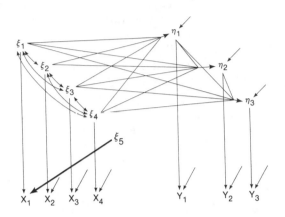

```
LISREL ESTIMATES (MAXIMUM LIKELIHOOD)

        LAMBDA X
              KSI 1      KSI 2      KSI 3      KSI 4      KSI 5
sex           1.000      0.0        0.0        0.0        1.000
age           0.0        1.000      0.0        0.0        0.0
educatio      0.0        0.0        1.000      0.0        0.0
attend        0.0        0.0        0.0        1.000      0.0

        BETA
              smoking1   asviews    asacts
smoking1      0.0        0.0        0.0
asviews      -0.879      0.0        0.0
asacts       -0.312      0.068      0.0

        GAMMA
              KSI 1      KSI 2      KSI 3      KSI 4      KSI 5
smoking1     -0.015     -0.004     -0.037     -0.036      0.0
asviews       0.545      0.017      0.047      0.042      0.0
asacts        0.161      0.0        0.046     -0.068

        PHI
              KSI 1      KSI 2      KSI 3      KSI 4      KSI 5
KSI 1         0.248
KSI 2         0.166    234.553
KSI 3         0.016     -7.282      6.218
KSI 4         0.139      7.269      0.233      4.191
KSI 5         0.0        0.0        0.0        0.0        0.002

        PSI
              smoking1   asviews    asacts
              0.204      2.736      1.156

        THETA EPS
              smoking1   asviews    asacts
              0.025      0.166      0.065

        THETA DELTA
              sex        age        educatio   attend
              0.0        12.345     0.691      0.466

CHI-SQUARE WITH   1 DEGREES OF FREEDOM IS      2.28 (PROB. LEVEL = 0.131)
```

all the other variables? The measurement error for sex ($\Theta_{\delta1}$), for example, can be replaced by setting $\Theta_{\delta1}$ to zero and creating a ξ_5, as in Figure 7.2. The additional column in Γ has values fixed at zero, and the additional row of Φ has fixed zero covariances and a variance fixed at the former error variance (.0025). Λ_x contains an extra column containing 0's except for a 1 linking ξ_5 to X_1 to transmit all the "error" variance of ξ_5 to X_1. Note that the same parameter estimates and model fit reappear. Work out equations paralleling Eqs. 7.1–7.3 if the implications of these changes seem unclear.

To solidify your understanding of these procedures, consider what each of the matrices $\mathbf{B}$, Γ, Λ_y, Λ_x, Φ, and Ψ would look like if all the Y error variances, or all the X error variances, or both the X and Y error variances were replaced with independent unnamed concepts.

7.1.2 Replacing Structural Disturbances (ζ's) with Concepts

Having replaced the measurement error variables ϵ and δ with independent unnamed concepts, we might consider a similar replacement for the ζ error variables in Eq. 4.1. Paralleling Section 7.1.1, we select an arbitrary ζ (in this case ζ_3, providing the error variance ψ_3 for the prediction of AS acts) and replace this first with an η and then with a ξ.

To replace ζ_3 by an η, we first fix ψ_3 to zero (creating the need for a replacement) and then specify a new concept η_4 that has no indicators (hence the new "fourth" column in Λ_y contains only 0's), that is influenced by no other concepts (the new "fourth" rows of $\mathbf{B}$ and Γ contain only 0's), and that influences only η_3 (the new "fourth" column of $\mathbf{B}$ contains a 1 in row 3 and 0's elsewhere). See Figure 7.3. The only variable influencing η_4 is its error variable ζ_4, whose variance ψ_4 is to be estimated. This revised model provides estimates identical to those obtained previously (again see Figure 7.3) except that ψ_3 is now zero and ψ_4 contains precisely the error variance ψ_3 had in the original model (1.156). Thus inserting the new η_4 merely transports the original error variance ψ_3 to another error variance ψ_4. Convince yourself (by using Eq. 1.23 twice) that the variance of η_4 and the variance η_4 contributes to η_3 are both 1.156.

One unanticipated, yet retrospectively reasonable, side effect of these changes is that the program output claims that η_3 is now perfectly predicted (its multiple R^2 is 1.0). From the program's perspective, η_3 is fully accounted for by η_1, η_2, and η_4. The program has no way of knowing that η_4 is not a real substantively meaningfully variable, so it treats the variance in η_3 arising from η_4 as explained rather than error variance.

Figure 7.4 depicts the replacement of the error term for η_3 with the artificial unnamed variable ξ_5. The same estimates again appear for all the original model parameters. Note the following features in the LISREL

Figure 7.3   Replacing a ζ with an η.

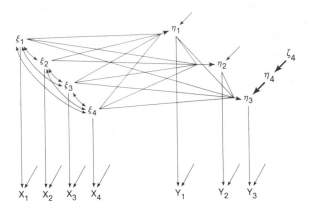

LISREL ESTIMATES (MAXIMUM LIKELIHOOD)

LAMBDA Y

	ETA 1	ETA 2	ETA 3	ETA 4
smoking1	1.000	0.0	0.0	0.0
asviews	0.0	1.000	0.0	0.0
asacts	0.0	0.0	1.000	0.0

BETA

	ETA 1	ETA 2	ETA 3	ETA 4
ETA 1	0.0	0.0	0.0	0.0
ETA 2	-0.879	0.0	0.0	0.0
ETA 3	-0.312	0.068	0.0	1.000
ETA 4	0.0	0.0	0.0	0.0

GAMMA

	sex	age	educatio	attend
ETA 1	-0.015	-0.004	-0.037	-0.036
ETA 2	0.545	0.017	0.047	0.042
ETA 3	0.161	0.0	0.046	-0.068
ETA 4	0.0	0.0	0.0	0.0

PHI

	sex	age	educatio	attend
sex	0.248			
age	0.166	234.553		
educatio	0.016	-7.282	6.218	
attend	0.139	7.269	0.233	4.191

PSI

ETA 1	ETA 2	ETA 3	ETA 4
0.204	2.736	0.0	1.156

THETA EPS

smoking1	asviews	asacts
0.025	0.166	0.065

THETA DELTA

sex	age	educatio	attend
0.002	12.345	0.691	0.466

SQUARED MULTIPLE CORRELATIONS FOR STRUCTURAL EQUATIONS

ETA 1	ETA 2	ETA 3	ETA 4
0.082	0.131	1.000	0.0

CHI-SQUARE WITH   1 DEGREES OF FREEDOM IS   2.28 (PROB. LEVEL = 0.131)

Figure 7.4   Replacing a ζ with a ξ.

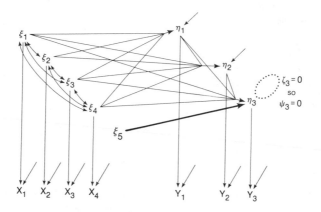

LISREL ESTIMATES (MAXIMUM LIKELIHOOD)

    LAMBDA X

	KSI 1	KSI 2	KSI 3	KSI 4	KSI 5
sex	1.000	0.0	0.0	0.0	0.0
age	0.0	1.000	0.0	0.0	0.0
educatio	0.0	0.0	1.000	0.0	0.0
attend	0.0	0.0	0.0	1.000	0.0

    BETA

	smoking1	asviews	asacts
smoking1	0.0	0.0	0.0
asviews	-0.879	0.0	0.0
asacts	-0.312	0.068	0.0

    GAMMA

	KSI 1	KSI 2	KSI 3	KSI 4	KSI 5
smoking1	-0.015	-0.004	-0.037	-0.036	0.0
asviews	0.545	0.017	0.047	0.042	0.0
asacts	0.161	0.0	0.046	-0.068	1.000

    PHI

	KSI 1	KSI 2	KSI 3	KSI 4	KSI 5
KSI 1	0.248				
KSI 2	0.166	234.553			
KSI 3	0.016	-7.282	6.218		
KSI 4	0.139	7.280	0.233	4.191	
KSI 5	0.0	0.0	0.0	0.0	1.156

    PSI

smoking1	asviews	asacts
0.204	2.736	0.0

    THETA EPS

smoking1	asviews	asacts
0.025	0.166	0.065

    THETA DELTA

sex	age	educatio	attend
0.002	12.345	0.691	0.466

    SQUARED MULTIPLE CORRELATIONS FOR STRUCTURAL EQUATIONS

smoking1	asviews	asacts
0.082	0.131	1.000

CHI-SQUARE WITH   1 DEGREES OF FREEDOM IS   2.28 (PROB. LEVEL = 0.131)

output: ξ_5 has no indicators, it has no effects on any η's (except η_3, whose error variable is being replaced), it has zero covariance with all the other ξ's, its variance is estimated (and ultimately equals the original error variance 1.156), and the multiple R^2 for η_3 is again 1.0 because LISREL "believes" ξ_5 is a real concept of known identity and hence classifies the variance in η_3 accounted for by ξ_5 as explained rather than unexplained variance. Actually, in terms of LISREL's internal calculations, the proportion of explained variance is calculated as the total variance in η_3 (given by Eq. 4.49) minus the variance of δ_3 (which is ψ_3), all divided by the total variance in η_3. Hence, fixing ψ_3 to 0 implies that LISREL will report 1.00 explained variance.

In summary, the preceding sections have replaced the error variables ϵ, δ, and ζ (and their variances Θ_ϵ, Θ_δ and Ψ) with either a ξ or an η (and a Φ or another Ψ variance). Since ϵ, δ, and ζ are error variables, we can represent them any way we like as long as we preserve the independence of these variables from all the other error variables and causal variables in the model. Converting error variables (ζ, ϵ, and δ) into error variances (Ψ, Θ_ϵ, and Θ_δ) (Sections 4.3 and 4.4), with subsequent references to only error variances in the model, frequently leads LISREL users to assume that the underlying error variables are fundamentally different from the other variables. The preceding "replacements" should force you to mentally reintegrate the error variables with the other variables in the model. The only differences are that the error variables are unnamed and independent of specific other variables (as emphasized by the lack of causal connections and noncausal associations between the replacement variables and the other variables in the model).

We can now appreciate why calling these variables errors is unnecessarily restrictive, in that the connotation of mistakes is stronger than is mathematically required. These variables contribute only to one specific variable, and hence they make unique contributions independent of all other contributing variables, but this need not make them errors. These variables may contribute real and important substantive unique effects, as opposed to reflecting limits on the sensitivity of measurement instruments. As Bentler puts it: "Only if truth is defined as a particular list of variables (namely as the list of predictor variables for any endogenous variable) is residual the same as error" (Bookstein, 1982:319). The notion of random errors is so entrenched that ϵ, δ, and ζ will continue to be called error variables, but this name should no longer deter analytical consideration of the sources of these uniquely effective forces.

7.1.3 Replacing One Coefficient with Two

7.1.3.1 Replacing a free coefficient with a fixed-free coefficient pair

Figure 7.5 eliminates the effect of sex (ξ_1) on AS views (η_2) in the smoking model by fixing γ_{21} at zero and replacing this eliminated effect by creating a new concept η_4 that has no indicators, is caused by sex, and causes AS views. Further, the effect of sex on the new η_4 is fixed at 3.0 (any nonzero value would work equally well), and ψ_4 is fixed at 0 so η_4 is perfectly predicted by sex.

Thus the eliminated γ_{21} has been replaced with a pathway from sex though the unnamed η_4 to AS views, where the first link in the pathway (γ_{41}) is fixed at 3.0 and the second link β_{24} is free to be estimated. The variance of the new η_4 is 3^2 times the variance in sex, because sex is the sole predictor of this variable (recall that ψ_4 was set to zero, and no other variables are allowed to influence η_4). Equivalently, η_4 is merely a rescaled sex variable taking on values 3 and 6 instead of 1 and 2.

Estimating this model (the lower portion of Figure 7.5) provides identical values to those obtained for the original model for all the model coefficients except the effect of η_4 on AS views which is one-third the effect of sex on AS views in the original model $[.182 = (.3333)(.545)]$. Thus, introducing η_4 provided an effect routing that replaces the original γ_{21}. The effect of sex on AS views via η_4 is the product of the effect coefficients composing the chain $3.0(.182) = .546$. (See Section 8.1 for further discussion of indirect effects.)

To summarize, this section demonstrates that *structural coefficients such as* Γ *and* B *can be replaced by a chain composed of one fixed and one free coefficient if the intervening variable is specified to have no indicators and no error variance.* The usefulness of this replacement is shown next.

7.1.3.2 Constraining effects to be multiples of other effects

Figure 7.6 takes the model developed in the preceding section and adds an additional constraint, forcing the effect of sex on AS acts (γ_{31}) to equal the effect of η_4 on AS views (β_{24}), thereby adding the hypothesis that sex influences AS "words" three times as much as AS "deeds." This forces the effect of sex on AS views to be three times the effect of sex on AS acts, because the effect of sex on AS views is multiplied threefold by the coefficient linking sex to η_4.

Figure 7.5   Replacing One Coefficient with Two.

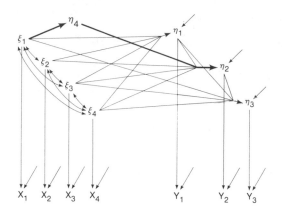

LISREL ESTIMATES (MAXIMUM LIKELIHOOD)

LAMBDA Y

	ETA 1	ETA 2	ETA 3	ETA 4
smoking1	1.000	0.0	0.0	0.0
asviews	0.0	1.000	0.0	0.0
asacts	0.0	0.0	1.000	0.0

BETA

	ETA 1	ETA 2	ETA 3	ETA 4
ETA 1	0.0	0.0	0.0	0.0
ETA 2	-0.879	0.0	0.0	0.182
ETA 3	-0.312	0.068	0.0	0.0
ETA 4	0.0	0.0	0.0	0.0

GAMMA

	sex	age	educatio	attend
ETA 1	-0.015	-0.004	-0.037	-0.036
ETA 2	0.0	0.017	0.047	0.042
ETA 3	0.161	0.0	0.046	-0.068
ETA 4	3.000	0.0	0.0	0.0

PHI

	sex	age	educatio	attend
sex	0.248			
age	0.166	234.553		
educatio	0.016	-7.282	6.218	
attend	0.139	7.269	0.233	4.191

PSI

ETA 1	ETA 2	ETA 3	ETA 4
0.204	2.736	1.156	0.0

THETA EPS

smoking1	asviews	asacts
0.025	0.166	0.065

THETA DELTA

sex	age	educatio	attend
0.002	12.345	0.691	0.466

SQUARED MULTIPLE CORRELATIONS FOR STRUCTURAL EQUATIONS

ETA 1	ETA 2	ETA 3	ETA 4
0 082	0.131	0.065	1.000

CHI-SQUARE WITH   1 DEGREES OF FREEDOM IS   2.28 (PROB. LEVEL = 0.131)

Figure 7.6   Making One Effect Three Times Another.

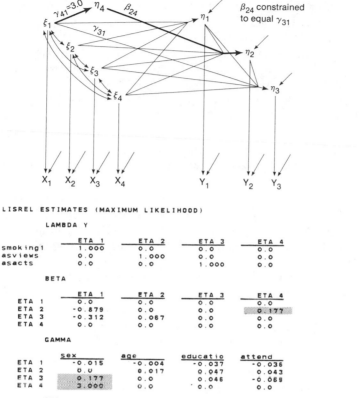

```
LISREL  ESTIMATES  (MAXIMUM  LIKELIHOOD)

        LAMBDA  Y

                  ETA 1         ETA 2         ETA 3         ETA 4
smoking1          1.000         0.0           0.0           0.0
asviews           0.0           1.000         0.0           0.0
asacts            0.0           0.0           1.000         0.0

        BETA

                  ETA 1         ETA 2         ETA 3         ETA 4
ETA 1             0.0           0.0           0.0           0.0
ETA 2            -0.879         0.0           0.0           0.177
ETA 3            -0.312         0.067         0.0           0.0
ETA 4             0.0           0.0           0.0           0.0

        GAMMA

                  sex           age           educatio      attend
ETA 1            -0.015        -0.004        -0.037        -0.036
ETA 2             0.0           0.017         0.947         0.043
ETA 3             0.177         0.0           0.046        -0.069
ETA 4             3.000         0.0           0.0           0.0

        PHI

                  sex           age           educatio      attend
sex               0.248
age               0.166         234.553
educatio          0.016        -7.282         6.218
attend            0.139         7.269         0.233         4.191

        PSI

                  ETA 1         ETA 2         ETA 3         ETA 4
                  0.204         2.736         1.156         0.0

THETA  EPS

                  smoking1      asviews       asacts
                  0.025         0.166         0.065

THETA  DELTA

                  sex           age           educatio      attend
                  0.002         12.345        0.691         0.466


SQUARED  MULTIPLE  CORRELATIONS  FOR  STRUCTURAL  EQUATIONS

                  ETA 1         ETA 2         ETA 3         ETA 4
                  0.082         0.130         0.066         1.000


CHI-SQUARE  WITH    2  DEGREES  OF  FREEDOM  IS        2.30 (PROB. LEVEL = 0.316)
```

Since the original effect of sex on AS views is not precisely three times the effect of sex on AS acts, the additional equality constraint prevents the same set of estimates reappearing for this model as have appeared for prior models. LISREL is now forced to find the best (maximum likelihood) compromise coefficient by balancing the seriousness (degree of ill fit of Σ to S) of overestimating one effect against underestimating the other. If the hypothesis that AS words are three times as responsive to sex differences as are AS deeds is correct, the ultimate fit of the constrained model should not be much worse than the unconstrained model. Indeed, the χ^2 for the constrained model is 2.30 with 2 $d.f.$, which implies a difference χ^2 (Section 6.1.2) of .02 with 1 $d.f.$ Thus there is an insignificant reduction in the fit of the model, or, equivalently, there is no evidence challenging the hypothesis that words are three times as responsive to sex differences as are deeds.

In general, the *replacement of one coefficient by one fixed and one free coefficient allows us to constrain any coefficient in the model to be a multiple of another coefficient.* Or if the fixed effect is set at -1.0 instead of 3.0, an effect can be constrained to be the negative of another effect.

The strategy of inserting a well-behaved η between two other variables provides the first way of avoiding the seeming constraint that exogenous concepts (ξ's) are not allowed to influence endogenous indicators (Y's) (Section 4.6). If the effect of the desired ξ on the new unnamed η is fixed at 1.0 (i.e., the relevant γ is fixed at 1.0), if the ψ for the unnamed η is fixed at 0.0, and if no other variables influence this η or are influenced by the η, then this η is precisely the same as the corresponding ξ (it has the same variance as ξ and is perfectly correlated with it). Either freeing or fixing (at a nonzero value) the λ_y linking the unnamed η to the appropriate endogenous indicator (Y) gives the second link in a chain providing an indirect routing via which the ξ can influence any Y without disturbing other aspects of the model.

7.1.3.3 Replacing a coefficient with its square root squared

Figure 7.7 presents the same model as Figure 7.5 except that the effect of sex on the newly created and perfectly predicted η_4 is not fixed at 3.0. Both the effect of sex on η_4 (i.e., γ_{41}) and the effect of η_4 on AS views (i.e., β_{24}) are set free, but they are constrained to be equal, so only a single coefficient is estimated.

It is not difficult to see that LISREL can duplicate the fit achieved for the original model (or any of its equivalent forms) by estimating the constrained γ_{41} and β_{24} to equal the square root of the original γ_{21} effect

Figure 7.7   Creating Square Roots of Effect Coefficients.

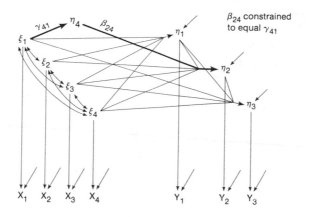

LISREL ESTIMATES (MAXIMUM LIKELIHOOD)

LAMBDA Y

	ETA 1	ETA 2	ETA 3	ETA 4
smoking1	1.000	0.0	0.0	0.0
asviews	0.0	1.000	0.0	0.0
asacts	0.0	0.0	1.000	0.0

BETA

	ETA 1	ETA 2	ETA 3	ETA 4
ETA 1	0.0	0.0	0.0	0.0
ETA 2	-0.879	0.0	0.0	-0.738
ETA 3	-0.312	0.068	0.0	0.0
ETA 4	0.0	0.0	0.0	0.0

GAMMA

	sex	age	educatio	attend
ETA 1	-0.015	-0.004	-0.037	-0.036
ETA 2	0.0	0.017	0.047	0.042
ETA 3	0.161	0.0	0.046	-0.068
ETA 4	-0.738	0.0	0.0	0.0

PHI

	sex	age	educatio	attend
sex	0.248			
age	0.166	234.553		
educatio	0.018	-7.282	6.218	
attend	0.139	7.269	0.233	4.191

PSI

ETA 1	ETA 2	ETA 3	ETA 4
0.204	2.736	1.156	0.0

THETA EPS

smoking1	asviews	asacts
0.025	0.166	0.065

THETA DELTA

sex	age	educatio	attend
0.002	12.345	0.691	0.466

SQUARED MULTIPLE CORRELATIONS FOR STRUCTURAL EQUATIONS

ETA 1	ETA 2	ETA 3	ETA 4
0.082	0.131	0.065	1.000

CHI-SQUARE WITH   1 DEGREES OF FREEDOM IS   2.28 (PROB. LEVEL = 0.131)

($\sqrt{.545}$ = .738). LISREL's actual estimate is $-.738$ or the negative square root of .545. The $-.738$ estimate provides .545 as the effect of sex on AS views $[(-.738)(-.738)$ = .545], with the negative sign merely informing us that the scale of η_4 is reversed in comparison to the scale of sex. The absence of indicator(s) for η_4 (a convenient fiction created in our attempt to undetectably deceive LISREL) implies that all properties of η_4 (including direction of the scale, size of scale units, and variance of η_4) are to be determined by a combination of model specification and estimation. In this instance model specification alone is insufficient to fix the scale direction, unit size or variance of η_4. It is only upon model estimation that the scale direction is specified as reversed [large values on the sex scale (female) correspond to low values on the η_4 scale], that the only values on the η_4 scale are set as $1(-.738)$ for males and $2(-.738)$ for females, and that the variance of η_4 is determined as $(-.738)^2\text{Var(sex)}$ = .545(.250) = .136.

The preceding liberates us from the formal LISREL model by introducing the possibility of using estimated square root constraints within our model. If, for example, theory suggests that the effect of sex on AS acts is the square root of the effect of sex on AS views, we could model this by constraining γ_{31} to equal γ_{41} and β_{24}. (Since the γ_{31} effect is positive in the basic model, the constraint to the square root coefficients makes the choice between a positive square root and a negative one no longer arbitrary, so a positive root would be expected.) Naturally, if the theory suggested that one effect was the negative of the square root of the other, we could combine this procedure for creating square roots with the procedure for reversing the sign of an effect (creating another η with one fixed effect of -1), as discussed in Section 7.1.3.2.

A further use of this square root procedure is to *constrain effect estimates to be positive*. With γ_{41} and β_{24} constrained to be equal, both effects must have the same sign; hence the effect of sex on AS views, which is the product of these effects, must be positive.

7.1.3.4 Modeling roots of variances

Occasionally we want to model the standard deviations of variables whose variances appear in the model as $\mathbf{\Psi}$, $\mathbf{\Phi}$, $\mathbf{\Theta}_\epsilon$, or $\mathbf{\Theta}_\delta$. Figure 7.8 shows how to create an estimated coefficient that is the square root of the error variance (ψ_3) for AS acts (η_3). The procedure is similar to that used in Figure 7.3. A new η_4 is created whose only causal impact is on η_3, but we switch the placement of the fixed and free cofficients accompanying η_4. In Figure 7.3 the error variance for η_4 (ψ_4) was free, and β_{34}, linking η_4 to η_3,

Figure 7.8   Square Roots of Variances.

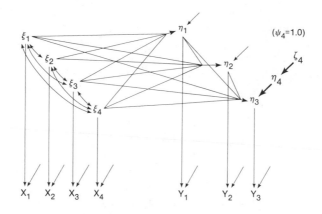

LISREL ESTIMATES (MAXIMUM LIKELIHOOD)

LAMBDA Y

	ETA 1	ETA 2	ETA 3	ETA 4
smoking1	1.000	0.0	0.0	0.0
asviews	0.0	1.000	0.0	0.0
asacts	0.0	0.0	1.000	0.0

BETA

	ETA 1	ETA 2	ETA 3	ETA 4
ETA 1	0.0	0.0	0.0	0.0
ETA 2	-0.879	0.0	0.0	0.0
ETA 3	-0.312	0.068	0.0	1.075
ETA 4	0.0	0.0	0.0	0.0

GAMMA

	sex	age	educatio	attend
ETA 1	-0.015	-0.004	-0.037	-0.036
ETA 2	0.545	0.017	0.047	0.042
ETA 3	0.161	0.0	0.046	-0.068
ETA 4	0.0	0.0	0.0	0.0

PHI

	sex	age	educatio	attend
sex	0.248			
age	0.166	234.553		
educatio	0.016	-7.282	6.218	
attend	0.139	7.269	0.233	4.191

PSI

ETA 1	ETA 2	ETA 3	ETA 4
0.204	2.736	0.0	1.000

THETA EPS

smoking1	asviews	asacts
0.025	0.166	0.065

THETA DELTA

sex	age	educatio	attend
0.002	12.345	0.691	0.466

SQUARED MULTIPLE CORRELATIONS FOR STRUCTURAL EQUATIONS

ETA 1	ETA 2	ETA 3	ETA 4
0.082	0.131	1.000	0.0

CHI-SQUARE WITH   1 DEGREES OF FREEDOM IS   2.28 (PROB. LEVEL = 0.131)

was fixed at 1.0. Here we fix ψ_4 at 1.0 and free β_{34}. This implies that the variance of η_4 is known to be 1.0 (since ψ_4 provides the only input into η_4), but the magnitude of the effect of η_4 on η_3 is unknown. Thus we are postulating errors of known magnitude whose transmission to the variable of concern is unknown, in contrast to errors of unknown size, which are assumed to be perfectly transmitted to the variable of concern.

How can η_4 provide an amount of error variance identical to the error variance estimated for the well-fitting original model? Since the variance η_4 contributes to η_3 will be β_{34} squared times the 1.0 variance of η_4, β_{34}^2 will have to equal the ψ_3 of the original model. Hence the estimated β_{34} should equal the square root of the original variance, as it does in Figure 7.8 ($1.075^2 = 1.156$).

Estimated measurement error variances (Θ_ϵ's and Θ_δ's) can be handled analogously, but Φ variances cannot. No variables are allowed to influence the exogenous ξ variables, and such influences are an integral part of this strategy. The covariances accompanying ξ variables (in contrast to the independence of error variables) provide an additional stumbling block to modeling ξ standard deviations. Section 7.2 demonstrates how to convert ξ's to η's, in which form the variances of ξ variables that are independent of all the other ξ's can be handled as in Figure 7.8, but I know of no procedure for obtaining standard deviations corresponding to Φ's describing variables correlated with other ξ's.

There are three primary uses of modeling error variances with models requiring estimation of "error standard deviations." First, this allows us to constrain other model coefficients to equal those standard deviations, thereby regaining or preserving degrees of freedom. Second, this provides a means for avoiding estimates of negative error variances (Heywood cases) (cf. Dillon, Kumar, and Mulani, 1987; Rindskopf, 1983a,b). LISREL does not constrain variance estimates to be positive (or for that matter implied correlations to be less than 1) (Joreskog, 1981b), so we occasionally encounter estimates falling outside the boundaries of the admissible parameter space (i.e., a negative variance or correlation exceeding 1.0). Though inadmissible estimates are rare, they do require correction. Combining the discussion in this section with the idea that squared coefficients must be positive (Section 7.1.3.3) provides an effective remedy for negative variance estimates.

A third use arises in conjunction with the creation of variables having variances that are specifiable functions of the variances of two or more other variables. We demonstrate this use in Sections 7.4 and 7.5.

7.1.4 Constraining One Effect to Equal or Exceed Another

Disregard the smoking model for a moment and consider a simple two-concept model for which theory dictates that concept η_1 influences concept η_2 with an effect of magnitude k or greater. Figure 7.9A illustrates a procedure for modeling this by fixing the direct effect of η_1 on η_2 at k and then providing an alternative effect routing through the phantom variable η_3. Constraining the effects in this indirect routing to be equal constrains the indirect effect to be positive. The "total" effect of η_1 on η_2 will be the sum of the direct and indirect effects, as is easily seen by examining the structural equations implied by Figure 7.9A:

$$\eta_2 = k\eta_1 + \beta_{23}\eta_3 \quad \text{and} \quad \eta_3 = \beta_{23}\eta_1 \qquad 7.4$$

since β_{31} is constrained to equal β_{23}. Substituting the second equation in the first gives

$$\eta_2 = k\eta_1 + \beta_{23}\beta_{23}\eta_1 = (k + \beta_{23}^2)\eta_1 \qquad 7.5$$

Thus, as required the effect of η_1 on η_2 equals k plus some amount (β_{23}^2) that is zero or some positive value depending on the estimated magnitude of β_{23}.

Constraining the effect of η_1 on η_2 to be of magnitude k or less is illustrated in Figure 7.9B and is a combination of the foregoing procedure with that of providing a negative coefficient according to Section 7.1.3.2. As an exercise, write out the equations for η_2, η_3, and η_4 in this model and solve for η_2 in terms of η_1, as we did for Figure 7.9A.

Note further that if we wished to constrain the effect of η_1 on η_2 to be equal to or greater than some other modeled effect, we could do it by constraining the effect labeled k to equal another effect rather than fixing its value at k, as in Figure 7.9A. Similarly, constraining rather than fixing k in Figure 7.9B provides a way of making the effect of η_1 on η_2 equal to or less than another effect (cf. Rindskopf, 1984b).

How would we have created a model forcing the effect of sex on AS views to equal or exceed the effect of sex on AS acts in the smoking model? Constrain γ_{21} to equal γ_{31}, and create an η_4 having γ_{41} constrained to equal β_{24}.

Note that, since there is only one unknown in either Figure 7.9A or B, if the original effect of η_1 on η_2 was identified, then the constrained effects are also identified. One interesting implication of this is that the unconstrained and constrained models would have the same degrees of freedom. That is, Eq. 6.3 would be the same for the two models, because the input data matrix is the same and the same number of coefficients are estimated. But note that the constrained model might fit more poorly than the unconstrained model, which happens if the constraint is unreasonable. The problem is that one model is clearly a more constrained version of the

Figure 7.9   Forcing Effects to be Larger or Smaller than a Specific Value k.

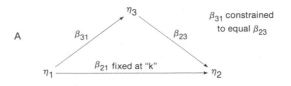

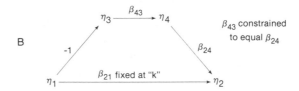

other, even though a direct transition from one to the other is precluded by the necessity of a new hypothetical and well-behaved variable. Thus the models *are* nested in the sense that one is more constrained than the other, but they are *not* nested in the sense of being able to obtain one from the other by constraining or fixing a coefficient.

The problem is further complicated if we consider multiple runs in which a single effect is constrained to equal or exceed .01, .1, 1.0, and 5.0. Clearly the magnitude of the fixed effect (i.e., the size of the effect that must be exceeded) is central to the issue of how constraining the extra constraint actually is, and this is not reflected in any count of how many free parameters remain to be estimated. This disjuncture between degrees of freedom and equal or exceeding constraints is one of several outstanding issues in LISREL modeling that point to unresolved issues in theoretical statistics.

7.2 Moving Φ into Ψ and Rethinking the Distinction between Exogenous/Endogenous

Figure 7.10 shows how we can respecify the basic smoking model to avoid using any ξ variables and hence avoid use of Φ, Λ_x, and Θ_δ. The original variables ξ_1, ξ_2, ξ_3, and ξ_4 have been replaced by η_4, η_5, η_6, and η_7. These new η variables display the same pattern of effects on η_1, η_2, and η_3 as did the original ξ variables. Like the original ξ variables, no causal effects reach the new η's from η_1, η_2 or η_3. The variances and covariances of the ξ's formerly appearing in Φ now appear in Ψ.

Examining the estimates accompanying Figure 7.10 should convince you that this is indeed the umpteenth way to get the same old estimates out of LISREL, but we should pause to examine the mechanics of the translation. That the last four rows of the $\mathbf{B}$ matrix in Figure 7.10 are fixed zeros implies that no modeled variables influence the new η's (except for the ζ "errors" associated with each of the new η's). Combining this with the implicit 1.0 values linking the ζ's to the new η's implies that the ζ's are identical to the corresponding η's. That is, writing the equations for the four new η's using the last four rows of the $\mathbf{B}$ matrix gives

$$\eta_4 = \zeta_4 \qquad \eta_5 = \zeta_5 \qquad \eta_6 = \zeta_6 \qquad \eta_7 = \zeta_7 \qquad\qquad 7.6$$

Thus the new ζ's and η's are identical, and both these sets of variables correspond to the original set of ξ's being replaced. Since the ζ's and η's are identical, the variances and covariances among the ζ's (namely Ψ's) give the variances and covariances among the new η's. Furthermore, since the η's are replacing the original ξ's, these Ψ variances and covariances must equal the original Φ matrix containing the variances and covariances among the original ξ's, which is easily verified by comparing the lower right portion of the new Ψ matrix to the original Φ matrix.

When the ξ's are converted to η's, the indicators of these concepts must be renamed Y's, which eliminates any need for Λ_x and Θ_δ, and which results in Θ_ϵ being partitioned to include the original Θ_ϵ and Θ_δ fixed measurement error variances.

Making all the concepts endogenous is a second way to allow exogenous concepts (ξ's) to influence endogenous indicators (Y's) or endogenous concepts (η's) to influence exogenous indicators (X's). Any element in the new Λ_y can be freed, so any concept can be allowed to influence any indicator. Since the Λ_y of the smoking model is an identity matrix, it is not printed by LISREL, but you should be able to reconstruct it as a 7 × 7 identity matrix, whose elements could be altered.

Previously we saw that the designation of errors (ϵ, δ, and ζ) as separate classes of variables was mere convention. The ability to replace all

Figure 7.10   Making ξ's into η's and Hence Φ's into Ψ's.

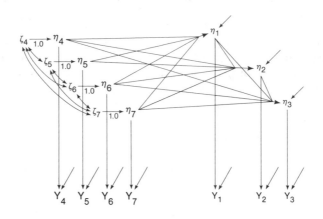

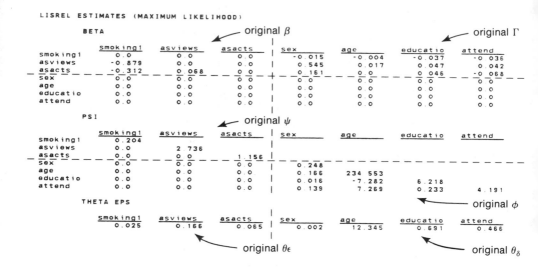

LISREL ESTIMATES (MAXIMUM LIKELIHOOD)

BETA   ← original β ← original Γ

	smoking1	asviews	asacts	sex	age	educatio	attend
smoking1	0.0	0.0	0.0	-0.015	-0.004	-0.037	-0.036
asviews	-0.879	0.0	0.0	0.545	0.017	0.047	0.042
asacts	-0.312	0.068	0.0	0.161	0.0	0.046	-0.068
sex	0.0	0.0	0.0	0.0	0.0	0.0	0.0
age	0.0	0.0	0.0	0.0	0.0	0.0	0.0
educatio	0.0	0.0	0.0	0.0	0.0	0.0	0.0
attend	0.0	0.0	0.0	0.0	0.0	0.0	0.0

PSI   ← original ψ

	smoking1	asviews	asacts	sex	age	educatio	attend
smoking1	0.204						
asviews	0.0	2.736					
asacts	0.0	0.0	1.156				
sex	0.0	0.0	0.0	0.248			
age	0.0	0.0	0.0	0.166	234.553		
educatio	0.0	0.0	0.0	0.016	-7.282	6.218	
attend	0.0	0.0	0.0	0.139	7.269	0.233	4.191

THETA EPS   ← original ϕ

smoking1	asviews	asacts	sex	age	educatio	attend
0.025	0.166	0.065	0.002	12.345	0.691	0.466

↖ original θ_ϵ ↖ original θ_δ

CHI-SQUARE WITH   1 DEGREES OF FREEDOM IS   2.28 (PROB. LEVEL = 0.131)

the ξ concepts with η concepts, and the corresponding movement of Φ into Ψ, demonstrate two further conventions. The choice between ξ and η labels is mere convention, and the restrictions on which concepts can influence which indicators (ξ's to only X's, and η's to only Y's) are conventional artifacts created by arbitrarily adopting the ξ labeling.

A more fundamental and unaltered aspect of the model becomes clear if we note that all of the variables previously listed as exogenous ξ's exhibit only mere correlations with one another (the portion of Ψ mirroring the original Φ). If a variable enters into even one correlation, it receives no direct effects from any other variables; and if a variable receives even one direct effect, it does not enter into any correlations. Thus there are two fundamental classes of η variable in the model with the ξ's replaced by η's and these classes preserve the fundamental commitment to explain some variables while leaving others unexplained. Thus *the fundamental distinction between exogenous and endogenous variables is preserved* even though the variables now appear on a single list (the vector η).

The simplicity gained by having fewer matrices in the η-only representation is offset by additional distinctions among the elements within the matrices (the partitioning of **B**, **Ψ**, and **Θ$_e$**). That is, **Ψ** contains some elements we wish to interpret as error variances, whereas other elements are to be interpreted as variances/covariances among the formerly exogenous variables. It is up to you as a researcher to decide whether you prefer fewer complex matrices (partitioning) or more homogeneous matrices. This choice can be influenced by decisions about the simplicity of presentation (where the nonpartitioned conventional form wins) or other modeling requirements, *such as avoiding the constraints of the basic version of the LISREL model, where no effects from exogenous concepts to endogenous indicators are allowed* (where the all-η partitioned matrix form wins). Among the authors pursuing the fewer partitioned matrices approach are Bentler (1982), Bentler and Weeks (1985), Graff and Schmidt (1982), McArdle (1980), and McDonald (1978). Bentler (1983c:38) and Bentler and Weeks (1985) use partitioned matrices to emphasize the distinction between exogenous and endogenous variables. Rindskopf (1984b) is strongly recommended to those wishing to pursue model reparameterizations or tricks of the trade. His sections on constraining "reliabilities" (1984b:41) and "imaginary" variables with fixed negative variances (1984b:43) are especially interesting. Joreskog (1977b: Sec. 3.5; 1978b; 1981b) provide interesting examples of reparameterizations.

All the preceding reexpressions of the basic smoking model (except the "constrained to be three times" model in Section 7.1.3.2) involve precisely the same number of unknowns as the original model and provide precisely the same estimates of the estimated coefficients and overall χ^2. The advantage of any of these alternative representations is not that they improve the model but merely that they highlight representation styles that might do what we want, whether circumventing the seeming inability of ξ's to directly influence Y's, allowing covariances among measurement errors of the X's and Y's (possible if Θ_δ is moved into Θ_ϵ), or constraining a Ψ variance to be positive (according to Section 7.1.3.4).

We agree with Rindskopf (1984b:46–47) when he states

> . . . it would be preferable to be able to make a direct statement of the desired constraints, and let the computer program implement them, instead of using 'tricks' to implement the constraints. The application of methods such as the penalty function technique (Lee, 1980) and the multiplier method (Bentler and Lee, 1983) in commonly available computer programs will eventually make the need for the techniques reported here disappear. (The "tricks"). . . will remain interesting on theoretical grounds, since they demonstrate the hidden properties of many models, and the equivalence among (seemingly different ways). . . of specifying a model.

7.3 Multiple Indicators

7.3.1 Defining Concepts as Composites of Indicators

Occasionally we encounter concepts that are created from, as opposed to measured by, some observable indicators. A classic example is socioeconomic status (SES), which is frequently defined as the sum of an individual's scores on education, income, and occupational prestige. It is not uncommon for researchers to model the SES variable as illustrated in Figure 7.11A, where education, income, and occupation are taken as indicators of the underlying concept SES in a fictitious model. This way of modeling SES alters the meaning of SES rather dramatically, because SES now appears as the "cause", "predictor," or "underlying fount" of education, income, and occupation, as opposed to being a composite created from these variables. That is, SES becomes the factor common to education, income, and occupation, as opposed to a concept integrating potentially diverse standings on the component dimensions. If SES is a common cause, then education, income, and occupation should be substantially correlated (Section 1.4.1), but if SES is an effect, then education, income, and occupation may display no, or even negative,

Figure 7.11   SES as a Composite.

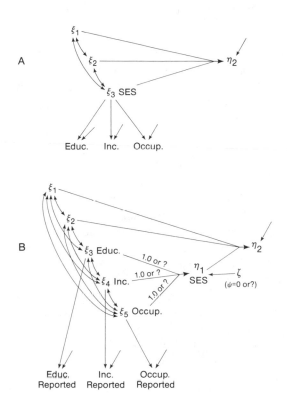

correlations with one another (cf. Bollen, 1984; Heise, 1972; Marsden, 1983).

A representation preserving the original conceptualization of SES as a composite of three components is presented in Figure 7.11B. In this formulation the concepts education, income, and occupation (ξ_3, ξ_4, and ξ_5) are scaled in the same units as the corresponding observed indicators (X variables) via specification of 1.0 Λ_x values. Measurement error is acknowledged through potentially nonzero (fixed or free) values in Θ_δ.

SES is then created from the education, income, and occupation concepts by inserting fixed 1.0 Γ values for the effects of these variables on the endogenous SES concept. The Ψ corresponding to SES might be fixed at zero if we believe that education, income, and occupation are the only concepts necessary for the definition of SES, or it might be fixed at some nonzero value or set free if we postulate specific or unknown contributions of other variables. SES has no indicators, and hence the meaning of SES arises from its creation as a composite of education, income, and occupation—a direct parallel to the original conceptualization of SES.

Figures 7.11A and B allow a single effect of SES on the dependent η_2. In Figure 7.11A education, income, and occupation do not directly affect η_2, though they do indirectly affect η_2 in Figure 7.11B. Also note that model 7.11B has fewer degrees of freedom because it has more Φ coefficients to be estimated.

The requirement of "equal" effects of education, income, and occupation on SES implied by fixing the appropriate Γ values at 1.0 forces us to consider the scales on which conceptual level education, income, and occupation are measured. The 1.0 Λ_x values imply the concepts are measured on the same scales as observed education, income, and occupation (namely in real units, such as "number of educational levels passed," "dollars," and "units of prestige"), but does it make sense to force the magnitudes of the effects of education, income, and occupation to be equal if the variables are measured on different scales? That is, is it reasonable to expect that one extra educational level has as much effect on SES as one extra dollar? Clearly it does not. One response to this problem is to standardize education, income, and occupational scores before the composite SES is created, thereby making SES a composite of the respondent's relative standing on three similarly scaled variables. Though the intent is to provide education, income, and occupation equal contributions to the variance in SES, you should see that since the equation for SES is of the form of Eq. 1.24, the variance of SES will be of the form of Eq. 1.26, which involves the covariances among the contributing variables (as well as their variances). Thus, even standardization is insufficient to guarantee an equal contribution to the variance of SES. [Burt (1973, 1976, and 1981) discuss how concepts derive their meanings from their links to observed indicators.]

In Figure 7.11B we consider standardizing the conceptual level variables, not the observed education, income, and occupation variables. Fixing the Φ variances for these concepts at 1.0 gives the appropriate variances, but it creates another problem because the links between conceptual education, income, and occupation and their respective indicators require reexamination. Whether the observed indicators are standardized or not, the Λ_x values linking the concepts to the indicators should no longer be 1.0 but should be fixed at the square root of the nonerror variance in the indicator [$\sqrt{\mathrm{Var}(X) - \mathrm{Var}(\text{meas. error})}$]. If the error variances are known, this is easy. If the error variances are unknown, there is no apparent way to do this, so we are forced to make a reasoned guess at the error variances and then see if there are any signs that these guesses hinder model estimation (recall the partial derivatives discussed in Chapter 6).

7.3.2 Multiple Indicators/Components and Proportional Effects

Though the preceding discussion highlights the possibilities and pitfalls of creating concepts as composites, we have yet to point out the most restrictive aspect of modeling SES in either Figure 7.11A or B and, by extension, the most restrictive aspect of modeling any composite of multiple variables or concept with multiple indicators. The single effect of SES on η_2 in 7.11B forces the indirect effects of conceptual education, income, and occupation to be equal (or proportional if their effects on SES are not all 1.0), because each indirect effect is the product of the SES-η_2 effect and the effect linking one of the three concepts to SES. A similar equality (proportionality) of the covariances between η_2 and observed education, income, and occupation is implied by Figure 7.11A. If SES is viewed as a common cause of η_2 and one of the three indicators, Eq. 1.67 describes the implied covariance resulting from the common cause SES. Using another indicator implies a second use of Eq. 1.67. The effect of SES on η_2 appears in both these equations and forces the equality (or proportionality) of covariances depending on the magnitude of the Λ values. Thus *both models imply a very strict pattern (either equality or proportionality) in the covariances between the three SES indicators and any indicator of η_2. The absence of a strict equality (proportionality) in the observed data implies that both models will be unable to fit the data.*[1]

One can view Figures 7.11A and B as forcing the three indicator variables to have effects (or display covariances) via a single causal mechanism—the mechanism implicating SES. Figure 7.11B provides a means for testing this "unitary causal pathway" assertion. The concepts education, income, and occupation can be allowed direct effects on the dependent η_2 by freeing any of the three appropriate Γ values. If education, income, and occupation influence η_2 by a variety of mechanisms, these direct or alternative paths to η_2 will carry substantial effects (or at least effects varying in magnitude) while the unifying pathway through SES remains unused and insignificant. Put another way, we must really believe education, income, and occupation are sufficiently coordinated by the abstract entity SES to provide equal or proportional effects if we are to entertain either model 7.11A or 7.11B. We must really believe the variables are precisely and fully channeled through SES en route to influencing η_2 before either the indicator model (7.11A) or the composite model (7.11B) can be taken as embodying our theoretical views.

It has been my experience that the particular composite called SES does not operate in this fashion. Any time I or my students have offered education, income, and occupation unique causal pathways simultaneously with a common pathway through SES, the results have unambiguously supported the unique effects at the expense of the common pathway. As a result, I routinely suggest that my students model education, income, and occupation as separate variables or as in Figure 7.11B, where the LISREL output can aid our decision about the unitary effect by requesting (through the partial derivatives) direct and unique effects.

Whether your experience with SES confirms or contrasts with mine will depend on the dependent variables that interest you, but this should not detract from the central point. *Whenever a concept has multiple indicators or is created as a composite of multiple dimensions, an equality (or proportionality) of effects/covariances occurs because these representations imply a unitary mechanism linking the concept to subsequent endogenous concepts.* If the multiple indicators or multiple components of the concept do not all share a single, uniformly effective, mechanism effecting subsequent model concepts, the model will fail to fit the data.[2]

7.3.3 Multiple Indicators in the Smoking Model

To illustrate the use of multiple indicators, we expand the basic smoking model to include double indicators of both smoking and religion. The concept smoking continues to have the current smoker/nonsmoker indicator. For a second indicator, we add the "physiological strain" measure, which scales the respondents' amount of smoking as light, moderate, and heavy and splits the current nonsmokers into those who have never smoked and those enduring the residual effects of former smoking (see Figure 4.4).

The two indicators of religion are the attendance of religious services (as routinely used before) and the respondent's rating of the truth/falsity of the statement "There is no definite proof that God exits" (see Figure 4.4). Although these indicators might report on the religiosity of the respondents, it is also possible that they do not function as multiple indicators of a single concept having a unitary set of causal connections. The behavioral slant of the first indicator contrasts markedly with the philosophical/scientific orientation of the second, and that may be sufficient to imply that these indicators are imbedded in essentially different causal webs.

The LISREL program and full results for this model are presented in Appendix C. You might examine the Λ_y and Λ_x matrices to see how the multiple indicators were specified. We highlight only the key results here.

First, the model fails to fit the data, $\chi^2 = 686$ with 16 *d.f.* and $p < .00$. Despite the poor fit the estimates in **B**, **Γ**, and **Ψ** differ minimally from the estimates in the basic model. The large λ estimate (2.856) for the supplemental smoking indicator appears because the physiological strain smoking indicator has considerably more variance than the original dichotomous smoker/nonsmoker indicator (see Figure 4.5 or Appendix C). Similarly, the small λ for the second religiosity indicator (.416) reflects the small variance of the God-exists indicator in comparision to the variance of the attendance indicator. The sizes of these λ values do not report on the quality of the multiple indicators; they merely make the variances of the corresponding concepts (which are still well specified by the fixed 1 λ values and the fixed Θ_ϵ and Θ_δ values as discussed in Chapter 4) as consistent as possible with the variances of the additional indicators. We might also note that the covariance between attendance and God exists implies a correlation of only .38, whereas the covariance between the smoking indicators implies a correlation of .90 (Eq. 1.54).

Two kinds of information suggest which of the new indicators are problematic. First, the partial derivatives are large (and the modification indices are huge) for numerous coefficients associated with the religiosity indicators (specificially the Θ_δ's and the Λ's linking these indicators to the other concepts). Second, the standardized residuals for the God-exists indicator are consistently large and in marked contrast to the small residuals for the smoking indicators. Indeed, one key aspect to the problematic fit of the religiosity indicators is apparent from the two elements at the lower right of the residual matrix: the variance of the God-exists indicator is underestimated, and the covariance between the religiosity indicators is overestimated. Increasing the value for the estimate of the λ linking religiosity to God exists would provide a larger variance estimate and a better fit, but a larger λ would also imply a greater covariance between the indicators (Eq. 1.67), which would provide a worse fit of the covariance. Or, in reverse, lowering the λ estimate to better fit the covariance would imply an even worse fit for the variance. Thus, there is sufficient inconsistency between the variances of the religiosity indicators and the covariance between these indicators to lead us to be suspicious of any common factor underlying these indicators, even without considering the covariances between these indicators and all the other indicators.

These observations suggest that the multiple smoking indicators are behaving very well: the proportionality of covariances implied by the multiple indicators of the concept smoking seems consistent with the data. In stark contrast, the data seem at odds with the proportionality of covariances implied by attempting to model God exists and attendance as multiple indicators of religiosity. These suspicions are reconfirmed by observing that a model with two smoking indicators (but only the attendance indicator of religion) fits well, whereas a model with the two religiosity indicators (but only the usual single indicator of smoking) fails dismally.

Clearly, the concept smoking can give a unitary causal mechanism consistent with both smoking indicators, and the concept religiosity cannot capture the diverse patterns of covariance (and implicitly diverse causal connections) of the behavioral attendance and the philosophical/scientific God-exists indicators. A model containing the two indicators of smoking and both God exists and attendance could be created by modeling smoking with multiple indicators and two separate religion concepts, reflecting the behavioral and philosophical aspects of religion, each with a single indicator. In such a model, the differences in the effects of these separate concepts (which would appear in an expanded Γ matrix) would provide direct evidence on which specific effects of these concepts are sufficiently different to hinder their modeling as proportional effects/covariances.[3]

If you are considering building a model with multiple indicators, I suggest beginning with the single best indicator of each concept and incrementally adding more indicators as necessary or informative. This simplifies the initial models by beginning with the fewest variables, for which the most is known. It also provides the clearest conceptual definitions and subsequently eliminates many sources of problems and indeterminacies.

If you are confronting a scale traditionally created as the sum of six or eight indicators, I would recommend you enter the indicators as single variables, use the best single indicator and a 1.0 λ to scale the concept, and fix the measurement error variances for the two or three best indicators. You might also adopt the strategy of first diagnosing indicator ills (by estimating a model with no η's, so all the concepts are merely correlated) and only subsequently introducing the postulated causal structure among the concepts (cf. Herting and Costner, 1985). This focuses LISREL's diagnostic power first on the measurement structure and subsequently on the conceptual model. Review the critique of this procedure at the end of Section 6.5.

I specifically discourage the use of composite scales created before calculating the data matrix S unless the creation of the composite is supported by extensive prior investigation (cf. Duncan, Sloane, and Brody, 1982). The smoking model, for example, would probably have been substantially more informative had we replaced the indicators of AS views and AS acts (which are counts of actions engaged in or supported) with single items reflecting agreement or disagreement with clear yet controversial issues. Unfortunately this type of data was unavailable.

7.4 Nonlinearity among the Concepts

Sections 7.4 and 7.5 illustrate how the tricks of the trade developed earlier can be adapted to address the issues of nonlinearity and interaction among conceptual level variables. Even if you have no specific interest in these topics, these sections should be read once (not skimmed and not studied) because considerable material of general interest is introduced.

In Section 2.6 we discussed the strategy of regressing a dependent variable on an independent variable and the square of that independent variable as a means of modeling many (but not necessarily all) of the types of nonlinear effects of one variable on another. This section adopts this conceptualization and draws heavily on the pathbreaking work of David Kenny and Charles Judd (1984), who extend the mathematics for obtaining the covariances among product random variables (Bohrnstedt and Goldberger, 1969) in developing a procedure for modeling nonlinear and interactive relations among concepts with multiple indicators. To test their procedures, Kenny and Judd created an artificial data set with known structural parameters that were hidden behind a mask of measurement error. They demonstrate the effectiveness of their procedures by recovering close estimates of the known underlying parameters.

We begin this section with a brief summary of the logic behind nonlinear effects in concepts and illustrate [using the Kenny and Judd (1984) model] the nature of the fundamental problem introduced by measurement error. We then express the Kenny and Judd model in LISREL notation, using many of the tricks of the trade already developed. [Kenny and Judd (1984) do not use LISREL or even matrix algebra.] We close with a discussion of some current issues in the modeling of nonlinearity.

We will be unable to provide the full details of the mathematics underlying the Kenny and Judd model (specifically of the statements preceding Eq. 7.15 and embodied in Eqs. 7.18–7.23). Kenny and Judd (1984), Bohrnstedt and Goldberger (1969), Bohrnstedt and Marwell (1978), and the related pages in Kendall and Stuart (1958) are mandatory reading for readers intending to model nonlinearities in new data sets. Anyone seeking a general idea of how this can be done, or wishing to see an example of how it is done, or even merely wishing to expand his or her knowledge of LISREL through a more sophisticated use of the tricks of the trade can take the nonproven points on faith and read the prerequisites only if a real need arises.

Figure 7.12   Nonlinear Effects of η_1 on η_3.

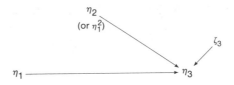

$$\eta = \beta\eta + \Gamma\xi + \zeta$$

$$\begin{bmatrix} \eta_1 \\ \eta_2 \\ \eta_3 \end{bmatrix} = \begin{bmatrix} 0 & & \\ 0 & 0 & \\ \beta_{31} & \beta_{32} & 0 \end{bmatrix} \begin{bmatrix} \eta_1 \\ \eta_2 \\ \eta_3 \end{bmatrix} + 0 + \begin{bmatrix} 0 \\ 0 \\ \zeta_3 \end{bmatrix}$$

Imagine a concept η_3 suspected to be nonlinearly dependent on a concept η_1. Adopting the regression strategy for investigating nonlinear effects, we express η_3 as a function of both η_1 and the square of η_1, which we will call η_2. This model is depicted in Figure 7.12. Imagine further that the causal variable η_1 has two indicators (Y_1 and Y_2), with the first being used to scale η_1 so that its λ is fixed at 1.0. The square of η_1 (namely η_2) has three indicators: Y_1 squared, Y_2 squared, and the product $Y_1 Y_2$, which we label Y_3, Y_4, and Y_5, respectively. The dependent variable η_3 is assumed to have a single error-free indicator Y_6. Figure 7.13 summarizes these concept–indicator relationships. The dotted portions of Figure 7.13 indicate portions of the model where we ultimately do some creative modeling after considering some implications of the steps already introduced.

7.4.1 The Implications of Squaring Variables Containing Measurement Error

Most of the creative modeling arises because squaring the observed indicators Y_1 and Y_2 or taking their product to obtain the observed

Figure 7.13   Indicators in the Nonlinear Model.

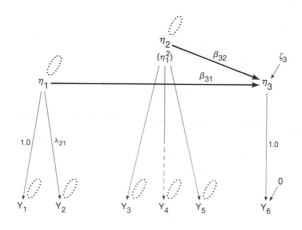

Y_1 = first indicator of η_1 (λ_{11}=1.0)

Y_2 = second indicator of η_1 (λ_{21}=?)

Y_3 = $(Y_1)^2$ = first indicator of the square of η_1 (η_2)

Y_4 = $(Y_2)^2$ = second indicator of the square of η_1 (η_2)

Y_5 = (Y_1Y_2) = third indicator of the square of η_1 (η_2)

Y_6 = indicator of the "dependent" concept η_3 (λ_{63}=1.0; $\psi_{6,6}$=0)

variables Y_3, Y_4, and Y_5 implies these variables are complex composites of measurement errors and the squared concept η_2. The nature of these complexities becomes clear if we first express Y_1 and Y_2 as indicators of η_1

$$Y_1 = 1.0\eta_1 + \epsilon_1 \qquad 7.7$$

$$Y_2 = \lambda_{21}\eta_1 + \epsilon_2 \qquad 7.8$$

We can now express the observed squared and product variables as

$$Y_1^2 = Y_3 = (\eta_1 + \epsilon_1)^2 \qquad 7.9$$

$$Y_3 = \eta_1^2 + 2\eta_1\epsilon_1 + \epsilon_1^2 \qquad 7.10$$

$$Y_2^2 = Y_4 = (\lambda \eta_1 + \epsilon_2)^2 \tag{7.11}$$

$$Y_4 = \lambda^2 \eta_1^2 + 2\lambda \eta_1 \epsilon_2 + \epsilon_2^2 \tag{7.12}$$

$$Y_1 Y_2 = Y_5 = (\eta_1 + \epsilon_1)(\lambda \eta_1 + \epsilon_2) \tag{7.13}$$

$$Y_5 = \lambda \eta_1^2 + \eta_1 \epsilon_2 + \lambda \eta_1 \epsilon_1 + \epsilon_1 \epsilon_2 \tag{7.14}$$

where the subscript 21 on λ has been omitted for simplicity.

The first terms on the right of the equals sign for Eqs. 7.10, 7.12, and 7.14 confirm that Y_3, Y_4, and Y_5 reflect the squared concept η_1^2 or η_2, but the remaining terms indicate that these squared and product variables also contain error components composed of variables created by squaring the original error variables or by multiplying the original error variables with each other or with η_1. The presence of the product variables $\eta_1 \epsilon_1$ and $\eta_1 \epsilon_2$ as the error terms for more than one equation creates the further problem that the complex error terms will be correlated due to the presence of the these "common causes."

If we do not know the individuals' real scores on η_1, ϵ_1, and ϵ_2, the individuals' scores on variables $\eta_1 \epsilon_1$, $\eta_1 \epsilon_2$, ϵ_1^2, and ϵ_2^2 are unknown. These complex and unknown error variables provide the fundamental stumbling block that Kenny and Judd (1984) circumvented. Specifically, following Bohrnstedt and Goldberger (1969), they (1984:202) demonstrate that if η_1, ϵ_1, and ϵ_2 are independent of one another, expressed as mean deviations, and multivariate normally distributed, then the covariances between all the variables η_1, ϵ_1, ϵ_2, η_2, $\eta_1 \epsilon_1$, $\eta_1 \epsilon_2$, and $\epsilon_1 \epsilon_2$ are zero and *the variances of the product variables are simple composites of the variances of* η_1, ϵ_1, *and* ϵ_2 (which we present in Eqs. 7.18–7.23).

This implies that Eqs. 7.10, 7.12, and 7.14 can be used to partition the variances in Y_3, Y_4, and Y_5 into components originating in the squared conceptual variable η_2 and other complex error variances, which can be expressed as simple functions of the variances of η_1, ϵ_1, and ϵ_2. For example, the variance of Y_3 can be expressed as follows if we treat Eq. 7.10 as a special instance of Eq. 1.24 and use 1.33 because the basic and product variables are independent.

$$\text{Var}(Y_3) = \text{Var}(\eta_1^2) + 4\text{Var}(\eta_1 \epsilon_1) + \text{Var}(\epsilon_1^2) \tag{7.15}$$

Similarly, the variances of Y_4 and Y_5 are

$$\text{Var}(Y_4) = (\lambda^2)^2 \text{Var}(\eta_1^2) + (2\lambda)^2 \text{Var}(\eta_1 \epsilon_2) + \text{Var}(\epsilon_2^2) \tag{7.16}$$

$$\text{Var}(Y_5) = \lambda^2 \text{Var}(\eta_1^2) + \text{Var}(\eta_1 \epsilon_2) + \lambda^2 \text{Var}(\eta_1 \epsilon_1) + \text{Var}(\epsilon_1 \epsilon_2) \tag{7.17}$$

The simple relationships between the variances of the product and squared variables proven by Kenny and Judd (1984:202 and appendix) are as follows if, to be consistent with LISREL notation, we represent the variance of measurement errors as Θ_ϵ's.

$$\text{Var}(\epsilon_1\epsilon_2) = \Theta_{\epsilon1}\Theta_{\epsilon2} \qquad\qquad 7.18$$

That is, the variance of the error product variable equals the product of the variances of the basic measurement error variables.

$$\text{Var}(\eta_1\epsilon_1) = \text{Var}(\eta_1)\Theta_{\epsilon1} \qquad\qquad 7.19$$

$$\text{Var}(\eta_1\epsilon_2) = \text{Var}(\eta_1)\Theta_{\epsilon2} \qquad\qquad 7.20$$

$$\text{Var}(\epsilon_1^2) = 2\Theta_{\epsilon1}^2 \qquad\qquad 7.21$$

That is, the variance of the variable created by squaring the original ϵ_1 variable equals twice the square of the variance of the original error variable.

$$\text{Var}(\epsilon_2^2) = 2\Theta_{\epsilon2}^2 \qquad\qquad 7.22$$

$$\text{Var}(\eta_1^2) = \text{Var}(\eta_2) = 2(\text{Var}(\eta_1))^2 \qquad\qquad 7.23$$

We can now rewrite Eqs. 7.15–7.17 as

$$\text{Var}(Y_3) - \text{Var}(\eta_2) + 4\text{Var}(\eta_1)\Theta_{\epsilon1} + 2\Theta_{\epsilon1}^2 \qquad\qquad 7.24$$

$$\text{Var}(Y_4) = (\lambda^2)^2\text{Var}(\eta_2) + (2\lambda)^2\text{Var}(\eta_1)\Theta_{\epsilon2} + 2\Theta_{\epsilon2}^2 \qquad\qquad 7.25$$

$$\text{Var}(Y_5) = \lambda^2\text{Var}(\eta_2) + \text{Var}(\eta_1)\Theta_{\epsilon2} + \lambda^2\text{Var}(\eta_1)\Theta_{\epsilon1} + \Theta_{\epsilon1}\Theta_{\epsilon2} \qquad\qquad 7.26$$

Though it is far from obvious, the complex error terms in these equations can be represented within LISREL by using simple extensions and combinations of the tricks discussed earlier. That is, $\Theta_{\epsilon1}$, $\Theta_{\epsilon2}$, and λ are coefficients appearing within the LISREL model, so we should be able to multiply these coefficients by one another or by constants, or to square or obtain the roots of the coefficients as necessary. Even the $\text{Var}(\eta_1)$, which does not normally appear as a specific numerical estimate, can be obtained through creative use of Ψ. The next few pages explicate the details of representing the foregoing as a LISREL model.

The strategy is to use the tricks from Section 7.1 to create "convenience variables" that mimic the complex variables $\eta_1\epsilon_1$, $\eta_1\epsilon_2$, $\epsilon_1\epsilon_1$, $\epsilon_2\epsilon_2$, and $\epsilon_1\epsilon_2$ in 7.10–7.14 by giving the convenience variables the "proper variances" according to Eqs. 7.18–7.23. We then constrain the convenience variables to have effects on the observed Y_3, Y_4, and Y_5 as required by Eqs. 7.10, 7.12, and 7.14. This accomplishes the variance partitioning required by 7.15–7.17 (or equivalently 7.24–7.26). Letting the convenience variables have effects on more than one of Y_3, Y_4, and Y_5 in precisely the same way as the corresponding complex variables $\eta_1\epsilon_1$ and $\eta_1\epsilon_2$ builds in the proper "common causal" structure and hence results in the proper amount of covariance between the complex error terms on the right sides of Eqs. 7.10, 7.12, and 7.14, even though we never actually calculate these covariances.

7.4.2 Representing the Nonlinear Model in LISREL

Only seven coefficients are to be estimated for the nonlinear model. Locate these coefficients in Figure 7.13 before tackling the more complex versions of the model that follow. The coefficients are

β_{31} and β_{32}: the effects of the concept and its square on the dependent concept η_3

ψ_3: the error variance in predicting η_3

λ_{21}: the loading of the second indicator on the concept η_1

$\Theta_{\epsilon1}$ and $\Theta_{\epsilon2}$: the measurement error variances for the two indicators

the variance of η_1: which we creatively model as ψ_1.

Convince yourself that only these seven coefficients are required to model the variances of the square and product indicators Y_3, Y_4, and Y_5, according to Eqs. 7.24–7.26 (recalling that $\text{Var}(\eta_2)$ is given by 7.23).

We now tackle the full version of the interaction model as depicted in Figure 7.14, represented in matrix form in Figure 7.15, and estimated in Figure 7.16. Consider the portion of Figure 7.14 near η_1. The only variable influencing η_1 is the error variable ζ_1, which has variance ψ_1. Since all the variance in η_1 originates in the error variable, $\text{Var}(\eta_1) = \psi_1$.

According to Eq. 7.23 the variance of the squared concept η_2 must equal $2(\text{Var}(\eta_1))^2$. This variance can be given to η_2 by first creating a new variable, η_4, whose variance is fixed at 2.0 (by fixing $\psi_4 = 2.0$), which influences η_2 by a coefficient constrained to equal ψ_1 (the variance of η_1), and simultaneously making η_4 the only contributor to the variance of η_2 by fixing ψ_2 to have zero variance. Thus, using Eq. 1.23, $\text{Var}(\eta_2) = (\psi_1)^2(2.0) = 2.0(\text{Var}(\eta_1))^2$, as required.

Moving to Y_1 and Y_2, we fix the measurement error variances for Y_1 and Y_2 to zero and replace these error terms with two new variables, η_5 and η_6, each having variance 1.0 ($\psi_5 = 1.0$, $\psi_6 = 1.0$), and affecting Y_1 and Y_2 with free Λ coefficients (cf. Sections 7.1.1.1 and 7.1.3.4). Since the variance contributed to Y_1 by η_5 is $\lambda_{15}^2\text{Var}(\eta_5) = \lambda_{15}^2$, λ_{15}^2 will equal the

Figure 7.14   The Full Nonlinear Model.[1]

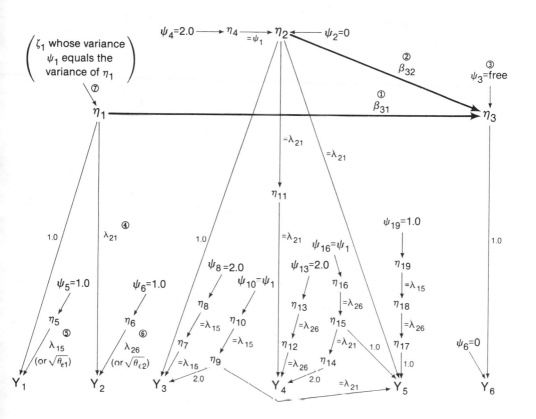

[1]Caution: Instead of plotting ζ variables, this figure plots the ψ providing the variance of the corresponding error variable. Coefficients constrained to equal another coefficient are indicated with the value they are constrained to have. Thus the effect of η_4 on η_2 is designated as $= \psi_1$ to indicate β_{24} is constrained to equal ψ_1. The seven coefficients to be estimated are indicated with circled numbers.

Figure 7.15   The Matrix Form of the Nonlinear Model.[1]

$$\eta = \beta\eta + \Gamma\xi + \zeta$$

$$Y = \Lambda_Y \eta + \epsilon$$

$$\left[\theta_\epsilon\right] = \left[\begin{matrix} 0 & 0 & 0 & 0 & 0 & 0 \end{matrix}\right]$$

[1]Constrained coefficients are indicated by entry of the coefficient to which the value is constrained. The seven estimated coefficients are circled. Omitted entries in β and Λ_Y are all zeroes.

Figure 7.16   LISREL MLE Estimates of the Nonlinear Model.

LISREL ESTIMATES (MAXIMUM LIKELIHOOD)

LAMBDA Y

	ETA 1	ETA 2	ETA 3	ETA 4	ETA 5	ETA 6	ETA 7	ETA 8	ETA 9	ETA 10
Y1	1.000	0.0	0.0	0.0	0.404	0.0	0.0	0.0	0.0	0.0
Y2	0.626	0.0	0.0	0.0	0.0	0.744	0.0	0.0	0.0	0.0
Y1SQ	0.0	1.000	0.0	0.0	0.0	0.0	0.404	0.0	2.000	0.0
Y2SQ	0.0	0.0	0.0	0.0	0.0	0.0	0.0	0.0	0.0	0.0
Y1Y2	0.0	0.626	0.0	0.0	0.0	0.0	0.0	0.0	0.626	0.0
Y6DEP	0.0	0.0	1.000	0.0	0.0	0.0	0.0	0.0	0.0	0.0

LAMBDA Y

	ETA 11	ETA 12	ETA 13	ETA 14	ETA 15	ETA 16	ETA 17	ETA 18	ETA 19
Y1	0.0	0.0	0.0	0.0	0.0	0.0	0.0	0.0	0.0
Y2	0.0	0.0	0.0	0.0	0.0	0.0	0.0	0.0	0.0
Y1SQ	0.0	0.0	0.0	0.0	0.0	0.0	0.0	0.0	0.0
Y2SQ	0.626	0.744	0.0	2.000	0.0	0.0	0.0	0.0	0.0
Y1Y2	0.0	0.0	0.0	0.0	1.000	0.0	1.000	0.0	0.0
Y6DEP	0.0	0.0	0.0	0.0	0.0	0.0	0.0	0.0	0.0

BETA

	ETA 1	ETA 2	ETA 3	ETA 4	ETA 5	ETA 6	ETA 7	ETA 8	ETA 9	ETA 10
ETA 1	0.0	0.0	0.0	0.0	0.0	0.0	0.0	0.0	0.0	0.0
ETA 2	0.0	0.0	0.0	0.995	0.0	0.0	0.0	0.0	0.0	0.0
ETA 3	0.248	-0.502	0.0	0.0	0.0	0.0	0.0	0.0	0.0	0.0
ETA 4	0.0	0.0	0.0	0.0	0.0	0.0	0.0	0.0	0.0	0.0
ETA 5	0.0	0.0	0.0	0.0	0.0	0.0	0.0	0.0	0.0	0.0
ETA 6	0.0	0.0	0.0	0.0	0.0	0.0	0.0	0.0	0.0	0.0
ETA 7	0.0	0.0	0.0	0.0	0.0	0.0	0.0	0.404	0.0	0.0
ETA 8	0.0	0.0	0.0	0.0	0.0	0.0	0.0	0.0	0.0	0.0
ETA 9	0.0	0.0	0.0	0.0	0.0	0.0	0.0	0.0	0.0	0.404
ETA 10	0.0	0.0	0.0	0.0	0.0	0.0	0.0	0.0	0.0	0.0
ETA 11	0.0	0.626	0.0	0.0	0.0	0.0	0.0	0.0	0.0	0.0
ETA 12	0.0	0.0	0.0	0.0	0.0	0.0	0.0	0.0	0.0	0.0
ETA 13	0.0	0.0	0.0	0.0	0.0	0.0	0.0	0.0	0.0	0.0
ETA 14	0.0	0.0	0.0	0.0	0.0	0.0	0.0	0.0	0.0	0.0
ETA 15	0.0	0.0	0.0	0.0	0.0	0.0	0.0	0.0	0.0	0.0
ETA 16	0.0	0.0	0.0	0.0	0.0	0.0	0.0	0.0	0.0	0.0
ETA 17	0.0	0.0	0.0	0.0	0.0	0.0	0.0	0.0	0.0	0.0
ETA 18	0.0	0.0	0.0	0.0	0.0	0.0	0.0	0.0	0.0	0.0
ETA 19	0.0	0.0	0.0	0.0	0.0	0.0	0.0	0.0	0.0	0.0

BETA

	ETA 11	ETA 12	ETA 13	ETA 14	ETA 15	ETA 16	ETA 17	ETA 18	ETA 19
ETA 1	0.0	0.0	0.0	0.0	0.0	0.0	0.0	0.0	0.0
ETA 2	0.0	0.0	0.0	0.0	0.0	0.0	0.0	0.0	0.0
ETA 3	0.0	0.0	0.0	0.0	0.0	0.0	0.0	0.0	0.0
ETA 4	0.0	0.0	0.0	0.0	0.0	0.0	0.0	0.0	0.0
ETA 5	0.0	0.0	0.0	0.0	0.0	0.0	0.0	0.0	0.0
ETA 6	0.0	0.0	0.0	0.0	0.0	0.0	0.0	0.0	0.0
ETA 7	0.0	0.0	0.0	0.0	0.0	0.0	0.0	0.0	0.0
ETA 8	0.0	0.0	0.0	0.0	0.0	0.0	0.0	0.0	0.0
ETA 9	0.0	0.0	0.0	0.0	0.0	0.0	0.0	0.0	0.0
ETA 10	0.0	0.0	0.0	0.0	0.0	0.0	0.0	0.0	0.0
ETA 11	0.0	0.0	0.0	0.0	0.0	0.0	0.0	0.0	0.0
ETA 12	0.0	0.0	0.744	0.0	0.0	0.0	0.0	0.0	0.0
ETA 13	0.0	0.0	0.0	0.0	0.0	0.0	0.0	0.0	0.0
ETA 14	0.0	0.0	0.0	0.0	0.626	0.0	0.0	0.0	0.0
ETA 15	0.0	0.0	0.0	0.0	0.0	0.744	0.0	0.0	0.0
ETA 16	0.0	0.0	0.0	0.0	0.0	0.0	0.0	0.0	0.0
ETA 17	0.0	0.0	0.0	0.0	0.0	0.0	0.744	0.0	0.0
ETA 18	0.0	0.0	0.0	0.0	0.0	0.0	0.0	0.0	0.404
ETA 19	0.0	0.0	0.0	0.0	0.0	0.0	0.0	0.0	0.0

PSI

ETA 1	ETA 2	ETA 3	ETA 4	ETA 5	ETA 6	ETA 7	ETA 8	ETA 9	ETA 10
0.995	0.0	0.196	2.000	1.000	1.000	0.0	2.000	0.0	0.995

PSI

ETA 11	ETA 12	ETA 13	ETA 14	ETA 15	ETA 16	ETA 17	ETA 18	ETA 19
0.0	0.0	2.000	0.0	0.0	0.995	0.0	0.0	1.000

CHI-SQUARE WITH 14 DEGREES OF FREEDOM IS 19.13 (PROB. LEVEL = 0.160)

Coefficient	Real Value (Kenny and Judd, 1984)	Maximum Likelihood Estimate	GLS Estimate (LISREL "GL" option or Kenny and Judd, 1984)
β_{31}	0.25	0.248	0.247
β_{32}	-0.50	-0.502	-0.500
ψ_3	0.20	0.196	0.199
ψ_1	1.00	0.995	0.989
λ_{21}	0.60	0.626	0.624
λ_{15}	$\sqrt{0.15}=0.387$	0.404	0.400
λ_{26}	$\sqrt{0.55}=0.742$	0.744	0.735

error variance that would have appeared if $\Theta_{\epsilon 1}$ had been free, which is another way of saying λ_{15} will equal the square root of $\Theta_{\epsilon 1}$. By a parallel argument, $\lambda_{26} = \sqrt{\Theta_{\epsilon 2}}$, or the square root of the error variance that would have appeared had we chosen to represent the measurement error in the traditional form.

We have now encountered all seven coefficients to be estimated [β_{31}, β_{32}, ψ_3, ψ_1 [or Var(η_1)], λ_{21}, λ_{15} (or $\sqrt{\Theta_{\epsilon 1}}$), and λ_{26} (or $\sqrt{\Theta_{\epsilon 2}}$)]. Only four of these coefficients (in various combinations) are required to create the complex error components influencing Y_3, Y_4, and Y_5, as indicated in Eqs. 7.24–7.26. The error component for Y_3 requires four convenience variables (η_7, η_8, η_9, and η_{10}) to be created. Variable η_8 is fixed to have variance 2.0 (by fixing its $\psi_8 = 2.0$), and η_{10} is given a variance equal to the variance of η_1 by constraining $\psi_{10} = \psi_1$. We further constrain the effects of η_8 on η_7, η_7 on Y_3, and η_{10} on η_9 to *all* equal λ_{15} (or $\sqrt{\Theta_{\epsilon 1}}$), while the effect of η_9 on Y_3 is fixed at 2.0. Following the chain of effects from η_8 to η_7 to Y_3 and using Eq. 1.23, we see that the variance of η_8 equals 2.0, so the variance of η_7 is $\lambda_{15}^2(2.0)$ or $2.0\Theta_{\epsilon 1}$. Subsequently, the variance contributed to Y_3 from this chain (using 1.23 again) is $\lambda_{15}^2(2.0\Theta_{\epsilon 1}) = \Theta_{\epsilon 1}2.0\Theta_{\epsilon 1} = 2\Theta_{\epsilon 1}^2$, as required by the last term in Eq. 7.24.

The chain from η_{10} to η_9 to Y_3 starts with the variance of η_{10} equaling the variance of η_1 (via ψ_1). Then the variance of η_9 is λ_{15}Var(η_1) = $\Theta_{\epsilon 1}$Var(η_1). Thus the variance contributed to Y_3 is $2^2\Theta_{\epsilon 1}$Var(η_1) = 4Var(η_1)$\Theta_{\epsilon 1}$, as required by the second-to-last term in Eq. 7.24. Thus the two chains of effects provide an error variance as required by Eq. 7.24, and the effect of η_2 on Y_3 increments the variance of Y_3 by 1.0^2Var(η_2), as required by the first term on the right of Eq. 7.24.

The setup of the error component of Y_4 parallels the error component of Y_3. Five new convenience variables are defined (η_{12}, η_{13}, η_{14}, η_{15}, η_{16}), and two are given specific variances. Then the variance of η_{13} is set at 2.0 by fixing $\psi_{13} = 2.0$, and the variance of η_{16} is set equal to Var(η_1) by constraining $\psi_{16} = \psi_1$. The effects of η_{13} on η_{12}, η_{12} on Y_4, and η_{16} on η_{15} are all constrained to equal λ_{26} (or $\sqrt{\Theta_{\epsilon 2}}$). The effect of η_{16} on η_{15} is constrained to equal λ_{21}, and the effect of η_{14} on Y_4 is fixed at 2.0.

Thus the variance of $\eta_{13} = 2.0$ (from ψ_{13}), the variance of η_{12} is $\lambda_{26}^2(2.0) = 2.0\Theta_{\epsilon 2}$, and, subsequently, the variance contributed to Y_4 by this chain of effects is $\lambda_{26}^2(2.0\Theta_{\epsilon 2}) = \Theta_{\epsilon 2}2.0\Theta_{\epsilon 2} = 2\Theta_{\epsilon 2}^2$, as required by the last term in Eq. 7.25. Similarly, the variance of η_{16} equals Var(η_1) by the constraint $\psi_{16} = \psi_1$, the variance of η_{15} is λ_{26}^2Var(η_1) = $\Theta_{\epsilon 2}$Var(η_1), the variance of η_{14} is $\lambda_{21}^2[\Theta_{\epsilon 2}$Var($\eta_1$)], and the variance contributed to Y_4 is $2^2\lambda_{21}^2\Theta_{\epsilon 2}$Var($\eta_1$), which is the second-to-last term in Eq. 7.25.

The first term on the right of Eq. 7.12 encapsulates the dependence of Y_4 on η_2. To provide the λ_{21}^2 prefix to this term, we introduce yet another convenience variable, η_{11}, which responds to η_2 according to a coefficient constrained to equal λ_{21}, and η_{11} influences Y_4 with a similarly constrained coefficient. Thus the variance of η_{11} is $\lambda_{21}^2 \text{Var}(\eta_2)$, and, the variance contributed to Y_4 from the chain originating in η_2 is $\lambda_{21}^2[\lambda_{21}^2 \text{Var}(\eta_2)]$ or $(\lambda_{21}^2)^2 \text{Var}(\eta_2)$, as required by Eq. 7.25.

To give Y_5 the appropriate variance and partitioning of variance as indicated in Eq. 7.26, we constrain the effect of η_2 on Y_5 to equal λ_{21} (cf. Eq. 7.14). This gives the $\lambda_{21}^2 \text{Var}(\eta_2)$, or $\lambda_{21}^2 \text{Var}(\eta_1^2)$, term in Eq. 7.26. Creating three more η variables for convenience and setting $\psi_{19} = 1.0$, the effect of η_{19} on η_{18} as λ_{15}, the effect of η_{18} on η_{17} as λ_{26}, and the effect of η_{17} on Y_5 as 1.0 give a variance contribution of $\lambda_{26}^2 \lambda_{15}^2 1.0 = \Theta_{\epsilon 1}\Theta_{\epsilon 2}$, which is the last term in Eq. 7.26.

Note that the product variable $\eta_1\epsilon_1$ appears in both Eqs. 7.10 and 7.14 (and hence in Eqs. 7.15 and 7.17, as well as 7.24 and 7.26), and $\eta_1\epsilon_2$ appears in Eqs. 7.12 and 7.14 (and hence in Eqs. 7.16, 7.17 and 7.25, 7.26). This allows us to capitalize on the creation of variance components for Y_3 and Y_4 in creating the necessary components for Y_5. Specifically, letting η_{15} influence Y_5 with coefficient 1.0 gives a variance of $\lambda_{26}^2\psi_1 = \Theta_{\epsilon 2}\text{Var}(\eta_1)$, and letting η_9 influence Y_5 with coefficient λ_{21} contributes the final variance required by Eq. 7.26, $\lambda_{21}^2\Theta_{\epsilon 1}\text{Var}(\eta_1)$.

Actually, the presence of the same product variable in the error terms for two equations makes the links from η_9 and η_{15} to Y_5 mandatory, not merely a matter of convenience. The common causal product variables in the error terms for two equations imply that the error terms are correlated. By creating variables with the proper variances and giving them the proper influence on the Y's for the equations in which the error terms appear, we have modeled the implied covariance.

To summarize, the previous steps first use the tricks of the trade to create a variable whose variance equals the variance of the complex product and squared variables on the right of Eqs. 7.10, 7.12, and 7.14 by using Eqs. 7.18–7.23. These variables are then allowed to influence *all* the indicators in whose equations they appear (7.10, 7.12, and 7.14) via effects constrained to equal the proper magnitudes. The contribution of the complex convenience variables to more than one equation automatically models the implicit covariances among the complex error terms.

LISREL structural equations equivalent to Figure 7.14 appear in Figure 7.15 with equality constraints indicated by multiple insertions of specific coefficient values. The placement of the constrained coefficients indicates which particular $\mathbf{B}$, $\mathbf{\Lambda}$, or $\mathbf{\Psi}$ value is being constrained.

7.4.3 Estimating the Nonlinear Model

Estimating this model[4] with MLE provides the estimates in Figure 7.16, which are presented in comparison to the actual values for the corresponding population parameters built in by Kenny and Judd (1984). Clearly, the estimates are sufficiently close to the known true values to claim a successful recovery of the coefficients involved in the nonlinear relationship between the concepts η_1 and η_3.

The equation linking η_3 to η_1 and η_2 is

$$\eta_3 = .248\eta_1 - .502\eta_1^2 + \zeta_3 \qquad\qquad 7.27$$

As in multiple regression, the coefficients .248 and $-.502$ should not be interpreted separately, because both coefficients link η_1 to η_3. The preferable strategy is to take the derivative of η_3 with respect to η_1, to obtain the "slopes" of the nonlinear function linking η_3 to η_1. From Section 3.2 $d\eta_3/d\eta_1 = .248 - 1.004\eta_1$, which highlights how strongly the slope, the effect of η_1 on η_2, depends on the particular value of η_1 that is considered. Since η_1 has mean zero and variance 1.0 (Kenny and Judd, 1984:203), η_1 values of -1, 0, and $+1$ represent realistic values, and the slopes at these values are 1.252, .248, and $-.756$, respectively.

7.4.4 Some Links to the Literature

Naturally, when results turn out this well, there is bound to be some hidden catch. Kenny and Judd (1984) point out that products of multinormally distributed variables cannot be themselves multinormally distributed, so MLE for this model is formally unjustified because the assumption of multinormality is known to be untenable for the product variables. Kenny and Judd (1984) used "generalized least squares" and another program to obtain their estimates. Using the generalized least squares estimation option within LISREL and the foregoing model give precisely the same values obtained by Kenny and Judd (1984), as indicated in Figure 7.16. Though Kenny and Judd did not report any maximum likelihood results, they did acknowledge having tried MLE and having found that "the parameter estimates were in most cases not appreciably different from the generalized least squares estimates" (1984:208). Indeed, since the maximum likelihood estimates come closer to the true values in three of seven instances, it seems the maximum likelihood procedure may be robust with respect to violation of the multivariate normality assumption. Though we cannot unconditionally recommend the use of MLE in the face of demonstrable violations of the multivariate normality assumption, the closeness of the estimates in this instance certainly does little to discourage its use, given the other benefits of this estimation strategy. We note that Kenny and Judd (1984) created their product variables from other variables known to be distributed multivariate normally. Taking the product of strongly nonnormal initial variables might give stronger violations of multivariate normality for the product variables, so extra caution is advised under these conditions.

In another discussion of the implications of measurement error in product variables, Busemeyer and Jones (1983) point out that covariances alone are insufficient information for deciding whether we are confronting a nonlinear measurement model appended to a linear conceptual level model, or a linear measurement model reflecting nonlinear relations among the concepts (as discussed earlier). The only current solution to this "placement of nonlinearity" problem is an appeal to substantive theory for guidance.

Busemeyer and Jones (1983:558) further contrast two recent approaches to modeling nonlinearity or product variables. One side of the contrast is illustrated by Bagozzi (1981), who created multiple product indicators of a product concept and proceeded to model this just as we would model ordinary concepts with indicators, paying no special attention to the implications of complex error terms (in stark contrast to what we have done). Naturally, Busemeyer and Jones (1983) do not recommend this. On the other side of the contrast are Bohrnstedt and Marwell (1978), who are characterized as detailing the implications of the error terms but requiring "prior knowledge of the reliabilities" of the concepts to be multiplicatively combined. The procedures in Sections 7.4 and 7.5 are in the tradition of precise calculation of complex errors (à la Bohrnstedt and Goldberger, 1969; Bohrnstedt and Marwell, 1978; Kenny and Judd, 1984), but it should be clear that this position does not require that one have "prior knowledge of reliabilities." The preceding example circumvents this condition on prior knowledge by using constrained estimates to simultaneously estimate the error in the basic and multiplicative indicators.

Also note that including product predictor variables is not the only traditional way of handling nonlinearities. We can occasionally transform the observed variables nonlinearly before modeling them. For example, if Y is a nonlinear function of X, it is possible that log Y is a linear function of X, or even that Y is a linear function of log X. Similarly, if our desired model is $Z = XY$, we could model it as the additive model log $Z =$ log $X +$ log Y. We see no inherent hindrance to using logarithmic or other nonlinearly transformed variables, just as they create no inherent problems in multiple regression. See Walling, Hotchkiss, and Curry (1984) for an introduction to the problem of specifying error terms in alternative formulations of nonlinear models.

Bentler (1986) reviews several recent advances in the statistics of "moment structures" that may provide more direct procedures for estimating nonlinear effects. The complexity of model estimation will also be reduced if current plans to allow user-specified arbitrary constraints between model coefficients are instituted in LISREL VII, much as they are in COSAN (McDonald, 1985). Heise (1986) gives a procedure for preadjusting for measurement error in nonlinear models, but this is not LISREL-oriented.

7.5 Interaction among the Concepts

There are two essentially different procedures for handling interactions in LISREL. If the modeled concepts are suspected of being involved in many interactions the procedures for stacking multiple groups of data in Chapter 9 are appropriate. If only a few interactions are suspected, we can mimic multiple regression procedures for handling interactions, just as Section 7.4 mimicked multiple regression procedures for handling nonlinearity. This section develops a LISREL model for the interaction between two concepts, each having two indicators. We again illustrate our discussion with an artificial data set created by Kenny and Judd (1984), so the population parameters are known. Since the representation of interactions within LISREL requires the same tricks of the trade we used with nonlinearities, we focus on the logic of the analysis and rely on you to provide some details.

7.5.1 The Kenny and Judd Interaction Model

Kenny and Judd (1984) postulate that two concepts (η_1 and η_2) interact in their production of another concept (η_4). The interaction is to be modeled by the concept η_3, which is the product of η_1 and η_2 (i.e., $\eta_3 = \eta_1\eta_2$). Taking the products of the indicators of η_1 and η_2 to create indicators of the product concept η_3 results in complex error terms depending on newly arising variables that are the products of the basic measurement error variables and the basic concepts η_1 and η_2. Though the specific values of these variables cannot be obtained (because the true values of the concepts and error variables remain unknown for the cases), it is of no concern because only the variances of these complex variables are ultimately required. Kenny and Judd (1984) provide the mathematical foundation for expressing the variances of these complex variables as functions of the variances of the basic concepts and measurement error variances.

The fundamental modeling strategy is to create convenience variables with precisely the proper variances (through creative LISREL modeling) and then allow them to influence all the product indicators in whose equations they appear. Equations 7.28–7.39 give the composition of the basic and product indicator variables as functions of the concepts and measurement error variables. Equations 7.44 to 7.52 give the variances of the complex conceptual and measurement error product variables in terms of simpler variances of the concepts and measurement error variances separately. These sets of equations are sufficient for LISREL modeling, but we take the process one step further and actually create and partition the variances of the indicator product variables in Eqs. 7.40–7.43 and 7.53–7.56. This maintains the parallel to the discussion of nonlinearity in the preceding section and verifies that the magnitudes of effects are those providing the proper effect sizes and variance partitionings. The covariances among the complex error terms are again automatically handled by this procedure, because we literally create a convenience variable whose variance *and effects* are precisely those of the corresponding concept–error product variables. Thus any covariances among the complex error terms created by the concept–error products influencing more than one indicator product variable are faithfully modeled through a direct insertion of convenience variables with identical variances and patterns of effects.

The Kenny and Judd (1984) interaction model appears in Figure 7.17. The equations for the indicators of η_1 are

$$Y_1 = 1.0\eta_1 + \epsilon_1 \tag{7.28}$$

$$Y_2 = \lambda_{21}\eta_1 + \epsilon_2 \tag{7.29}$$

For the indicators of η_2 the equations are

$$Y_3 = 1.0\eta_2 + \epsilon_3 \tag{7.30}$$

$$Y_4 = \lambda_{42}\eta_2 + \epsilon_4 \tag{7.31}$$

The product indicator variables are therefore

$$Y_5 = Y_1 Y_3 = (\eta_1 + \epsilon_1)(\eta_2 + \epsilon_3) \tag{7.32}$$

$$Y_5 = \eta_1\eta_2 + \eta_1\epsilon_3 + \eta_2\epsilon_1 + \epsilon_1\epsilon_3 \tag{7.33}$$

$$Y_6 = Y_1 Y_4 = (\eta_1 + \epsilon_1)(\lambda_{42}\eta_2 + \epsilon_4) \tag{7.34}$$

$$Y_6 = \lambda_{42}\eta_1\eta_2 + \eta_1\epsilon_4 + \lambda_{42}\eta_2\epsilon_1 + \epsilon_1\epsilon_4 \tag{7.35}$$

$$Y_7 = Y_2 Y_3 = (\lambda_{21}\eta_1 + \epsilon_2)(\eta_2 + \epsilon_3) \tag{7.36}$$

$$Y_7 = \lambda_{21}\eta_1\eta_2 + \lambda_{21}\eta_1\epsilon_3 + \eta_2\epsilon_2 + \epsilon_2\epsilon_3 \tag{7.37}$$

$$Y_8 = Y_2 Y_4 = (\lambda_{21}\eta_1 + \epsilon_2)(\lambda_{42}\eta_2 + \epsilon_4) \qquad 7.38$$

$$Y_8 = \lambda_{21}\lambda_{42}\eta_1\eta_2 + \lambda_{21}\eta_1\epsilon_4 + \lambda_{42}\eta_2\epsilon_2 + \epsilon_2\epsilon_4 \qquad 7.39$$

Note the complex product variables of the form $\eta\epsilon$ and $\epsilon\epsilon$ in addition to the product variable $\eta_1\eta_2$, which we call η_3.

Since the measurement error variables and the product variables are all independent (as proven by Kenny and Judd, 1984), we see that, using Eq. 1.33, the preceding equations imply that the variances of the product indicators are

$$\text{Var}(Y_5) = \text{Var}(\eta_1\eta_2) + \text{Var}(\eta_1\epsilon_3) + \text{Var}(\eta_2\epsilon_1) + \text{Var}(\epsilon_1\epsilon_3) \qquad 7.40$$

$$\text{Var}(Y_6) = \lambda_{42}^2\text{Var}(\eta_1\eta_2) + \text{Var}(\eta_1\epsilon_4) + \lambda_{42}^2\text{Var}(\eta_2\epsilon_1) + \text{Var}(\epsilon_1\epsilon_4) \qquad 7.41$$

$$\text{Var}(Y_7) = \lambda_{21}^2\text{Var}(\eta_1\eta_2) + \lambda_{21}^2\text{Var}(\eta_1\epsilon_3) + \text{Var}(\eta_2\epsilon_2) + \text{Var}(\epsilon_2\epsilon_3) \qquad 7.42$$

$$\text{Var}(Y_8) = \lambda_{21}^2\lambda_{42}^2\text{Var}(\eta_1\eta_2) + \lambda_{21}^2\text{Var}(\eta_1\epsilon_4) + \lambda_{42}^2\text{Var}(\eta_2\epsilon_2) + \text{Var}(\epsilon_2\epsilon_4) \quad 7.43$$

Figure 7.17   Simple Version of the Interaction Model.

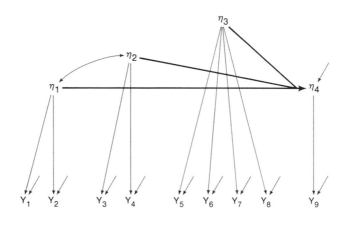

Kenny and Judd (1984:204) give the following relationships between the variances of the product variables and the variances of the single variables.

$$\text{Var}(\eta_1\eta_2) = \text{Var}(\eta_3) = \text{Var}(\eta_1)\text{Var}(\eta_2) + \text{Cov}(\eta_1\eta_2)^2 \qquad 7.44$$

$$\text{Var}(\epsilon_1\epsilon_3) = \Theta_{\epsilon1}\Theta_{\epsilon3} \qquad 7.45$$

$$\text{Var}(\epsilon_1\epsilon_4) = \Theta_{\epsilon1}\Theta_{\epsilon4} \qquad 7.46$$

$$\text{Var}(\epsilon_2\epsilon_3) = \Theta_{\epsilon2}\Theta_{\epsilon3} \qquad 7.47$$

$$\text{Var}(\epsilon_2\epsilon_4) = \Theta_{\epsilon2}\Theta_{\epsilon4} \qquad 7.48$$

$$\text{Var}(\eta_1\epsilon_3) = \text{Var}(\eta_1)\Theta_{\epsilon3} \qquad 7.49$$

$$\text{Var}(\eta_1\epsilon_4) = \text{Var}(\eta_1)\Theta_{\epsilon4} \qquad 7.50$$

$$\text{Var}(\eta_2\epsilon_1) = \text{Var}(\eta_2)\Theta_{\epsilon1} \qquad 7.51$$

$$\text{Var}(\eta_2\epsilon_2) = \text{Var}(\eta_2)\Theta_{\epsilon2} \qquad 7.52$$

Using these equations, we write the variance partitionings in Eqs. 7.40–7.43 as follows:

$$\begin{aligned}\text{Var}(Y_5) = &\, [\text{Var}(\eta_1)\text{Var}(\eta_2) + \text{Cov}(\eta_1\eta_2)^2] + \text{Var}(\eta_1)\Theta_{\epsilon3} \\ &+ \text{Var}(\eta_2)\Theta_{\epsilon1} + \Theta_{\epsilon1}\Theta_{\epsilon3} \qquad 7.53\end{aligned}$$

$$\begin{aligned}\text{Var}(Y_6) = &\, \lambda_{42}^2[\text{Var}(\eta_1)\text{Var}(\eta_2) + \text{Cov}(\eta_1\eta_2)^2] \\ &+ \text{Var}(\eta_1)\Theta_{\epsilon4} + \lambda_{42}^2\text{Var}(\eta_2)\Theta_{\epsilon1} + \Theta_{\epsilon1}\Theta_{\epsilon4} \qquad 7.54\end{aligned}$$

$$\begin{aligned}\text{Var}(Y_7) = &\, \lambda_{21}^2[\text{Var}(\eta_1)\text{Var}(\eta_2) + \text{Cov}(\eta_1\eta_2)^2] \\ &+ \lambda_{21}^2\text{Var}(\eta_1)\Theta_{\epsilon3} + \text{Var}(\eta_2)\Theta_{\epsilon2} + \Theta_{\epsilon2}\Theta_{\epsilon3} \qquad 7.55\end{aligned}$$

$$\begin{aligned}\text{Var}(Y_8) = &\, \lambda_{21}^2\lambda_{42}^2[\text{Var}(\eta_1)\text{Var}(\eta_2) + \text{Cov}(\eta_1\eta_2)^2] \\ &+ \lambda_{21}^2\text{Var}(\eta_1)\Theta_{\epsilon4} + \lambda_{42}^2\text{Var}(\eta_2)\Theta_{\epsilon2} + \Theta_{\epsilon2}\Theta_{\epsilon4} \qquad 7.56\end{aligned}$$

Figure 7.18 gives the full version of the LISREL model maintaining these variance partitionings. Figure 7.19 gives the actual LISREL estimates for the coefficients in the model using MLE. Comparing the MLEs with the true values of these coefficients (Figure 7.20) verifies that

Figure 7.18   The Full Interaction Model.[1]

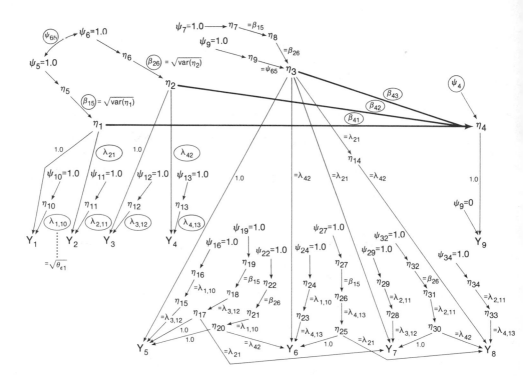

[1] ψ variances rather than ζ variables are plotted. The thirteen coefficients to be estimated are circled.

this procedure adequately recovers the coefficients hidden by Kenny and Judd (1984).

We again caution that this use of maximum likelihood knowingly violates the assumptions of multivariate normality, but we also note that the maximum likelihood procedure, even with the violated assumption, gives slightly closer estimates of the true values than those provided by generalized least squares. Using generalized least squares in LISREL (using the foregoing model but specifying GL on LISREL's output card) identically reproduced Kenny and Judd's results, but the GLS estimates are actually slightly worse than the maximum likelihood estimates.

Kenny and Judd (1984) started with truly multivariate normal variables η_1 and η_2 in creating the product variable η_3. Starting from strongly nonnormal variables could give more serious violations of the assumption of multivariate normality for the product variable.

This discussion of interaction among concepts directly parallels traditional multiple regression interaction procedures and hence shares the limitations of those procedures. The most fundamental limitation is that only regular or smooth (as opposed to erratic) interactions between the values of the concepts η_1 and η_2 can be modeled in this way.

The interpretation of interactions, like nonlinearities, is best attacked by using the partial derivative of the equation for η_4 with respect to one concept, say η_1, as the effect of η_1 on η_4. The presence of the product $\eta_1\eta_2$ forces the effect of η_1 to be conditional on the actual value of η_2 we considered, and similarly the effect of η_2 on η_4 is conditional on the specific value of η_1 considered. The equation creating η_4 from η_1, η_2, and $\eta_1\eta_2$ is

$$\eta_4 = -.17\eta_1 + .322\eta_2 + .714\eta_1\eta_2 + \zeta_4 \qquad 7.57$$

The effect of η_1 on η_4 is

$$\frac{\partial \eta_4}{\partial \eta_1} = -.17 + .714\eta_2 \qquad 7.58$$

and the effect of η_2 on η_4 is

$$\frac{\partial \eta_4}{\partial \eta_2} = .322 + .714\eta_1 \qquad 7.59$$

The variable $\eta_1\eta_2$ should never be interpreted independently of η_1 and η_2, and the effects of η_1 and η_2 should never be interpreted without considering $\eta_1\eta_2$.

Hayduk and Wonnacott (1980) discuss in more detail the interpretation of interactions, and Stolzenberg (1980) uses partial derivatives to interpret nonlinear and interactive structural equation models. Those intending to pursue the modeling of interactions, or nonlinearities might review Marsden (1983) on interactions with composite constructs and Allison (1977) on the role played by scale zero points.

Figure 7.19 LISREL MLE Estimates for the Interaction Model.

19 SEP 86 LISREL-INTERACTION PAGE 26
17:17:56 University of Alberta

KENNY+J84,ALL ETAS,TWO IND,LINTER.1x1 fromscratch

LISREL ESTIMATES (MAXIMUM LIKELIHOOD)

LAMBDA Y

	ETA 1	ETA 2	ETA 3	ETA 4	ETA 5	ETA 6	ETA 7	ETA 8	ETA 9	ETA 10
Y1	1.000	0.0	0.0	0.0	0.0	0.0	0.0	0.0	0.0	0.675
Y2	0.638	0.0	0.0	0.0	0.0	0.0	0.0	0.0	0.0	0.0
Y3	0.0	1.000	0.0	0.0	0.0	0.0	0.0	0.0	0.0	0.0
Y4	0.0	0.691	0.0	0.0	0.0	0.0	0.0	0.0	0.0	0.0
Y5	0.0	0.0	1.000	0.0	0.0	0.0	0.0	0.0	0.0	0.0
Y6	0.0	0.0	0.691	0.0	0.0	0.0	0.0	0.0	0.0	0.0
Y7	0.0	0.0	0.638	0.0	0.0	0.0	0.0	0.0	0.0	0.0
Y8	0.0	0.0	0.0	0.0	0.0	0.0	0.0	0.0	0.0	0.0
Y9	0.0	0.0	0.0	1.000	0.0	0.0	0.0	0.0	0.0	0.0

LAMBDA Y

	ETA 11	ETA 12	ETA 13	ETA 14	ETA 15	ETA 16	ETA 17	ETA 18	ETA 19	ETA 20
Y1	0.0	0.0	0.0	0.0	0.0	0.0	0.0	0.0	0.0	0.0
Y2	0.861	0.0	0.0	0.0	0.0	0.0	0.0	0.0	0.0	0.0
Y3	0.0	0.671	0.0	0.0	0.0	0.0	0.0	0.0	0.0	0.0
Y4	0.0	0.0	0.751	0.0	0.0	0.0	0.0	0.0	0.0	0.0
Y5	0.0	0.0	0.0	0.0	0.671	0.0	1.000	0.0	0.0	1.000
Y6	0.0	0.0	0.0	0.0	0.0	0.0	0.0	0.0	0.0	0.691
Y7	0.0	0.0	0.0	0.0	0.0	0.0	0.638	0.0	0.0	0.0
Y8	0.0	0.0	0.0	0.691	0.0	0.0	0.0	0.0	0.0	0.0
Y9	0.0	0.0	0.0	0.0	0.0	0.0	0.0	0.0	0.0	0.0

LAMBDA Y

	ETA 21	ETA 22	ETA 23	ETA 24	ETA 25	ETA 26	ETA 27	ETA 28	ETA 29	ETA 30
Y1	0.0	0.0	0.0	0.0	0.0	0.0	0.0	0.0	0.0	0.0
Y2	0.0	0.0	0.0	0.0	0.0	0.0	0.0	0.0	0.0	0.0
Y3	0.0	0.0	0.0	0.0	0.0	0.0	0.0	0.0	0.0	0.0
Y4	0.0	0.0	0.0	0.0	0.0	0.0	0.0	0.0	0.0	0.0
Y5	0.0	0.0	0.0	0.0	0.0	0.0	0.0	0.0	0.0	0.0
Y6	0.0	0.0	0.751	0.0	1.000	0.0	0.0	0.0	0.0	0.0
Y7	0.0	0.0	0.0	0.0	0.0	0.0	0.0	0.671	0.0	1.000
Y8	0.0	0.0	0.0	0.0	0.638	0.0	0.0	0.0	0.0	0.691
Y9	0.0	0.0	0.0	0.0	0.0	0.0	0.0	0.0	0.0	0.0

19 SEP 86 LISREL-INTERACTION PAGE 27
17:17:56 University of Alberta

LAMBDA Y

	ETA 31	ETA 32	ETA 33	ETA 34
Y1	0.0	0.0	0.0	0.0
Y2	0.0	0.0	0.0	0.0
Y3	0.0	0.0	0.0	0.0
Y4	0.0	0.0	0.0	0.0
Y5	0.0	0.0	0.0	0.0
Y6	0.0	0.0	0.0	0.0
Y7	0.0	0.0	0.0	0.0
Y8	0.0	0.0	0.751	0.0
Y9	0.0	0.0	0.0	0.0

BETA

	ETA 1	ETA 2	ETA 3	ETA 4	ETA 5	ETA 6	ETA 7	ETA 8	ETA 9	ETA 10
ETA 1	0.0	0.0	0.0	0.0	1.396	0.0	0.0	0.0	0.0	0.0
ETA 2	0.0	0.0	0.0	0.0	0.0	1.308	0.0	0.0	0.0	0.0
ETA 3	0.0	0.0	0.0	0.0	0.0	0.0	0.0	1.308	0.222	0.0
ETA 4	-0.170	0.322	0.714	0.0	0.0	0.0	0.0	0.0	0.0	0.0
ETA 5	0.0	0.0	0.0	0.0	0.0	0.0	0.0	0.0	0.0	0.0
ETA 6	0.0	0.0	0.0	0.0	0.0	0.0	0.0	0.0	0.0	0.0
ETA 7	0.0	0.0	0.0	0.0	0.0	0.0	0.0	0.0	0.0	0.0
ETA 8	0.0	0.0	0.0	0.0	0.0	0.0	1.396	0.0	0.0	0.0
ETA 9	0.0	0.0	0.0	0.0	0.0	0.0	0.0	0.0	0.0	0.0
ETA 10	0.0	0.0	0.0	0.0	0.0	0.0	0.0	0.0	0.0	0.0
ETA 11	0.0	0.0	0.0	0.0	0.0	0.0	0.0	0.0	0.0	0.0
ETA 12	0.0	0.0	0.0	0.0	0.0	0.0	0.0	0.0	0.0	0.0
ETA 13	0.0	0.0	0.0	0.0	0.0	0.0	0.0	0.0	0.0	0.0
ETA 14	0.0	0.0	0.638	0.0	0.0	0.0	0.0	0.0	0.0	0.0
ETA 15	0.0	0.0	0.0	0.0	0.0	0.0	0.0	0.0	0.0	0.0
ETA 16	0.0	0.0	0.0	0.0	0.0	0.0	0.0	0.0	0.0	0.0
ETA 17	0.0	0.0	0.0	0.0	0.0	0.0	0.0	0.0	0.0	0.0
ETA 18	0.0	0.0	0.0	0.0	0.0	0.0	0.0	0.0	0.0	0.0
ETA 19	0.0	0.0	0.0	0.0	0.0	0.0	0.0	0.0	0.0	0.0
ETA 20	0.0	0.0	0.0	0.0	0.0	0.0	0.0	0.0	0.0	0.0
ETA 21	0.0	0.0	0.0	0.0	0.0	0.0	0.0	0.0	0.0	0.0
ETA 22	0.0	0.0	0.0	0.0	0.0	0.0	0.0	0.0	0.0	0.0
ETA 23	0.0	0.0	0.0	0.0	0.0	0.0	0.0	0.0	0.0	0.0
ETA 24	0.0	0.0	0.0	0.0	0.0	0.0	0.0	0.0	0.0	0.0
ETA 25	0.0	0.0	0.0	0.0	0.0	0.0	0.0	0.0	0.0	0.0
ETA 26	0.0	0.0	0.0	0.0	0.0	0.0	0.0	0.0	0.0	0.0
ETA 27	0.0	0.0	0.0	0.0	0.0	0.0	0.0	0.0	0.0	0.0
ETA 28	0.0	0.0	0.0	0.0	0.0	0.0	0.0	0.0	0.0	0.0
ETA 29	0.0	0.0	0.0	0.0	0.0	0.0	0.0	0.0	0.0	0.0
ETA 30	0.0	0.0	0.0	0.0	0.0	0.0	0.0	0.0	0.0	0.0
ETA 31	0.0	0.0	0.0	0.0	0.0	0.0	0.0	0.0	0.0	0.0
ETA 32	0.0	0.0	0.0	0.0	0.0	0.0	0.0	0.0	0.0	0.0
ETA 33	0.0	0.0	0.0	0.0	0.0	0.0	0.0	0.0	0.0	0.0
ETA 34	0.0	0.0	0.0	0.0	0.0	0.0	0.0	0.0	0.0	0.0

Figure 7.19   Continued.

BETA

	ETA 11	ETA 12	ETA 13	ETA 14	ETA 15	ETA 16	ETA 17	ETA 18	ETA 19	ETA 20
ETA 1	0.0	0.0	0.0	0.0	0.0	0.0	0.0	0.0	0.0	0.0
ETA 2	0.0	0.0	0.0	0.0	0.0	0.0	0.0	0.0	0.0	0.0
ETA 3	0.0	0.0	0.0	0.0	0.0	0.0	0.0	0.0	0.0	0.0
ETA 4	0.0	0.0	0.0	0.0	0.0	0.0	0.0	0.0	0.0	0.0
ETA 5	0.0	0.0	0.0	0.0	0.0	0.0	0.0	0.0	0.0	0.0
ETA 6	0.0	0.0	0.0	0.0	0.0	0.0	0.0	0.0	0.0	0.0
ETA 7	0.0	0.0	0.0	0.0	0.0	0.0	0.0	0.0	0.0	0.0
ETA 8	0.0	0.0	0.0	0.0	0.0	0.0	0.0	0.0	0.0	0.0
ETA 9	0.0	0.0	0.0	0.0	0.0	0.0	0.0	0.0	0.0	0.0
ETA 10	0.0	0.0	0.0	0.0	0.0	0.0	0.0	0.0	0.0	0.0
ETA 11	0.0	0.0	0.0	0.0	0.0	0.0	0.0	0.0	0.0	0.0
ETA 12	0.0	0.0	0.0	0.0	0.0	0.0	0.0	0.0	0.0	0.0
ETA 13	0.0	0.0	0.0	0.0	0.0	0.0	0.0	0.0	0.0	0.0
ETA 14	0.0	0.0	0.0	0.0	0.0	0.0	0.0	0.0	0.0	0.0
ETA 15	0.0	0.0	0.0	0.0	0.0	0.675	0.0	0.0	0.0	0.0
ETA 16	0.0	0.0	0.0	0.0	0.0	0.0	0.0	0.0	0.0	0.0
ETA 17	0.0	0.0	0.0	0.0	0.0	0.0	0.0	0.671	0.0	0.0
ETA 18	0.0	0.0	0.0	0.0	0.0	0.0	0.0	0.0	1.396	0.0
ETA 19	0.0	0.0	0.0	0.0	0.0	0.0	0.0	0.0	0.0	0.0
ETA 20	0.0	0.0	0.0	0.0	0.0	0.0	0.0	0.0	0.0	0.0
ETA 21	0.0	0.0	0.0	0.0	0.0	0.0	0.0	0.0	0.0	0.0
ETA 22	0.0	0.0	0.0	0.0	0.0	0.0	0.0	0.0	0.0	0.0
ETA 23	0.0	0.0	0.0	0.0	0.0	0.0	0.0	0.0	0.0	0.0
ETA 24	0.0	0.0	0.0	0.0	0.0	0.0	0.0	0.0	0.0	0.0
ETA 25	0.0	0.0	0.0	0.0	0.0	0.0	0.0	0.0	0.0	0.0
ETA 26	0.0	0.0	0.0	0.0	0.0	0.0	0.0	0.0	0.0	0.0
ETA 27	0.0	0.0	0.0	0.0	0.0	0.0	0.0	0.0	0.0	0.0
ETA 28	0.0	0.0	0.0	0.0	0.0	0.0	0.0	0.0	0.0	0.0
ETA 29	0.0	0.0	0.0	0.0	0.0	0.0	0.0	0.0	0.0	0.0
ETA 30	0.0	0.0	0.0	0.0	0.0	0.0	0.0	0.0	0.0	0.0
ETA 31	0.0	0.0	0.0	0.0	0.0	0.0	0.0	0.0	0.0	0.0
ETA 32	0.0	0.0	0.0	0.0	0.0	0.0	0.0	0.0	0.0	0.0
ETA 33	0.0	0.0	0.0	0.0	0.0	0.0	0.0	0.0	0.0	0.0
ETA 34	0.0	0.0	0.0	0.0	0.0	0.0	0.0	0.0	0.0	0.0

BETA

	ETA 21	ETA 22	ETA 23	ETA 24	ETA 25	ETA 26	ETA 27	ETA 28	ETA 29	ETA 30
ETA 1	0.0	0.0	0.0	0.0	0.0	0.0	0.0	0.0	0.0	0.0
ETA 2	0.0	0.0	0.0	0.0	0.0	0.0	0.0	0.0	0.0	0.0
ETA 3	0.0	0.0	0.0	0.0	0.0	0.0	0.0	0.0	0.0	0.0
ETA 4	0.0	0.0	0.0	0.0	0.0	0.0	0.0	0.0	0.0	0.0
ETA 5	0.0	0.0	0.0	0.0	0.0	0.0	0.0	0.0	0.0	0.0
ETA 6	0.0	0.0	0.0	0.0	0.0	0.0	0.0	0.0	0.0	0.0
ETA 7	0.0	0.0	0.0	0.0	0.0	0.0	0.0	0.0	0.0	0.0
ETA 8	0.0	0.0	0.0	0.0	0.0	0.0	0.0	0.0	0.0	0.0
ETA 9	0.0	0.0	0.0	0.0	0.0	0.0	0.0	0.0	0.0	0.0
ETA 10	0.0	0.0	0.0	0.0	0.0	0.0	0.0	0.0	0.0	0.0
ETA 11	0.0	0.0	0.0	0.0	0.0	0.0	0.0	0.0	0.0	0.0
ETA 12	0.0	0.0	0.0	0.0	0.0	0.0	0.0	0.0	0.0	0.0
ETA 13	0.0	0.0	0.0	0.0	0.0	0.0	0.0	0.0	0.0	0.0
ETA 14	0.0	0.0	0.0	0.0	0.0	0.0	0.0	0.0	0.0	0.0
ETA 15	0.0	0.0	0.0	0.0	0.0	0.0	0.0	0.0	0.0	0.0
ETA 16	0.0	0.0	0.0	0.0	0.0	0.0	0.0	0.0	0.0	0.0
ETA 17	0.0	0.0	0.0	0.0	0.0	0.0	0.0	0.0	0.0	0.0
ETA 18	0.0	0.0	0.0	0.0	0.0	0.0	0.0	0.0	0.0	0.0
ETA 19	0.0	0.0	0.0	0.0	0.0	0.0	0.0	0.0	0.0	0.0
ETA 20	0.675	0.0	0.0	0.0	0.0	0.0	0.0	0.0	0.0	0.0
ETA 21	0.0	1.308	0.0	0.0	0.0	0.0	0.0	0.0	0.0	0.0
ETA 22	0.0	0.0	0.0	0.0	0.0	0.0	0.0	0.0	0.0	0.0
ETA 23	0.0	0.0	0.0	0.675	0.0	0.0	0.0	0.0	0.0	0.0
ETA 24	0.0	0.0	0.0	0.0	0.0	0.0	0.0	0.0	0.0	0.0
ETA 25	0.0	0.0	0.0	0.0	0.0	0.751	0.0	0.0	0.0	0.0
ETA 26	0.0	0.0	0.0	0.0	0.0	0.0	1.396	0.0	0.0	0.0
ETA 27	0.0	0.0	0.0	0.0	0.0	0.0	0.0	0.0	0.0	0.0
ETA 28	0.0	0.0	0.0	0.0	0.0	0.0	0.0	0.0	0.861	0.0
ETA 29	0.0	0.0	0.0	0.0	0.0	0.0	0.0	0.0	0.0	0.0
ETA 30	0.0	0.0	0.0	0.0	0.0	0.0	0.0	0.0	0.0	0.0
ETA 31	0.0	0.0	0.0	0.0	0.0	0.0	0.0	0.0	0.0	0.0
ETA 32	0.0	0.0	0.0	0.0	0.0	0.0	0.0	0.0	0.0	0.0
ETA 33	0.0	0.0	0.0	0.0	0.0	0.0	0.0	0.0	0.0	0.0
ETA 34	0.0	0.0	0.0	0.0	0.0	0.0	0.0	0.0	0.0	0.0

Figure 7.19   Continued.

```
19 SEP 86   LISREL-INTERACTION                                                              PAGE  30
17:17:57    University of Alberta

        BETA

               ETA 31      ETA 32      ETA 33      ETA 34
   ETA  1       0.0         0.0         0.0         0.0
   ETA  2       0.0         0.0         0.0         0.0
   ETA  3       0.0         0.0         0.0         0.0
   ETA  4       0.0         0.0         0.0         0.0
   ETA  5       0.0         0.0         0.0         0.0
   ETA  6       0.0         0.0         0.0         0.0
   ETA  7       0.0         0.0         0.0         0.0
   ETA  8       0.0         0.0         0.0         0.0
   ETA  9       0.0         0.0         0.0         0.0
   ETA 10       0.0         0.0         0.0         0.0
   ETA 11       0.0         0.0         0.0         0.0
   ETA 12       0.0         0.0         0.0         0.0
   ETA 13       0.0         0.0         0.0         0.0
   ETA 14       0.0         0.0         0.0         0.0
   ETA 15       0.0         0.0         0.0         0.0
   ETA 16       0.0         0.0         0.0         0.0
   ETA 17       0.0         0.0         0.0         0.0
   ETA 18       0.0         0.0         0.0         0.0
   ETA 19       0.0         0.0         0.0         0.0
   ETA 20       0.0         0.0         0.0         0.0
   ETA 21       0.0         0.0         0.0         0.0
   ETA 22       0.0         0.0         0.0         0.0
   ETA 23       0.0         0.0         0.0         0.0
   ETA 24       0.0         0.0         0.0         0.0
   ETA 25       0.0         0.0         0.0         0.0
   ETA 26       0.0         0.0         0.0         0.0
   ETA 27       0.0         0.0         0.0         0.0
   ETA 28       0.0         0.0         0.0         0.0
   ETA 29       0.0         0.0         0.0         0.0
   ETA 30       0.861       0.0         0.0         0.0
   ETA 31       0.0         1.308       0.0         0.0
   ETA 32       0.0         0.0         0.0         0.0
   ETA 33       0.0         0.0         0.0         0.861
   ETA 34       0.0         0.0         0.0         0.0
```

```
19 SEP 86   LISREL-INTERACTION                                                              PAGE  31
17:17:57    University of Alberta

        PSI

               ETA 1    ETA 2    ETA 3    ETA 4    ETA 5    ETA 6    ETA 7    ETA 8    ETA 9    ETA 10
   ETA  1       0.0
   ETA  2       0.0      0.0
   ETA  3       0.0      0.0      0.0
   ETA  4       0.0      0.0      0.0      0.230
   ETA  5       0.0      0.0      0.0      0.0      1.000
   ETA  6       0.0      0.0      0.0      0.0      0.222    1.000
   ETA  7       0.0      0.0      0.0      0.0      0.0      0.0      1.000
   ETA  8       0.0      0.0      0.0      0.0      0.0      0.0      0.0      0.0
   ETA  9       0.0      0.0      0.0      0.0      0.0      0.0      0.0      0.0      1.000
   ETA 10       0.0      0.0      0.0      0.0      0.0      0.0      0.0      0.0      0.0      1.000
   ETA 11       0.0      0.0      0.0      0.0      0.0      0.0      0.0      0.0      0.0      0.0
   ETA 12       0.0      0.0      0.0      0.0      0.0      0.0      0.0      0.0      0.0      0.0
   ETA 13       0.0      0.0      0.0      0.0      0.0      0.0      0.0      0.0      0.0      0.0
   ETA 14       0.0      0.0      0.0      0.0      0.0      0.0      0.0      0.0      0.0      0.0
   ETA 15       0.0      0.0      0.0      0.0      0.0      0.0      0.0      0.0      0.0      0.0
   ETA 16       0.0      0.0      0.0      0.0      0.0      0.0      0.0      0.0      0.0      0.0
   ETA 17       0.0      0.0      0.0      0.0      0.0      0.0      0.0      0.0      0.0      0.0
   ETA 18       0.0      0.0      0.0      0.0      0.0      0.0      0.0      0.0      0.0      0.0
   ETA 19       0.0      0.0      0.0      0.0      0.0      0.0      0.0      0.0      0.0      0.0
   ETA 20       0.0      0.0      0.0      0.0      0.0      0.0      0.0      0.0      0.0      0.0
   ETA 21       0.0      0.0      0.0      0.0      0.0      0.0      0.0      0.0      0.0      0.0
   ETA 22       0.0      0.0      0.0      0.0      0.0      0.0      0.0      0.0      0.0      0.0
   ETA 23       0.0      0.0      0.0      0.0      0.0      0.0      0.0      0.0      0.0      0.0
   ETA 24       0.0      0.0      0.0      0.0      0.0      0.0      0.0      0.0      0.0      0.0
   ETA 25       0.0      0.0      0.0      0.0      0.0      0.0      0.0      0.0      0.0      0.0
   ETA 26       0.0      0.0      0.0      0.0      0.0      0.0      0.0      0.0      0.0      0.0
   ETA 27       0.0      0.0      0.0      0.0      0.0      0.0      0.0      0.0      0.0      0.0
   ETA 28       0.0      0.0      0.0      0.0      0.0      0.0      0.0      0.0      0.0      0.0
   ETA 29       0.0      0.0      0.0      0.0      0.0      0.0      0.0      0.0      0.0      0.0
   ETA 30       0.0      0.0      0.0      0.0      0.0      0.0      0.0      0.0      0.0      0.0
   ETA 31       0.0      0.0      0.0      0.0      0.0      0.0      0.0      0.0      0.0      0.0
   ETA 32       0.0      0.0      0.0      0.0      0.0      0.0      0.0      0.0      0.0      0.0
   ETA 33       0.0      0.0      0.0      0.0      0.0      0.0      0.0      0.0      0.0      0.0
   ETA 34       0.0      0.0      0.0      0.0      0.0      0.0      0.0      0.0      0.0      0.0
```

Figure 7.19   Continued.

```
19 SEP 86   LISREL-INTERACTION                                                              PAGE  32
17:17:57    University of Alberta

        PSI

            ETA 11      ETA 12      ETA 13      ETA 14      ETA 15      ETA 16      ETA 17      ETA 18      ETA 19      ETA 20
ETA 11      1.000
ETA 12      0.0         1.000
ETA 13      0.0         0.0         1.000
ETA 14      0.0         0.0         0.0         0.0
ETA 15      0.0         0.0         0.0         0.0         0.0
ETA 16      0.0         0.0         0.0         0.0         0.0         1.000
ETA 17      0.0         0.0         0.0         0.0         0.0         0.0         0.0
ETA 18      0.0         0.0         0.0         0.0         0.0         0.0         0.0         0.0
ETA 19      0.0         0.0         0.0         0.0         0.0         0.0         0.0         0.0         1.000
ETA 20      0.0         0.0         0.0         0.0         0.0         0.0         0.0         0.0         0.0         0.0
ETA 21      0.0         0.0         0.0         0.0         0.0         0.0         0.0         0.0         0.0         0.0
ETA 22      0.0         0.0         0.0         0.0         0.0         0.0         0.0         0.0         0.0         0.0
ETA 23      0.0         0.0         0.0         0.0         0.0         0.0         0.0         0.0         0.0         0.0
ETA 24      0.0         0.0         0.0         0.0         0.0         0.0         0.0         0.0         0.0         0.0
ETA 25      0.0         0.0         0.0         0.0         0.0         0.0         0.0         0.0         0.0         0.0
ETA 26      0.0         0.0         0.0         0.0         0.0         0.0         0.0         0.0         0.0         0.0
ETA 27      0.0         0.0         0.0         0.0         0.0         0.0         0.0         0.0         0.0         0.0
ETA 28      0.0         0.0         0.0         0.0         0.0         0.0         0.0         0.0         0.0         0.0
ETA 29      0.0         0.0         0.0         0.0         0.0         0.0         0.0         0.0         0.0         0.0
ETA 30      0.0         0.0         0.0         0.0         0.0         0.0         0.0         0.0         0.0         0.0
ETA 31      0.0         0.0         0.0         0.0         0.0         0.0         0.0         0.0         0.0         0.0
ETA 32      0.0         0.0         0.0         0.0         0.0         0.0         0.0         0.0         0.0         0.0
ETA 33      0.0         0.0         0.0         0.0         0.0         0.0         0.0         0.0         0.0         0.0
ETA 34      0.0         0.0         0.0         0.0         0.0         0.0         0.0         0.0         0.0         0.0

        PSI

            ETA 21      ETA 22      ETA 23      ETA 24      ETA 25      ETA 26      ETA 27      ETA 28      ETA 29      ETA 30
ETA 21      0.0
ETA 22      0.0         1.000
ETA 23      0.0         0.0         0.0
ETA 24      0.0         0.0         0.0         1.000
ETA 25      0.0         0.0         0.0         0.0         0.0
ETA 26      0.0         0.0         0.0         0.0         0.0         0.0
ETA 27      0.0         0.0         0.0         0.0         0.0         0.0         1.000
ETA 28      0.0         0.0         0.0         0.0         0.0         0.0         0.0         0.0
ETA 29      0.0         0.0         0.0         0.0         0.0         0.0         0.0         0.0         1.000
ETA 30      0.0         0.0         0.0         0.0         0.0         0.0         0.0         0.0         0.0         0.0
ETA 31      0.0         0.0         0.0         0.0         0.0         0.0         0.0         0.0         0.0         0.0
ETA 32      0.0         0.0         0.0         0.0         0.0         0.0         0.0         0.0         0.0         0.0
ETA 33      0.0         0.0         0.0         0.0         0.0         0.0         0.0         0.0         0.0         0.0
ETA 34      0.0         0.0         0.0         0.0         0.0         0.0         0.0         0.0         0.0         0.0

        PSI

            ETA 31      ETA 32      ETA 33      ETA 34
ETA 31      0.0
ETA 32      0.0         1.000
ETA 33      0.0         0.0         0.0
ETA 34      0.0         0.0         0.0         1.000
```

```
19 SEP 86   LISREL-INTERACTION                                                              PAGE  33
17:17:57    University of Alberta


W_A_R_N_I_N_G :  THE MATRIX PSI          IS NOT POSITIVE DEFINITE

        SQUARED MULTIPLE CORRELATIONS FOR STRUCTURAL EQUATIONS

            ETA 1       ETA 2       ETA 3       ETA 4       ETA 5       ETA 6       ETA 7       ETA 8       ETA 9       ETA 10
            1.000       1.000       1.000       0.893       0.0         0.0         0.0         1.000       0.0         0.0

        SQUARED MULTIPLE CORRELATIONS FOR STRUCTURAL EQUATIONS

            ETA 11      ETA 12      ETA 13      ETA 14      ETA 15      ETA 16      ETA 17      ETA 18      ETA 19      ETA 20
            0.0         0.0         0.0         1.000       1.000       0.0         1.000       1.000       0.0         1.000

        SQUARED MULTIPLE CORRELATIONS FOR STRUCTURAL EQUATIONS

            ETA 21      ETA 22      ETA 23      ETA 24      ETA 25      ETA 26      ETA 27      ETA 28      ETA 29      ETA 30
            1.000       0.0         1.000       0.0         1.000       1.000       0.0         1.000       0.0         1.000

        SQUARED MULTIPLE CORRELATIONS FOR STRUCTURAL EQUATIONS

            ETA 31      ETA 32      ETA 33      ETA 34
            1.000       0.0         1.000       0.0


            MEASURES OF GOODNESS OF FIT FOR THE WHOLE MODEL :

    CHI-SQUARE WITH  32 DEGREES OF FREEDOM IS      44.98 (PROB. LEVEL = 0.064)

            GOODNESS OF FIT INDEX IS 0.980

            ADJUSTED GOODNESS OF FIT INDEX IS 0.972

            ROOT MEAN SQUARE RESIDUAL IS      0.145
```

Figure 7.20   True, MLE, and GLS Estimates of the Interaction Model.

Coefficient	True Value from Kenny and Judd, 1984	MLE Estimate	GLS Estimate Kenny and Judd, 1984
β_{41}	-0.15	-0.170	-0.169
β_{42}	0.35	0.322	0.321
β_{43}	0.70	0.714	0.710
β_{15}	$\sqrt{2.15} = 1.466$	1.396	1.372
β_{26}	$\sqrt{1.60} = 1.265$	1.308	1.286
ψ_4	0.16	0.230	0.265
ψ_{65}	0.20	0.222	0.369
λ_{21}	0.60	0.638	0.646
λ_{42}	0.70	0.691	0.685
$\lambda_{1,10}$	$\sqrt{0.36} = 0.60$	0.675	0.654
$\lambda_{2,11}$	$\sqrt{0.81} = 0.90$	0.861	0.849
$\lambda_{3,12}$	$\sqrt{0.49} = 0.70$	0.671	0.666
$\lambda_{4,13}$	$\sqrt{0.64} = 0.80$	0.751	0.743

Notes

1.　　That is, the concept must really function and not be merely specified as functioning, as what Burt (1976, 1981) calls a "point variable." Concepts labeled "needs" or "attitudes" are often problematic. These titles indicate the researcher is thinking about multiple entities and implicitly about multiple potential effect routings. Relabeling these concepts as "the need" or "the attitude" will not be resisted by a researcher who is indeed thinking of a single entity that could conceivably function to channel the effects of the multiple observed variables.

2.　　Alwin and Jackson (1980) provide a useful discussion of multiple indicators that demonstrates a nesting of various types of multiple-indicator models. From the most mathematically more general to the more specific, these models are the *common factor model*, the *congeneric measures model*, the *tau-equivalent model*, and the *parallel measures model*. Multiple indicators arise from a *common factor model* if any indicator can respond to any of the underlying factors (concepts) with a loading (structural coefficient) that can vary from indicator to indicator. In a *congeneric model*, each indicator is allowed to respond to only a single underlying concept (factor), though the loadings are free to vary from indicator to indicator. In *tau-equivalent models*, each indicator reflects only a single concept, and the loadings of the indicators on that concept are equal, so the concept contributes the same amount of variance into each indicator. In a *parallel measures model*, the loadings and the error variances of the multiple indicators are constrained to be equal between measures. Thus the total variances of the indicators are equal because the true (conceptual) variances and the error variances are equal. For further discussion of these terms see Alwin and Jackson (1980), Joreskog (1968, 1970a, 1971b, 1973b, 1978a), and Lord and Novick (1968).

These terms have traditionally been employed in models with a few correlated factors (concepts). No standard terminology has yet arisen for models allowing effects among the concepts and where mixtures of the preceding multiple-indicator specifications appear—that is, where some concepts have congeneric measures and others have tau-equivalent or parallel measures.

3.　　Attempting to fix the model by freeing the Θ_δ for God exists improves the model substantially, but it continues to have an unacceptable χ^2 ($\chi^2 = 42.22$ with 15 *d.f.*, $p < .00$), and the estimated Θ_δ for the God-exists indicator (1.805) implies that 84% of the variance in this indicator is error variance. Again, we are led to the conclusion that the attendance and God-exists indicators are measuring different concepts.

4.　　Attempting to estimate this model in November 1984 with the version of LISREL VI provided through SPSSX was unsuccessful until a LISREL program bug was eliminated. SPSS has been advised of the problem and presumably has been supplying corrected versions since that date. The bug was that too small a matrix was zeroed out during calculation of the SIG subroutine.

Chapter 8

Interpreting It All

One fundamental aspect of interpreting structural coefficients concerns the consistency of the implications of those estimates with the observed data. In this sense we have been "interpreting" the meanings of structural coefficients ever since Chapter 4, which demonstrated how model coefficients combine to imply a covariance matrix among the observed indicators, and Chapter 6, which discussed how the fit between the implied and observed covariances can be tested with a χ^2 test. As useful and fundamental as these strategies are for understanding model coefficients, they fail to capitalize on the human proclivity to interpret $\mathbf{B}$, $\boldsymbol{\Gamma}$, $\boldsymbol{\Lambda}_y$, and $\boldsymbol{\Lambda}_x$ coefficients as constituting "effects," that can combine to provide various indirect effects and total effects. We begin by reviewing the interpretation of regression coefficients as effect coefficients, and we then consider some simple non-LISREL models to illustrate the complexities that arise when we try to calculate indirect effects and the "effects" of feedback loops. Subsequently, we return to LISREL notation and develop parallel formulas for calculating the total, direct, and indirect effects between all pairs of variables in any model. Sections 8.3 and 8.4 discuss how these procedures can be used to obtain various effect decompositions and recompositions that can further assist interpretation. We conclude with a discussion of the total, direct, and indirect effects in a modified smoking model.

8.1 The Basics of Interpretation

The interpretation of structural coefficients as "effect coefficients" originates with ordinary regression equations like

$$X_0 = a + b_1 X_1 + b_2 X_2 + b_3 X_3 + e \qquad\qquad 8.1$$

for the effects of variables X_1, X_2, and X_3 on variable X_0. We can interpret the estimate of b_1 as *the magnitude of the change in X_0 that would be predicted to accompany a unit change in X_1 with X_2 and X_3 left untouched at their original values.* We avoid ending with the phrase "held constant" because this phrase must be abandoned for models containing multiple equations, as we shall later see. Parallel interpretations are appropriate for b_2 and b_3.

Before proceeding, we should be clear about the hypothetical nature of the "unit intervention" and "resultant change" in the preceding interpretation. The imagery of changing some cases' scores is merely a verbally efficient way of comparing the scores on the dependent variable (X_0) for two sets of cases having identical scores on all the independent variables, except for the variable whose slope is being interpreted (X_1). For b_1 the compared groups would be identical on all the predictor variables, except that one group has X_1 scores 1 unit higher than the X_1 scores possessed by the members of the other group. Figure 8.1 depicts three types of groupings we might be referring to if X_0 is regressed on X_1 and X_2. Interpretation A is the most frequently intended comparison (all the cases have the same X_2 score, and the groups have X_1 scores that are homogeneous though 1 unit apart), but comparisons of types B (matched X_2 scores, homogeneous clusters of X_1 scores) and C (identical X_2 scores, matched X_1 scores 1 unit apart) are equally adequate. The difference between the means (expected values) of the dependent variable X_0 for each pair of groups can be obtained by using the slope b_1 and interpreting this as a difference, just as slopes for dummy variables can be interpreted as differences (Section 2.3.1).

Though it is convenient to encapsulate the difference between the groups as the expected change (increase or decrease) in the dependent variable that would appear if each case in the group on the left had its score on X_1 increased 1 unit, the switch to the terminology of a change precipitating an effect, invokes a stronger dependence on cause-effect relations than is required by a simple comparison of two groups. For example, if individuals self-selected themselves into X_1 groupings on the basis of X_0 scores, the difference-in-expected-value interpretation remains valid while the change-resulting-in-an-effect interpretation is incorrect.

Naturally, the closer the match between the model specifications of dependent and independent variables and the real causes and effects, the more sound the change-resultant-difference terminology. We adopt the change-effect terminology for simplicity and because of the natural proclivity of researchers to feel they are creating models of a world that is inherently composed of causal forces. Despite this general position, we feel

Figure 8.1 Multiple Interpretations of b_1.

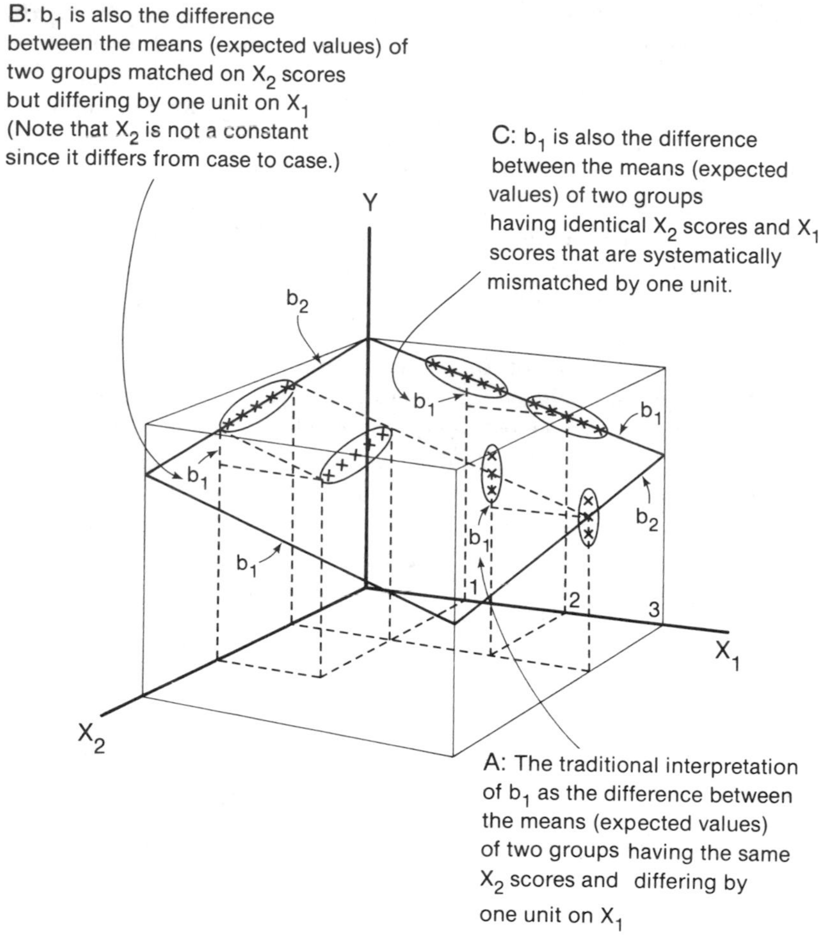
B: b1 is also the difference between the means (expected values) of two groups matched on X2 scores but differing by one unit on X1 (Note that X2 is not a constant since it differs from case to case.)
C: b1 is also the difference between the means (expected values) of two groups having identical X2 scores and X1 scores that are systematically mismatched by one unit.
Y
b2
b1
b1
b1
b1
b1
b2
1
2
3
X1
X2
A: The traditional interpretation of b1 as the difference between the means (expected values) of two groups having the same X2 scores and differing by one unit on X1

free to revert to the more fundamental comparison of differing groups whenever necessary—for example, when sex or age is the independent variable, and a unit change is beyond our control.

Another way to emphasize the hypothetical nature of the unit intervention, resultant change interpretational strategy is to note that if we were to enter the real world and artificially change the independent variable X_1 by 1 unit, there would be no guarantee that the dependent variable in the real world would respond with precisely b_1 units of change. The potential problem is that *artifical changes in an independent variable might not function the same way as natural changes in that same variable.* For example, even if naturally reading the Bible every day (high X_1) makes one a better citizen (i.e., a positive b_1 leads to high citizenship X_0), artifically forcing others to read the Bible daily may nonetheless engender resistance (a negative b_1), rather than improved citizenship.

8.1.1 Moving from Single Equations to Models

Consider the model in Figure 8.2A where X_3 responds to X_1, X_2, and X_4, and X_2 responds only to X_1. In this model, X_1 influences X_3 directly or indirectly through X_2. We can obtain an estimate of the *indirect effect* of X_1 on X_3 as $b_{32}b_{21}$—that is, *as the product of the coefficient estimates making up the string of paths* linking X_1 to X_3. You can think of this as an initial unit change in X_1 producing b_{21} units of change in X_2. These b_{21} units of change in X_2 lead to b_{21} times b_{32} units of change in X_3, namely some fraction or multiple of the amount of change 1 unit of change in X_2 would produce in X_3 (cf. Duncan, 1975; Alwin and Hauser, 1975). We interpret the product $b_{32}b_{21}$ as the magnitude of the change predicted to appear in X_3 if X_1 is changed 1 unit, if X_4 remains unchanged, if the mechanism providing the b_{31} effect is eliminated, and if X_2 is allowed no changes other than that transmitted from the unit change in X_1. That is, X_2 *is allowed to change in response to, but only in response to, the hypothetical unit change in X_1.*

Adding the direct (b_{31}) and indirect $(b_{21}b_{32})$ effects gives the **total effect** of X_1 on X_3, and is best interpreted as the change in X_3 predicted to follow a unit change in X_1 if all the other variables in the model are left untouched except for changes originating in the hypothetical unit change in X_1. Hence, X_4 remains constant (even if not forcibly held constant), and X_2 is allowed to change but only in response to the unit change in X_1. Interpretation of total effects in even recursive models (models with no loops, reciprocal causation, or unspecified causal orderings) requires that we imagine a system (the variables composing the model) resting in a stable state (possessing a set of nonchanging values) into which we insert a 1-unit change at a particular variable. We then wait for the system to

Figure 8.2   Some Illustrative Examples.[1]

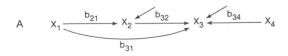

A $X_1 \xrightarrow{\ b_{21}\ } X_2 \xrightarrow{\ b_{32}\ } X_3 \xleftarrow{\ b_{34}\ } X_4$

b_{31}

B $X_1 \xrightarrow{\ b_{21}\ } X_2 \underset{b_{32}}{\overset{b_{23}}{\rightleftarrows}} X_3 \xleftarrow{\ b_{34}\ } X_4$

C $X_1 \xrightarrow{\ b_{21}\ } X_2 \underset{b_{32}}{\overset{b_{23}}{\rightleftarrows}} X_3 \xrightarrow{\ b_{43}\ } X_4$

[1]A correlation between X_1 and X_4 is allowed in A and B
but is not represented for diagrammatic simplicity.

restabilize as the effect is transmitted through the various routings connecting the hypothetically manipulated variable to the dependent variable before examining the total amount of change reaching the dependent variable.

Next consider the model in Figure 8.2B. Here, none of the structural coefficients b_{21}, b_{32}, b_{23}, or b_{34} reflect the amount of change in one variable that the model would predict to accompany a unit change in another variable. To see why, consider what happens to X_2 if X_1 changes by 1 unit: X_2 responds with a change of b_{21} units, *but the process does not stop here.* The change in X_2 causes a change in X_3, and this change in X_3 leads to a further change in X_2 (by the reciprocal effect b_{23}). This latest change in X_2 results in a further change in X_3, which in turn makes another change in X_2, and the cycle is set to repeat endlessly.[1]

Thus a unit change in X_1 produces a series of changes in X_2, only the first of which is reflected by the structural coefficient b_{21}. Hence, b_{21} can be interpreted as *the first of a series of changes* in X_2 resulting from the overall model's response to the unit change in X_1, or as the change in X_2 that would be predicted to follow a unit change in X_1 *with the X_2–X_3 loop rendered inoperative*, which would occur if the mechanisms linking X_2 to X_3 were rendered inoperative or if the value of X_3 could forcibly be held constant (e.g., by legislation or contractual agreement if X_3 happened to be a variable such as wage rates).

Typically, we are interested in the *overall model's prediction for the change in one variable following a unit change in another variable, not merely the first change, or the amount of change likely to appear with extra constraints rendering portions of the model inoperative.* That is, we are usually interested in the effects of a unit change in the independent variable with no other changes and with no additional restrictions on the model.

It should now be clear why we routinely employ the phrase "with the other variables left untouched" rather than "holding the other variables constant": "holding constant" implies restrictions over and above those inherent in the model itself. It should also be clear why the "held constant" phraseology has not been purged from the literature. For single regression equations, which constitute much social science literature, the other variables can be held constant without disrupting the effects of the unit change investigated, though even here the phrase "left untouched" is equally appropriate. In recursive models containing indirect effects and in nonrecursive models, however, "holding the other variables constant" does more than eliminate irrelevant variations in other portions of the model. It restricts the mechanisms by which a unit change can influence the dependent variable to only those routings that do not proceed through intervening variables or causal loops.

We will speak of the structural coefficient b_{21} in Figure 8.2B as the **basic direct** effect of X_1 on X_2, and if we wish to speak of the effect a unit change in X_1 has on X_2, considering the enhancement of this effect by repeated cyclings through the X_2–X_3 loop, we will speak of this as the **loop-enhanced direct effect** or simply the **enhanced direct effect** of X_1 on X_2.

It may not yet be clear that the effect of X_3 on X_4 in Figure 8.2C also depends on the X_2–X_3 loop. The b_{43} coefficient reflects the predicted change in X_4 for a unit change in X_3 with all the other variables forcibly held constant (by legislation or by removal of the mechanisms carrying effects to other variables), but if the other variables are left untouched yet free to vary, we observe a different response in X_4. An initial unit change in X_3 will influence X_2 (as well as X_4), and feedback from that unit change will subsequently alter X_3 itself and hence produce further changes in X_4.

Figure 8.2C provides further instruction if we consider the effect of X_1 on X_4. This effect would normally be expressed as the product of the coefficients $b_{43}b_{32}b_{21}$ (the chain of coefficients linking X_1 to X_4), but note that this is a basic or unenhanced indirect effect, because it does not consider the influence of the X_2-X_3 loop. The magnitude of $b_{43}b_{32}b_{21}$ is again the first of a series of impacts reaching X_4. The next in the series would result from the original change in X_3 cycling once through the X_2-X_3 loop and subsequently being passed on to X_4. Thus *the distinction between basic and loop-enhanced effects is as appropriate for indirect effects as it is for direct effects.*

8.1.2 Quantifying the Effects of Loops

Having observed that models containing loops imply repeated cyclings through those loops, we confront the issue of how to quantify the effects of such cyclings. The top portion of Figure 8.3 depicts X_1 as a cause of X_2, where X_2 is known to be one of the variables forming a causal loop. For simplicity we have not shown the variables in this loop, but we simply denote the product of all the structural coefficients constituting the loop as L.

There are two conceptually distinct ways to represent Figure 8.3A in equation form, both of which give the important conclusion that the effect of adding the loop L to the dependent variable increases the change in X_2 predicted to accompany a unit change in X_1 from b_{21} to an amount $1/(1 - L)$ times larger than the basic direct effect, namely $b_{21}/(1 - L)$.

The first way is to write the structural determinants of X_2 as including X_2 itself:

$$X_2 = b_{21}X_1 + LX_2 \qquad \qquad 8.2$$

Rearranging gives

$$X_2(1 - L) = b_{21}X_1$$
$$X_2 = [b_{21}/(1 - L)]X_1 \qquad \qquad 8.3$$

which is a reduced form equation providing $b_{21}/(1 - L)$ as the response of X_2 to a change in X_1.

Alternatively, consider the series of incremental changes in X_2 that follow a change in X_1. The initial change in X_2 that appears is b_{21} times the change in X_1 (as would be reflected in the equation $X_2 = b_{21}X_1$),[2] the next change in X_2 reflects the proportion (L) of the initial X_2 change ($b_{21}X_1$) that reenters X_2 after transmission through the variables in the loop (or $Lb_{21}X_1$). This latest change recycles through the loop to provide a

Figure 8.3   Simplified Loops.[1]

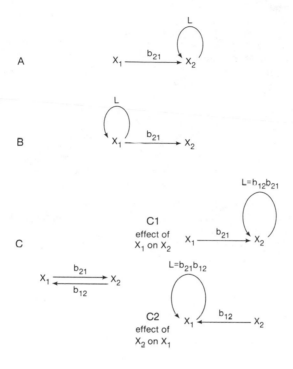

further effect of $LLb_{21}X_1$, and so on, ultimately providing the overall equation linking X_1 to X_2 as

$$X_2 = b_{21}X_1 + Lb_{21}X_1 + LLb_{21}X_1 + LLLb_{21}X_1 + \cdots \qquad 8.4$$

This same series is produced if the equation for X_2 (Eq. 8.2) is repeatedly substituted for the X_2 term on the right side of Eq. 8.2. By factoring out X_1 and b_{21} we can write this as

$$X_2 = b_{21}(1 + L + L^2 + L^3 + \cdots)X_1 \qquad 8.5$$

As long as the absolute value of L is less than 1—that is, as long as each cycle through the loop provides a change in X_2 that is smaller than the change created by the preceding cycle, the terms in the parentheses form a geometric series in L that has sum $1/(1 - L)$ (Protter and Morrey, 1964:74), and hence

$$X_2 = [b_{21}/(1 - L)]X_1 \qquad 8.6$$

Thus if we consider the loop as either the variable X_2 providing input into its own value or as a mechanism by which X_1 initiates a series of increments in X_2, we come to the same conclusion: the impact of a unit change in X_1 on X_2 is not b_{21} but $b_{21}/(1 - L)$ [i.e., b_{21} multiplicatively augmented by $1/(1 - L)$]. Note that information on the magnitude of the effect returning to a variable after a single pass through a loop (L) is sufficient to compute the overall influence provided by multiple passes through the loop. The mathematical equivalence of the self-influence or incremental representations (Eqs. 8.2 and 8.4, respectively) implies that the choice between these representations must be dictated by the utility of the resulting interpretations. The incremental representation seems superior in this respect (it complements the interpretive strategy of examining the results of a hypothetical unit intervention), and hence it dominates much of the following discussion.

Returning to Figure 8.3B, note that a unit change in X_1 would be passed along toward X_2 after each of zero, one, two, three, or more loopings through L, thus giving

$$X_2 = b_{21}X_1 + Lb_{21}X_1 + LLb_{21}X_1 + LLLb_{21}X_1 + \cdots \qquad 8.7$$

which reduces to

$$X_2 = [b_{21}/(1 - L)]X_1 \qquad 8.8$$

We again recognize this as a basic direct effect multiplicatively enhanced by $1/(1 - L)$.

The symmetry in Figure 8.3C implies that we can examine the effect of X_1 on X_2 or of X_2 on X_1. In either case, we can think of this as a direct effect plus a loop that just happens to include the other variable in the model. The loop takes on a value $L = b_{21}b_{12} = b_{12}b_{21}$ and provides a series whose enhancement is $1/(1 - b_{21}b_{12})$. Thus the enhanced effect of X_1 on X_2 is $b_{21}/(1 - b_{12}b_{21})$, and the enhanced effect of X_2 on X_1 is $b_{12}/(1 - b_{12}b_{21})$. If, for example, b_{12} and b_{21} were .6 and .7, respectively, the model predicts that a unit change in X_1 would produce a change of $.7/[1 - (.6)(.7)]$ or 1.21 units in X_2 after cyclings through the loop are considered.

Note that the value of the causal variable itself changes by more than 1 unit by the time the model returns to a stable state. If X_1 initially changed 1 unit, the overall change in X_1 would be about 1.72 units—1.0 unit from the initial intervention and .72 of a unit from the cyclings through the loop. [The value .72 is obtained as the total change in X_1 that would accompany a b_{21} (or .7 unit) change in X_2, which is .7 of $.6/(1 - (.6)(.7))$.] It is clear that we would *not* interpret b_{21} or b_{12} as the overall effects of X_1 and X_2 on each other. Although these are the proper structural coefficients, they are not interpretable as total effect coefficients, because they do not account for loop or feedback effects.

If the value of a loop (L) is negative, the enhancement provided by the loop is a reduction in total effect below the level of the initial increment. For example, if in Figure 8.3A the loop had value $L = -.2$ (as it would if a nondiagrammed X_3 variable responded to X_2 with a coefficient of $-.4$ and influenced X_2 with coefficient .5), the total effect of X_1 on X_2 would be $b_{21}/(1 - (-.2)) = .83b_{21}$. Hence, the eventual total effect would be only 83% of the initial impact of a unit change in X_1. Mentally tracing the series of changes resulting from a unit increase in X_1 uncovers an initial b_{21} increase in X_2 followed by alternating *decreases* and increases for subsequent cycles through the loop. The first cycle gives a decrease of $.2b_{21}$ unit, the next an increase (the negative of the preceding decrease) of $.2(.2b_{21})$ unit, and so on. Since the first and largest loop-induced change is a decrease, this series ultimately reduces the size of the initial impact. Hence, *negative loops buffer or cushion dependent variables against changes introduced from other variables, and positive loops magnify the responses of the dependent variables they touch.*

8.2 General Matrix Formulas

Discussions of matrix procedures for calculating the total effects between *all* pairs of variables in models have appeared in several substantive areas (cf. Fox, 1980,[3] 1985; Graff and Schmidt, 1982; Greene, 1977; Lewis-Beck and Mohr, 1976). We now demonstrate how matrix calculations can provide direct, indirect, and total effects through extension of the notion of "repeated cyclings through loops" or "enhancements" as discussed before. We ultimately obtain formulas identical to those provided in LISREL V and VI, p. III-39, which Joreskog and Sorbom provide without derivation.

8.2.1 Basic Direct Effects

We begin by returning to the basic structural equations linking the conceptual level variables $\eta = \mathbf{B}\eta + \Gamma\xi + \zeta$, as developed in Chapter 4. The estimates of the structural coefficients in the $m \times m$ matrix **B** contain the basic direct effects of the endogenous variables on one another. Specifically, the typical element β_{ij} represents the basic direct effect of the jth endogenous variable on the ith endogenous variable. Similarly, the $m \times n$ matrix Γ of structural coefficients with typical element γ_{ik} represents the basic direct effect of the kth exogenous variable on the ith endogenous variable. Thus the elements of $\mathbf{B}$ and Γ reflect the basic direct effect of a column variable on the row variable and can be interpreted as the amount of change in an η expected to accompany a unit change in the causal variable (η or ξ) with all the other variables "held constant," so there is no possibility of indirect or looped effects.

Indirect effects are most easily visualized in the context of a particular model, so you might examine Figure 8.4 and its basic direct effects before proceeding. The coefficient β_{31}, for example, is the change in η_3 predicted to accompany a unit change in η_1 if no cyclings through the η_1–η_2 loop are allowed.

8.2.2 Indirect Effects of Endogenous Concepts (η's) on One Another

In this section we develop a procedure for calculating the numerical magnitude of the indirect effect of any one endogenous concept on any other endogenous concept. We will record these indirect effect coefficients in an $m \times m$ matrix called $\mathbf{I}_{\eta\eta}$, whose elements are the coefficients reflecting the indirect effects of the endogenous concepts labeling the columns of this matrix on the endogenous concepts labeling the rows of the matrix. Recall that any particular routing between two variables transmits an effect equal to the product of the coefficients constituting that routing. Summing the indirect effects transmitted through all the routings of length 2 or more that link any two variables provides the overall indirect effect between those variables—the quantities we eventually record in $\mathbf{I}_{\eta\eta}$.

Consider the $m \times m$ matrix obtained by squaring the $\mathbf{B}$ matrix. This $\mathbf{B}^2$ matrix contains cell entries that are the sum of the products of the coefficients of all routings of length 2 connecting the variables specifying the columns and rows of the matrix. That is, the entries in the matrix $\mathbf{B}^2$ are the indirect effects transmitted through routings containing two paths (or one intervening variable). Since much of what follows is a direct extension of the idea presented in this paragraph, we strongly recommended that you verify how matrix multiplication gives the elements in row 1, column 1 and row 3, column 2 of $\mathbf{B}^2$ provided as the first matrix in the "Indirect Effects" section of Figure 8.4, and that you locate the indirect paths corresponding to these elements of $\mathbf{B}^2$ on the diagram in Figure 8.4. Now, reread the paragraph if you have lost track of what it said.

Figure 8.4   Illustration of Indirect and Total Effect Calculations.

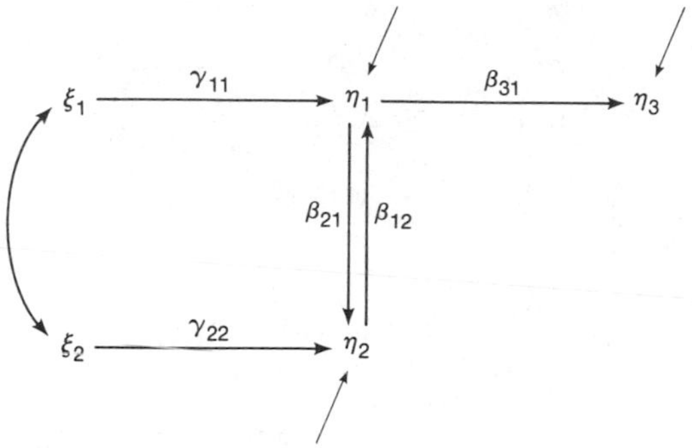

$$\eta \;=\; \beta\eta \;\;+\;\; \Gamma\xi \;\;+\;\; \zeta$$

$$\begin{bmatrix} \eta_1 \\ \eta_2 \\ \eta_3 \end{bmatrix} = \begin{bmatrix} 0 & \beta_{12} & 0 \\ \beta_{21} & 0 & 0 \\ \beta_{31} & 0 & 0 \end{bmatrix}\begin{bmatrix} \eta_1 \\ \eta_2 \\ \eta_3 \end{bmatrix} + \begin{bmatrix} \gamma_{11} & 0 \\ 0 & \gamma_{22} \\ 0 & 0 \end{bmatrix}\begin{bmatrix} \xi_1 \\ \xi_2 \end{bmatrix} + \begin{bmatrix} \zeta_1 \\ \zeta_2 \\ \zeta_3 \end{bmatrix}$$

<u>Direct Effects of η on η = β =</u> $\begin{bmatrix} 0 & \beta_{12} & 0 \\ \beta_{21} & 0 & 0 \\ \beta_{31} & 0 & 0 \end{bmatrix}$

Indirect Effects of η on η = $I_{\eta\eta} = \beta^2 + \beta^3 + \beta^4 + \beta^5 + \beta^6 + \ldots\ldots\ldots \beta^\infty$

$$= \begin{bmatrix} \beta_{12}\beta_{21} & 0 & 0 \\ 0 & \beta_{21}\beta_{12} & 0 \\ 0 & \beta_{31}\beta_{12} & 0 \end{bmatrix} + \begin{bmatrix} 0 & \beta_{12}\beta_{21}\beta_{12} & 0 \\ \beta_{21}\beta_{12}\beta_{21} & 0 & 0 \\ \beta_{31}\beta_{12}\beta_{21} & 0 & 0 \end{bmatrix} + \begin{bmatrix} (\beta_{12}\beta_{21})^2 & 0 & 0 \\ 0 & (\beta_{21}\beta_{12})^2 & 0 \\ 0 & (\beta_{31}\beta_{12})(\beta_{21}\beta_{12}) & 0 \end{bmatrix}$$

$$+ \begin{bmatrix} 0 & (\beta_{12}\beta_{21})^2\beta_{12} & 0 \\ (\beta_{21}\beta_{12})^2\beta_{21} & 0 & 0 \\ (\beta_{31}\beta_{12})(\beta_{21}\beta_{12})\beta_{21} & 0 & 0 \end{bmatrix} + \begin{bmatrix} (\beta_{12}\beta_{21})^3 & 0 & 0 \\ 0 & (\beta_{21}\beta_{12})^3 & 0 \\ 0 & (\beta_{31}\beta_{12})(\beta_{21}\beta_{12})^2 & 0 \end{bmatrix} + \ldots\ldots\ldots$$

or

<u>$I_{\eta\eta} = (I - \beta)^{-1} - I - \beta$</u>

$$= \begin{bmatrix} \dfrac{1}{1-\beta_{21}\beta_{12}} & \dfrac{\beta_{12}}{1-\beta_{21}\beta_{12}} & 0 \\[2ex] \dfrac{\beta_{21}}{1-\beta_{21}\beta_{12}} & \dfrac{1}{1-\beta_{21}\beta_{12}} & 0 \\[2ex] \dfrac{\beta_{31}}{1-\beta_{21}\beta_{12}} & \dfrac{\beta_{31}\beta_{12}}{1-\beta_{21}\beta_{12}} & 1 \end{bmatrix} - \begin{bmatrix} 1 & 0 & 0 \\ 0 & 1 & 0 \\ 0 & 0 & 1 \end{bmatrix} - \begin{bmatrix} 0 & \beta_{12} & 0 \\ \beta_{21} & 0 & 0 \\ \beta_{31} & 0 & 0 \end{bmatrix} = \begin{bmatrix} \dfrac{\beta_{21}\beta_{12}}{1-\beta_{21}\beta_{12}} & \dfrac{\beta_{21}\beta_{12}\beta_{12}}{1-\beta_{21}\beta_{12}} & 0 \\[2ex] \dfrac{\beta_{21}\beta_{12}\beta_{21}}{1-\beta_{21}\beta_{12}} & \dfrac{\beta_{21}\beta_{12}}{1-\beta_{21}\beta_{12}} & 0 \\[2ex] \dfrac{\beta_{21}\beta_{12}\beta_{31}}{1-\beta_{21}\beta_{12}} & \dfrac{\beta_{31}\beta_{12}}{1-\beta_{21}\beta_{12}} & 0 \end{bmatrix}$$

Figure 8.4   Continued.

Total Effects of η on $\eta = T_{\eta\eta} = \beta^1 + \beta^2 + \beta^3 + \beta^4 \ldots\ldots \beta^\infty = (I - \beta)^{-1} - I$

$$
= \begin{bmatrix} \dfrac{1}{1-\beta_{21}\beta_{12}} & \dfrac{\beta_{12}}{1-\beta_{21}\beta_{12}} & 0 \\[2ex] \dfrac{\beta_{21}}{1-\beta_{21}\beta_{12}} & \dfrac{1}{1-\beta_{21}\beta_{12}} & 0 \\[2ex] \dfrac{\beta_{31}}{1-\beta_{21}\beta_{12}} & \dfrac{\beta_{31}\beta_{12}}{1-\beta_{21}\beta_{12}} & 1 \end{bmatrix} - \begin{bmatrix} 1 & 0 & 0 \\ 0 & 1 & 0 \\ 0 & 0 & 1 \end{bmatrix} = \begin{bmatrix} \dfrac{\beta_{21}\beta_{12}}{1-\beta_{21}\beta_{12}} & \dfrac{\beta_{12}}{1-\beta_{21}\beta_{12}} & 0 \\[2ex] \dfrac{\beta_{21}}{1-\beta_{21}\beta_{12}} & \dfrac{\beta_{21}\beta_{12}}{1-\beta_{21}\beta_{12}} & 0 \\[2ex] \dfrac{\beta_{31}}{1-\beta_{21}\beta_{12}} & \dfrac{\beta_{31}\beta_{12}}{1-\beta_{21}\beta_{12}} & 0 \end{bmatrix}
$$

Direct Effects of ξ on $\eta = \Gamma = \begin{bmatrix} \gamma_{11} & 0 \\ 0 & \gamma_{22} \\ 0 & 0 \end{bmatrix}$

Indirect Effects of ξ on $\eta = I_{\eta\xi} = (\beta^1 + \beta^2 + \beta^3 + \ldots\ldots \beta^\infty)\Gamma = ((I - \beta)^{-1} - I)\Gamma$

$$
\begin{bmatrix} \dfrac{\beta_{21}\beta_{12}}{1-\beta_{21}\beta_{12}} & \dfrac{\beta_{12}}{1-\beta_{21}\beta_{12}} & 0 \\[2ex] \dfrac{\beta_{21}}{1-\beta_{21}\beta_{12}} & \dfrac{\beta_{21}\beta_{12}}{1-\beta_{21}\beta_{12}} & 0 \\[2ex] \dfrac{\beta_{31}}{1-\beta_{21}\beta_{12}} & \dfrac{\beta_{31}\beta_{12}}{1-\beta_{21}\beta_{12}} & 0 \end{bmatrix} \begin{bmatrix} \gamma_{11} & 0 \\ 0 & \gamma_{22} \\ 0 & 0 \end{bmatrix} = \begin{bmatrix} \dfrac{\gamma_{11}\beta_{21}\beta_{12}}{1-\beta_{21}\beta_{12}} & \dfrac{\gamma_{22}\beta_{12}}{1-\beta_{21}\beta_{12}} \\[2ex] \dfrac{\gamma_{11}\beta_{21}}{1-\beta_{21}\beta_{12}} & \dfrac{\gamma_{22}\beta_{21}\beta_{12}}{1-\beta_{21}\beta_{12}} \\[2ex] \dfrac{\gamma_{11}\beta_{31}}{1-\beta_{21}\beta_{12}} & \dfrac{\gamma_{22}\beta_{31}\beta_{12}}{1-\beta_{21}\beta_{12}} \end{bmatrix}
$$

Total Effects of ξ on $\eta = T_{\eta\xi} = (\beta^0 + \beta^1 + \beta^2 + \ldots\ldots \beta^\infty)\Gamma = (I - \beta)^{-1}\Gamma$

$$
\begin{bmatrix} \dfrac{1}{1-\beta_{21}\beta_{12}} & \dfrac{\beta_{12}}{1-\beta_{21}\beta_{12}} & 0 \\[2ex] \dfrac{\beta_{21}}{1-\beta_{21}\beta_{12}} & \dfrac{1}{1-\beta_{21}\beta_{12}} & 0 \\[2ex] \dfrac{\beta_{31}}{1-\beta_{21}\beta_{12}} & \dfrac{\beta_{31}\beta_{12}}{1-\beta_{21}\beta_{12}} & 1 \end{bmatrix} \begin{bmatrix} \gamma_{11} & 0 \\ 0 & \gamma_{22} \\ 0 & 0 \end{bmatrix} = \begin{bmatrix} \dfrac{\gamma_{11}}{1-\beta_{21}\beta_{12}} & \dfrac{\gamma_{22}\beta_{12}}{1-\beta_{21}\beta_{12}} \\[2ex] \dfrac{\gamma_{11}\beta_{21}}{1-\beta_{21}\beta_{12}} & \dfrac{\gamma_{22}}{1-\beta_{21}\beta_{12}} \\[2ex] \dfrac{\gamma_{11}\beta_{31}}{1-\beta_{21}\beta_{12}} & \dfrac{\gamma_{22}\beta_{31}\beta_{12}}{1-\beta_{21}\beta_{12}} \end{bmatrix}
$$

If a measurement structure is specified for the model, the effects of the concepts on the observed Y variables can be calculated as:

Direct Effects of η on $Y = D_{Y\eta} = \Lambda_Y$

Indirect Effects of η on $Y = I_{Y\eta} = \Lambda_Y T_{\eta\eta}$

Total Effects of η on $Y = T_{Y\eta} = \Lambda_Y + \Lambda_Y T_{\eta\eta}$

Total Effects of ξ on $Y = T_{Y\xi} = \Lambda_Y T_{\eta\xi}$

Similarly, the entries in the $m \times m$ matrix $\mathbf{B}^3$ are the sum of the indirect effects of the column variable on the row variable that are transmitted through paths of length 3 (two intervening variables), and so on, for $\mathbf{B}^4$, $\mathbf{B}^5$, and so on, for paths of lengths 4, 5, and so on. Examine a few of the entries in $\mathbf{B}^3$ in Figure 8.4 and locate the corresponding paths of length 3 on the diagram.

The overall indirect effect between any pair of endogenous variables $(\mathbf{I}_{\eta\eta})$ *is the sum of the indirect effects transmitted through all routings of length 2 or more.*

$$\mathbf{I}_{\eta\eta} = \mathbf{B}^2 + \mathbf{B}^3 + \mathbf{B}^4 + \cdots + \mathbf{B}^\infty \qquad 8.9$$

We will find a simpler expression for this sum of matrix powers later.[4]

In Figure 8.4 the only indirect effects of length 2 among the endogenous variables are the impact of η_2 on η_3, η_1 on itself through η_2, and η_2 on itself through η_1, as we see from the $\mathbf{B}^2$ matrix. Note that η_2 and η_1 influence "themselves" at even powers of $\mathbf{B}$, because these powers correspond to one, two, three, . . . cycles through the η_1–η_2 loop. Note the similar placement of nonzero elements in the second, fourth, and sixth powers of $\mathbf{B}$ and the appearance of successive powers of the loop effect $\beta_{12}\beta_{21}$ in these matrices.

η_1 will influence η_3 in the odd powers of $\mathbf{B}$, because the even-length paths provided by the multiple cyclings through the loop are followed by the single β_{31} effect to create an odd number of links in the effect chain.

In models containing no loops or reciprocal causes, the longest possible routing among the m endogenous variables is $m-1$ paths long—that is, a chain of paths linking all the m endogenous variables would contain only $m-1$ paths. This implies that no effects can be transmitted along routings longer than $m-1$ path lengths, and hence that all powers of $\mathbf{B}$ greater than $m-1$ should contain only zero entries for recursive models. In models containing loops, the higher powers will not necessarily be zero because routings of considerable length may result from repeated cyclings through the loops. *Even in models with loops, however, the numerical value of the entries in the higher powers of $\mathbf{B}$ typically approach zero because longer routings produce weaker effects due to the imperfect transmittance of effects through the many paths in long routings.* Typically, calculation of the first few powers of $\mathbf{B}$, both numerically and in symbolic form as in Figure 8.4, will be especially informative.

Even though the diagonal elements of $\mathbf{B}$ are zero by definition, the diagonals of the powers of $\mathbf{B}$ need not be zero (again see Figure 8.4). A nonzero diagonal element in a power of $\mathbf{B}$ indicates the presence of a loop, by which that variable can influence its own value, and the loop contains as many paths as the power of $\mathbf{B}$ in which the nonzero diagonal appears.[5,6]

8.2.3 Total Effects of Endogenous Concepts (η's) on One Another

We are now ready to express the *total effects* of the endogenous concepts on one another $(\mathbf{T}_{\eta\eta})$ as the sum of the basic direct effects $(\mathbf{B}^1)$ plus and indirect effects $(\mathbf{B}^2 + \mathbf{B}^3 + \cdots + \mathbf{B}^\infty)$:

$$\mathbf{T}_{\eta\eta} = \mathbf{B}^1 + \mathbf{B}^2 + \mathbf{B}^3 + \cdots + \mathbf{B}^\infty \qquad 8.10$$

We can simplify this formula by invoking a bit of matrix algebra that allows one to express the sum of a series of matrix powers as an inverse. Specifically,

$$\mathbf{B}^0 + \mathbf{B}^1 + \mathbf{B}^2 + \cdots + \mathbf{B}^\infty = (\mathbf{I} - \mathbf{B})^{-1} \qquad 8.11$$

if the model is of a stable or a nonexplosive system,[7] and where the unsubscripted $\mathbf{I}$ (here and after) is an $m \times m$ identity matrix.

$\mathbf{T}_{\eta\eta}$ is this series, minus $\mathbf{B}^0$ which is defined as an $m \times m$ identity matrix $\mathbf{I}$ (just as any scalar raised to the zero power is defined to equal 1), so we can rewrite the expression for total effects as

$$\mathbf{T}_{\eta\eta} = (\mathbf{I} - \mathbf{B})^{-1} - \mathbf{I} \qquad 8.12$$

Thus we can obtain total effects by calculating an inverse rather than by summing an infinite series of matrix powers. Note the parallel between equating the sum of a series of powers to an inverse here, and equating the sum of a series to a reciprocal in Section 8.1.2.

Noting that $\mathbf{I}_{\eta\eta}$ is $\mathbf{T}_{\eta\eta}$ minus $\mathbf{B}$ allows us to express $\mathbf{I}_{\eta\eta}$ as

$$\mathbf{I}_{\eta\eta} = (\mathbf{I} - \mathbf{B})^{-1} - \mathbf{I} - \mathbf{B} \qquad 8.13$$

and similarly remove the problem of summing an infinite power series when calculating indirect effects. The equations for the direct, indirect, and total effects of endogenous variables on one another are summarized in Figure 8.4.

8.2.4 Indirect Effects of Exogenous (ξ) Concepts on Endogenous (η) Concepts

We have already specified the direct effects of the exogenous concepts on the endogenous concepts as the elements of $\mathbf{\Gamma}$. Indirect effects leading from an exogenous concept to an endogenous concept can do so only by going through other intervening endogenous concepts. That is, no indirect effects can be transmitted through exogenous variables, because, by definition, exogenous variables are never causally dependent on any other variables. Thus indirect effects of length 2 originating from the exogenous concepts have as their first link an effect of a ξ on an η (a Γ coefficient), and the second link captures the effect of an η on another η (a B coefficient). Similarly, all routings three or more paths long have a first path whose value is a Γ coefficient, and all subsequent paths are B coefficients. And, as seen earlier, the paths of various lengths among the η's are provided by powers of B. This reasoning prepares us for the form of the series of matrices capturing the indirect effects of the exogenous concepts on the endogenous concepts via routings of all possible lengths, $I_{\eta\xi}$,

$$\underset{(m \times n)}{I_{\eta\xi}} = \underset{(m \times m)(m \times n)}{B^1\Gamma + B^2\Gamma + B^3\Gamma + B^4\Gamma + \cdots + B^\infty\Gamma} \qquad 8.14$$

You should calculate $B\Gamma$ for the model in Figure 8.4, paying special attention to how matrix multiplication supplies the nonzero entries in the product matrix and the diagrammatic location of the paths corresponding to the nonzero elements. Though it is not obvious from the example, you should convince yourself that each element in the matrix product $B\Gamma$ captures *all* the routings of length 2 linking an exogenous variable to an endogenous variable—that is, the sum of several routings of length 2 each going through a different intervening variable. (For example, if an effect of ξ_1 on η_2 and an effect of η_2 on η_3 are added to the model, there would be two routings of length 2 by which ξ_1 could influence η_3 and hence the 3,1 element of the product matrix would be the sum of the corresponding products of paths.)

By direct extension, the elements in the matrix $B^2\Gamma$ are the sums of the products of the paths comprising *all* of the routings of length 3 linking the column exogenous variable to the row endogenous variable, and so on, for the series created by postmultiplying higher powers of B by Γ.

We can calculate $I_{\eta\xi}$ by an inverse, rather than as the sum of a power series, if we factor out Γ,

$$I_{\eta\xi} = (B^1 + B^2 + B^3 + \cdots + B^\infty)\Gamma \qquad 8.15$$

and use Eqs. 8.10–8.12:

$$I_{\eta\xi} = ((I - B)^{-1} - I)\Gamma \qquad 8.16$$

8.2.5 Total Effects of Exogenous (ξ) Concepts on Endogenous (η) Concepts

Paralleling our previous discussion, we can express the total effects of the exogenous concepts on the endogenous concepts ($\mathbf{T}_{\eta\xi}$) as the sum of the direct effects ($\mathbf{\Gamma}$) and the indirect effects ($\mathbf{I}_{\eta\xi}$) from Eq. 8.14.

$$\mathbf{T}_{\eta\xi} = \mathbf{\Gamma} + \mathbf{B}^1\mathbf{\Gamma} + \mathbf{B}^2\mathbf{\Gamma} + \mathbf{B}^3\mathbf{\Gamma} + \cdots + \mathbf{B}^\infty\mathbf{\Gamma} \qquad 8.17$$

$$= (\mathbf{I} + \mathbf{B}^1 + \mathbf{B}^2 + \mathbf{B}^3 + \cdots + \mathbf{B}^\infty)\mathbf{\Gamma} \qquad 8.18$$

$$= (\mathbf{I} - \mathbf{B})^{-1}\mathbf{\Gamma} \qquad 8.19$$

In Figure 8.4 the basic direct effect of ξ_1 on η_1 is γ_{11}, and the total effect of ξ_1 on η_1 is $\gamma_{11}/(1 - \beta_{21}\beta_{12})$ from the row 1, column 1 element of $\mathbf{T}_{\eta\xi}$. Recognizing $\beta_{21}\beta_{12}$ as the product of the coefficients constituting the only loop in this model (called L in our discussion of Figure 8.3), we recognize this as the basic direct effect multiplied by the enhancing factor $1/(1 - L)$, as discussed in Section 8.1.2.

We note further that the only denominator present in any of the indirect or total effects matrices in the example is precisely this loop enhancement factor $1/(1 - \beta_{21}\beta_{12})$. Thus the total effect of ξ_1 on η_3 (the row 3, column 1 element of $\mathbf{T}_{\eta\xi}$) is the chain of effects linking ξ_1 to η_3 (i.e., $\gamma_{11}\beta_{31}$) multiplied by the loop enhancement $1/(1 - \beta_{11}\beta_{12})$. Similarly, the total effect of ξ_2 on η_3 is the chain of effects linking ξ_2 to η_3 multiplied by the same loop enhancement. Note that any effect that touches the loop, even the basic direct effect of η_1 on η_3, is enhanced because of that touching.

Another, more traditional, derivation of $\mathbf{T}_{\eta\xi}$ begins from the basic equation $\eta = \mathbf{B}\eta + \mathbf{\Gamma}\xi + \zeta$, which is rearranged to give

$$\eta - \mathbf{B}\eta = \mathbf{\Gamma}\xi + \zeta \qquad 8.20$$

$$(\mathbf{I} - \mathbf{B})\eta = \mathbf{\Gamma}\xi + \zeta \qquad 8.21$$

Premultiplying by $(\mathbf{I} - \mathbf{B})^{-1}$ gives what is called the reduced form equation

$$\eta = (\mathbf{I} - \mathbf{B})^{-1}\mathbf{\Gamma}\xi + (\mathbf{I} - \mathbf{B})^{-1}\zeta \qquad 8.22$$

in which we discover $\mathbf{T}_{\eta\xi}$ as the reduced form coefficients linking the exogenous variables to the endogenous variables. This derivation parallels the rearrangements in Eqs. 8.2 and 8.3, but it camouflages (within the inverse) the repeated cyclings highlighted by our previous derivation. It shows, however, that effects originating in unidentified or error sources are subject to the same cyclic enhancements as effects originating in the ξ or η variables.

8.2.6 Thinking about the Effects of Loops

The preceding section tells us how to calculate direct, indirect, and total effects, but the effects of loops are lumped together with all the other indirect effects. Although it seems desirable to decompose the indirect effects into single pass (or basic) indirect effects and multiple pass (or loop-enhancing) indirect effects, such a decomposition could not give a simple sum of basic plus loop-enhancing components. Recalling our discussion of Figure 8.3 and noting the form of the total effect of ξ_1 on η_1 in Figure 8.4, we see that loops *do not add an effect onto that provided by the basic direct effects, they multiply (or enhance) the basic effects by a factor of the form* $1/(1 - L)$. Hence, the effects of loops are only interpretable in conjunction with discussions of *basic* direct or indirect effects.

Consider the indirect effect of η_2 on η_3 in Figure 8.4 (the row 3, column 2 element of $\mathbf{I}_{\eta\eta}$), which is $\beta_{31}\beta_{12}(1/(1 - \beta_{21}\beta_{12}))$. We could partition this indirect effect by subtracting the single-pass indirect effect $\beta_{31}\beta_{12}$ to leave a numerical quantity that appears to be attributable only to looping, because it is neither a direct effect nor a single-pass (basic) indirect effect. This numerical quantity, however, is not interpretable as an effect of the loop on η_3. The loop per se cannot cause η_3. *Only variables have causal powers in structural equation models.* The interpretation of this loop-associated quantity requires that it be reunited with the basic routing from η_2 to η_3. The "difference" between the total and basic indirect effects is the cumulative change expected in η_3 *resulting from a unit change in* η_2 if the first (and only nonlooped) effect reaching η_3 is discounted.

Note that the coefficients on the diagonal of $\mathbf{I}_{\eta\eta}$ usually need no decomposition because they are composed entirely of loop effects. For such terms, the only thing decompositon could do is to specify which of several possible loops provide the greatest portion of the self-enhancement, but even this is obvious from the magnitudes of the products of the coefficients in the various loops (L's).

8.2.7 Total Effects of Concepts on Indicators

The previous section gives the direct, indirect, and total effects among the conceptual level variables. We may also be interested in the direct, indirect, and total effects of the η's or the ξ's on the Y variables. (The X variables receive direct effects from the ξ's, but they never receive any indirect effects because the ξ's can never "receive" any effects to "transmit indirectly" to the X's.)

Once the direct, indirect, and total effects have been obtained (as before), the only way these effects filter down to the observed Y indicators is through Λ_y. Indeed, the direct effects of the η's on the Y's are given by Λ_y, so we can immediately express $\mathbf{D}_{y\eta}$ as

$$\mathbf{D}_{y\eta} = \Lambda_y \qquad 8.23$$

The indirect effects of the η's on the Y's are given by the total effects of any one η on the other η's, which then filter down to the Y's:

$$\mathbf{I}_{y\eta} = \Lambda_y \mathbf{T}_{\eta\eta} = \Lambda_y((\mathbf{I} - \mathbf{B})^{-1} - \mathbf{I}) \qquad 8.24$$

The total effects of the η's on the Y's are their direct effects (Eq. 8.23) plus their indirect effects (Eq. 8.24):

$$\mathbf{T}_{y\eta} = \Lambda_y + \Lambda_y \mathbf{T}_{\eta\eta} = \Lambda_y + \Lambda_y((\mathbf{I} - \mathbf{B})^{-1} - \mathbf{I}) \qquad 8.25$$

$$= \Lambda_y(\mathbf{I} - \mathbf{B})^{-1} \qquad 8.26$$

All of the effects of the ξ's on the Y's must be indirect, because their effects must be transmitted through the η's in order to reach the Y's. The total effects of the ξ's on the Y's are

$$\mathbf{T}_{y\xi} = \Lambda_y \mathbf{T}_{\eta\xi} = \Lambda_y(\mathbf{I} - \mathbf{B})^{-1}\Gamma \qquad 8.27$$

These equations are given in the table on p. III-39 of the LISREL V and VI manuals.[8] LISREL's output provides only the basic direct effects and the total effects for η on η, ξ on η, η on Y, and ξ on Y. If needed, the corresponding indirect effects can be calculated by simple subtraction.

8.3 Two Extensions of Effect Decompositions

8.3.1 Modeling Covariance versus Accounting for Covariance

Indirect and total effects are model dependent in that practically any model revision implies different total and indirect effects. For example, if ξ_2 in Figure 8.4 is made endogenous by allowing ξ_1 to influence ξ_2, a whole new set of indirect routings is introduced between ξ_1 and the other endogenous concepts. It would also change the actual estimates of the existing coefficients. In both ways, changing the model changes the indirect and total effects.

At a more general level, we should always consider effects (total, indirect, or direct) as model specific, because all statements of model-implied effects are conditioned by the identity of the included variables and by the placement of directed paths and nondirected correlations among the variables. Indeed, the ability of a model to "account for" the covariance between variables must be thought of as model dependent. In Figure 8.4, for example, the covariance between ξ_1 and η_2 is partially a function of the indirect effect of ξ_1 on η_2, partially a function of the loop enhancement of this effect, and partially a function of ξ_1's correlation with another cause of η_2, namely ξ_2. Changing the model by replacing the ξ_1–ξ_2 correlation with an effect of ξ_1 on ξ_2, for example, inserts a whole new set of indirect effects of ξ_1 on η_2 and would eliminate the contribution of any correlational component to the covariance between ξ_1 and η_2, because only directed effects would remain in the model.

The extent to which mere correlations contribute to the modeled covariances reflects the extent to which a model is operating through unspecified causal structures. Hence, there is a clear conceptual distinction between how *consistent* the model is with the observed covariances between the variables, and how well the model *causally accounts* for the covariances between the variables. A good model will be consistent with the observed covariances (recall the discussion of χ^2), but a better model will provide that consistency as the result of directed causal effects, as opposed to mere correlational associations. Though LISREL provides the direct and total effects of each variable on the others (and hence the indirect effects via subtraction), it does not currently support decompositions that quantify how much of the association between two variables remains to be explained over and above the total effects implied by the model. Noncausal aspects of the decomposition of effects have been discussed by Alwin and Hauser (1975), Fox (1980[3]), Greene (1977), and Lewis-Beck and Mohr (1976), but none of these discussions use the general LISREL model or even differentiate between observed indicators and unobserved concepts.

8.3.2 Decomposition with a Variable Forcibly Held Constant

Another perspective on effect decomposition is provided by Greene (1977). He suggests eliminating the rows and columns of $\mathbf{B}$ and $\mathbf{\Gamma}$ that correspond to a variable (or group of variables) in the model, and then using *the original set of coefficient estimates* to calculate another set of indirect and total effect matrices using the shrunken matrices and the procedures summarized in Figure 8.4. This is equivalent to forcibly holding the eliminated variable(s) constant, and therefore it eliminates the possibility of causal effects originating in, or being transmitted through, the eliminated variable(s) while the remainder of the model functions as before. The interpretations developed earlier have routinely required that the other variables remain untouched yet free, but they are easily modified to routinely state that the indirect and total effects (as calculated after eliminating some rows and columns of $\mathbf{B}$ and $\boldsymbol{\Gamma}$) reflect the effects that would appear if the eliminated variables were forcibly held constant while all the other variables continued to be untouched yet free. Such side calculations are particularly appropriate where legislative interventions, contractual agreements, or experimental conditions might allow the corresponding variable(s) to be forcibly fixed in the real world.

Fox (1985) discusses how we can extend these ideas to calculate the effects that *go through a specific set variables but not through another set*. Essentially, we first obtain the matrix containing the total effect of any variable on another via all the routes that do not include a set of variables i by eliminating the columns and/or rows of $\mathbf{B}$ and $\boldsymbol{\Gamma}$ referring to the set i, as suggested by Greene (denote this matrix $\mathbf{T}_{-i}$). We next obtain a matrix containing the total effects of any variable on another via all routes that do not include a second (larger) set of variables. Denote this matrix $\mathbf{T}_{-ij}$ to indicate that both the original set i and a new set j of variables have been forcibly held constant. Obtaining the difference between the corresponding elements of $\mathbf{T}_{-i}$ and $\mathbf{T}_{-ij}$ (i.e., calculating $\mathbf{T}_{-i} - \mathbf{T}_{-ij}$ for the corresponding elements) gives the total effects of any variable on another that go through set j but not through set i.

The ability to calculate indirect effects transmitted through some variables while avoiding (holding constant) other variables has not been implemented in LISREL, so side calculations are required. The hand calculation of summing the products of the paths making up the indirect effect routings of interest obtains the same results, and it is likely to be simpler unless we have a particularly large model or wish to do many calculations of this type. Neither Fox (1985) nor Greene (1977) uses LISREL notation.

8.4 On Developing Equivalent Models (Effect Recomposition)

Another way to understand models involves recombining the basic effects in incremental steps, such that "equivalent" models with fewer variables but more complex coefficients are produced. We begin by choosing any two variables as the variables of interest. To investigate and understand the model-implied causal links between these variables, we pose the hypothetical problem of creating simpler models (i.e., models with fewer variables) that imply the same overall causal structure between the selected variables.

Consider the model in Figure 8.5A. Four models giving identical total effects between the focal variables X_1 on X_4 appear as Figure 8.5B, C, D, and E.   Figure 8.5B eliminates X_3 but maintains its modeled influence by recognizing its role in providing feedback to X_2 and in recognizing that the effect of X_2 on X_4 requires transmission through coefficients of value b_{32} and b_{43} (the usual multiplication of coefficients for indirect effects). Similarly, Figure 8.5C eliminates X_2 from the model, yet recognizes the role X_2 played in providing feedback to X_3 and transmitting any effects from X_1 to X_3. Figure 8.5D eliminates the loop in favor of a more complex coefficient, using the rule that a loop multiplicatively enhances effects by a factor $1/(1 - L)$. Finally, Figure 8.5E eliminates both X_2 and X_3 (again by the multiplication of coefficients providing an indirect effect) while maintaining our focus on the X_1–X_4 relationship. If the coefficients in the original model had been estimated, the complex coefficients in the restructured models could be obtained by direct substitution into the formulas in the figure.

Simplification proceeds by eliminating one variable at a time; different elimination series can produce different "partially reduced form" models. The basic rules for simplification are as follows. If a variable has no route by which it can effect either of the focal variables of interest and if the variable does not depend on the causally prior variable of interest, that variable can simply be eliminated from the model. If the variable to be eliminated provides only a loop, that variable is eliminated and replaced with a loop of the appropriate value. (Variable X_4 could have been eliminated if we were focusing on the X_1–X_3 relationship.) If a variable merely transmits basic indirect effects, it can be eliminated by using the usual coefficient multiplication to express indirect effects. (Several multiplications may be required if the variable is involved in several indirect routings.) If the variable to be eliminated functions as a part of a loop and as a mechanism transmitting basic indirect effects, both of these functions must be maintained in the simplified model (as in Figure 8.5B, C). It is often simplest to first eliminate the variables involved in the fewest relationships.

Although the preceding simplification strategies will not necessarily handle all the models we can create, they do cover most models, and they will not lead us astray, because more complex models will simply lead to situations not covered by the preceding rules of thumb and hence will lead us back to the original flow graph[9] sources from which these rules arise. These procedures are discussed in Heise (1975) and in Huggins and Entwisle (1968), where they are called "node absorption."

Figure 8.5   Model Simplification.[1]

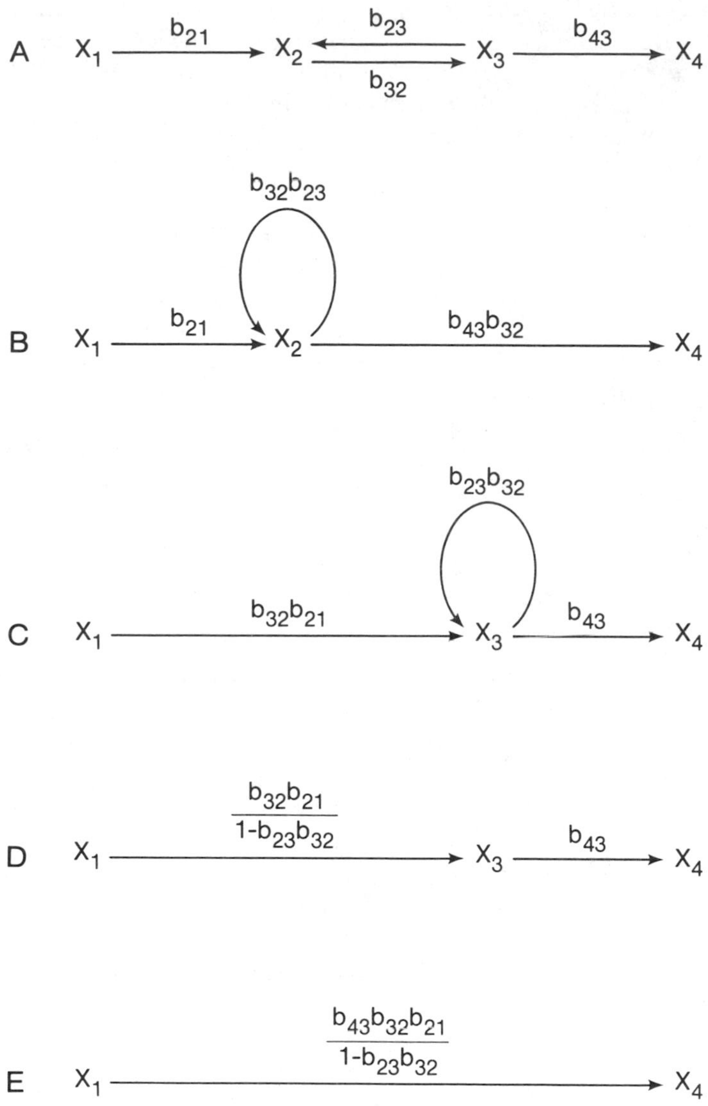

A $X_1 \xrightarrow{\quad b_{21} \quad} X_2 \underset{b_{32}}{\overset{b_{23}}{\rightleftarrows}} X_3 \xrightarrow{\quad b_{43} \quad} X_4$

B $X_1 \xrightarrow{\quad b_{21} \quad} X_2 \xrightarrow{\quad b_{43}b_{32} \quad} X_4$ (loop $b_{32}b_{23}$ on X_2)

C $X_1 \xrightarrow{\quad b_{32}b_{21} \quad} X_3 \xrightarrow{\quad b_{43} \quad} X_4$ (loop $b_{23}b_{32}$ on X_3)

D $X_1 \xrightarrow{\quad \frac{b_{32}b_{21}}{1-b_{23}b_{32}} \quad} X_3 \xrightarrow{\quad b_{43} \quad} X_4$

E $X_1 \xrightarrow{\quad \frac{b_{43}b_{32}b_{21}}{1-b_{23}b_{32}} \quad} X_4$

[1]Error terms have been omitted for diagrammatic simplicity.

We can always check the accuracy of our simplifications by verifying that the total effect implied by the simplest model equals the total effect obtained from the matrix procedures for the original model (as provided by LISREL). The coefficient in Figure 8.5E, for example, presents the same total effect of X_1 on X_4 as any of the other versions of this model. Though the total effects should be identical, slightly different time dynamics may be implied by the different models. Figure 8.5E, for example, might function by a single effect transmission, whereas 8.5B,C imply incremental effects.

The researcher trying to interpret model A now has several ways of describing and interpreting the basic model (models B–E). Note that some of these models avoid the need to talk about loops, and others let us choose precisely the loop we wish to talk about. Clearly, the availability of several equivalent models provides a flexibility in interpretation unimagined until the possibility of equivalent models is raised.

8.5 Interpreting the Smoking Model with an Inserted Loop

Figure 8.6 is a revised version of the smoking model in which a loop or reciprocal causal relation has been inserted between smoking (η_1) and antismoking views (η_2) by freeing β_{12}, and γ_{11} and γ_{24} have been fixed at zero to identify the reciprocal-effects coefficients. The resulting reciprocal-effects model fits acceptably $\chi^2 - 3.07$ with 2 $d.f.$ and $p = .22$. Figure 8.6 also contains some LISREL output (the matrices containing row/column headings) and some hand calculation of matrices that integrate the output with our preceding discussion of direct, indirect, and total effects.

Each coefficient estimate in **B** and **Γ** is a basic direct effect and can be interpreted as the amount of change in an η variable expected to follow a unit change in the causal variable (ξ or η) if no effects are transmitted through indirect routings or cycled though the loop (i.e., under the hypothetical condition that all other variables are forcibly held constant). Thus γ_{33} indicates that incrementing education by "one diploma" is expected to increase by .046 the number of AS bylaws with which one agrees (a trivial but significant increase), assuming no effects are transmitted through AS views (η_2) or smoking behavior (η_1).

The basic direct effect of smoking on AS acts (β_{31}) is $-.312$. Hence, converting a group of persons from nonsmokers to smokers leads us to expect that these persons on average would engage in .312 fewer AS acts than previously, again assuming the mechanisms providing indirect and

Figure 8.6   A Revised Smoking Model.

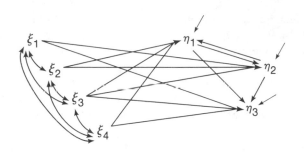

$$\eta = \beta\eta + \Gamma\xi + \zeta$$

Basic Direct Effects β and Γ
LISREL ESTIMATES (maximum likelihood)

| | BETA | | | | GAMMA | | | |
	SMOKING 1	AS-VIEWS	AS-ACTS		SEX	AGE	EDUCATION	ATTENDANCE
SMOKING 1	0.0	0.027	0.0	SMOKING 1	0.0	-0.004	-0.040	-0.039
AS-VIEWS	-1.258	0.0	0.0	AS-VIEWS	0.557	0.017	0.035	0.0
AS-ACTS	-0.312	0.067	0.0	AS-ACTS	0.162	0.0	0.046	-0.068

$$I_{\eta\eta} = \beta^2 + \beta^3 + \beta^4 + \cdots \beta^\infty = (I-\beta)^{-1} - I - \beta$$

$$= \begin{bmatrix} -0.034 & 0 & 0 \\ 0 & -0.034 & 0 \\ -0.084 & -0.008 & 0 \end{bmatrix} + \begin{bmatrix} 0 & -0.001 & 0 \\ 0.043 & 0 & 0 \\ 0.010 & -0.002 & 0 \end{bmatrix} + \cdots \cdots = \begin{bmatrix} -0.033 & -0.001 & 0 \\ 0.041 & -0.033 & 0 \\ -0.072 & -0.010 & 0 \end{bmatrix}$$

(by hand calculation)   (by subtracting direct
                                                                 from total effects)

$$T_{\eta\eta} = \beta^1 + \beta^2 + \beta^3 + \beta^4 + \cdots \beta^\infty = (I-\beta)^{-1} - I$$

	SMOKING 1	AS-VIEWS	AS ACTS
SMOKING 1	-0.033	0.026	0.0
AS-VIEWS	-1.217	-0.033	0.0
AS-ACTS	-0.384	0.057	0.0

Figure 8.6   Continued.

$$I_{\eta\xi} = (\beta^1 + \beta^2 + \beta^3 + \dots \beta^\infty)\Gamma = ((I - \beta)^{-1} - I)\Gamma = T_{\eta\xi} - \Gamma$$

$$= \begin{bmatrix} 0.015 & 0 & 0.003 & 0.001 \\ -0.018 & 0.004 & 0.047 & 0.048 \\ 0.032 & 0.003 & 0.017 & 0.015 \end{bmatrix}$$

(by subtracting direct from total effects)

$$T_{\eta\xi} = (\beta^0 + \beta^1 + \beta^2 + \beta^3 + \dots \beta^\infty)\Gamma = (I - \beta)^{-1}\Gamma$$

	SEX	AGE	EDUCATION	ATTENDANCE
= SMOKING 1	0.015	-0.004	-0.037	-0.038
AS-VIEWS	0.539	0.021	0.082	0.048
AS-ACTS	0.194	0.003	0.063	-0.053

$$\underline{T_{Y\xi} = \Lambda_Y T_{\eta\xi}}$$

since $\Lambda_Y - I$ for this model

$$T_{Y\xi} = T_{\eta\xi} = \text{as just above}$$

$$\underline{T_{Y\eta} = \Lambda_Y + \Lambda_Y T_{\eta\eta}}$$

since $\Lambda_Y = I$ for this model

$$T_{Y\eta} = I + T_{\eta\eta} = \begin{bmatrix} 0.967 & 0.026 & 0.0 \\ -1.217 & 0.967 & 0.0 \\ -0.384 & 0.057 & 1.0 \end{bmatrix}$$

looping effects have been rendered inoperative and, in addition, reminding ourselves of the possible enigmatic consequences of artifical interventions as discussed at the beginning of this chapter. The total effect of smoking status (η_1) on AS acts (η_3) involves this basic direct effect, a basic indirect effect (through η_2 views), and loop enhancements. The basic direct and indirect effects will be enhanced because both effect routings touch the loop between η_1 and η_2. The basic indirect effect of η_1 on η_3 is $\beta_{32}\beta_{21} = .067(-1.258) = -.084$. Hence, if the basic direct and indirect effects are allowed in the absence of looping (i.e., the mechanism providing the η_2 to η_1 effect is eliminated), a unit change in η_1 (smoking) on η_3 (acts) gives an effect that is the sum of the basic direct and basic indirect effects: $-.312 + (-.084) = -.396$.

If the loop is allowed to operate, it reduces the total expected effect below this value, because the loop's value is negative. Specifically, the product of the coefficients in the loop is $(-1.258)(.027) = -.0339$, so each cycle through the loop would return an effect that is $-.0339$ of the effect that entered. The total effect of η_1 on η_3 is the sum of the enhanced direct effect and the enhanced indirect effect [with the loop enhancement being $1/(1 - L) = 1/(1 + .0339)) = .967$, as discussed in Section 8.1.2]:

$$\text{total effect} = -.312\left(\frac{1}{1 + .0339}\right) + -.084\left(\frac{1}{1 + .0339}\right) \qquad 8.28$$

$$= -.383$$

Note that this is slightly less than the value obtained by summing the basic direct and indirect effects (with the loop rendered inoperative) and that it is within rounding error of the $-.384$ LISREL reports as the total effect of η_1 on η_3 from its matrix calculations. Thus, on average, smokers would be expected to take .383 fewer AS actions than nonsmokers.[10] This is substantial, given that the respondents overall reported having only taken .605 (of eight possible) AS actions (Figure 4.5).

If you are unsure about the preceding calculations, verify that the total effect of η_2 on η_3 as reported by LISREL can be obtained as the sum of the enhanced direct effect of η_2 on η_3 plus the enhanced indirect effect routed from η_2 to η_1 to η_3.

The diagonal of the matrix containing the total effects of the η's on the η's contains two values of $-.033$: one refers to the effect of η_1 on itself, and the other refers to the effect of η_2 on itself. Variable η_1 can influence itself through η_2 with a basic indirect effect of $\beta_{21}\beta_{12} = -1.258(.027)$, which when enhanced is $-1.258(.027)(.967) = -.033$. Convince yourself that the effect of η_2 on itself gives the product of the same numbers except that the order of the first two numbers is reversed.

The effect of education (ξ_3) on AS acts (η_3) is instructive because even though there are only three effects originating in education, there are five routings by which ξ_3 can influence η_3, and all five routings must be summed to obtain the total effect of ξ_3 on η_3. The routings are (1) the basic direct effect of ξ_3 on η_3 (which receives no enhancing), (2) the routing from ξ_3 to η_1 to η_3 (which will be enhanced), (3) the routing from ξ_3 to η_2 to η_3 (enhanced), (4) the routing from ξ_3 to η_2 to η_1 to η_3 (enhanced), and (5) the routing from ξ_3 to η_1 to η_2 to η_3 (enhanced). Summing these enhanced effects and the unenhanced direct effect gives the total effect .063.

Though γ_{11} was fixed at zero, sex does have an effect on smoking in this model—an indirect effect through η_2 that is enhanced. Specifically, $\gamma_{21}\beta_{12}(1/(1 - L)) = .557(.027).967 = .015$, the total effect reported for ξ_1 on η_1.

In concluding our discussion of the reciprocal-effects smoking model, we point out two problems with the model. First, the largest eigenvalue of $\mathbf{BB'}$, the stability index,[7] is 1.68. Hence, there is some vector of scores for which $\mathbf{B}$ could act to create an "explosive" feedback system. The second disquieting symptom is the $-.973$ correlation between the estimates of β_{12} and β_{21}, which indicates a strong colinearity between them. Though the fixing of two Γ values to zero, mentioned at the beginning of this section, was intended to create instrumental variables that would identify these reciprocal-effect coefficients, it seems that this was successful only to the point of avoiding severe estimation problems and totally unreasonable estimates (the β_{21} estimate of -1.258 does not differ markedly from the $-.879$ estimate for the basic model in Appendix B). Despite the larger magnitude of the β_{21} estimate, it has dropped below significance because the standard error for this coefficient is about four times as large as previously (.83 versus .19), which is another sign of colinearity problems.

These observations reduce our belief in the numerical estimates in the model and particularly in the estimates of the reciprocal effects. Since the loop provides enhancements to many of the total effects, these total effects are also dubious. Although we could have "improved" this model before using it as an example, we decided to use it to illustrate the point that *even unacceptable models can be interpreted*. There was nothing in this model that interfered with the mechanics of interpretation. Providing a clear and precise interpretation does not guarantee that a model is good or even acceptable.

8.6 Summary

Two concepts grounded our discussion of model interpretation. The first involves *routinely thinking of all the variables in a model (except the changed variable) as untouched yet free*. The notion that all variables in a model are free to respond to a hypothetical yet interpretively useful unit change is implicit in the calculation of indirect effects as the product of the effect coefficients constituting the indirect routing, and hence it underlies the calculation of total effects in recursive models as the sum of the direct and indirect effects. The notion of untouched yet free variables is even more fundamental to the calculation and interpretation of total effects in models containing loops, because of its implicit use when repeated cyclings through loops are considered.

The second conceptual foundation is that *loops multiplicatively enhance any basic direct or indirect effects they touch*. The enhancements may imply total effects that are larger or smaller than the basic effects, depending on whether the product of the coefficients constituting the loop is positive or negative. If the product of the coefficients constituting a loop is L, that loop will enhance any effects transmitted along paths touching that loop by a factor of $1/(1 - L)$. *Loops do not add in new effects but merely multiplicatively enlarge or shrink effects transmitted along direct or indirect paths.* Loops should be interpreted in conjunction with basic direct or indirect effects, and they should always be considered as multiplicative enhancements to those effects. In practice, a loop will not have much effect unless L is larger than about .1, because $1/(1 - L)$ is very near 1 for smaller values of L.

Total effect for all the pairs of variables in a model can be obtained with matrix algebra if the matrices of structural coefficients (the basic direct effects) are known (recall Section 8.2). Calculating the difference between a total effect and a basic direct effect provides an indirect effect that is composed of basic indirect effect routings and loop enhancements of both the direct and indirect routings (again recall Section 8.2).

These interpretation strategies *may be used no matter what estimation strategy has been used to obtain the estimates of the structural coefficients* (maximum likelihood, generalized least squares, ordinary least squares, etc.), and they are *appropriate for standardized and unstandardized coefficients*.[11]

Notes

1.   If simultaneous equation models are to be thought of as "causal models," with or without loops, this requires a coordination that precludes a variable from acting as a cause and as an effect at the same time. The problem with a simultaneous cause and effect is that we are unable to specify what the causal value of the variable is, because that value was changing at the time the variable was supposed to be acting as a cause. This problem is minimized if we think of causes as beginning instantaneously, even if they require some time increment before the effects arrive at the dependent variable. Fisher (1970) discusses simultaneous equation models as approximations to real systems with time lags that approach zero. See note 7 for the statistical condition that must be met for a simultaneous equation model to be a reasonable representation of the limiting case of a nonsimultaneous real world. See also Lewis-Beck and Mohr (1976), Heise (1975), and Huggins and Entwisle (1968).

Note specifically that although the time variable does not appear in structural equation models, it is implied if we are modeling a suspected causal world. This contradictory presence/absence of time arises in the movement from correlation to regression. If we decide to regress Y on X (rather than X on Y) because this meshes with the suspected causal sequence of X causing Y, this invokes time as a component of the real world problem founding the regression even though time per se is not in the regression equation. The statistician's insistence that time is not in the structural equations, as contrasted with the researcher's insistence that a suspected causal (time) order dictated the selection of this particular regression/structural equation, has been the fount of innumerable frictions. Those interested in pursuing the question of time and causality might begin by considering that causality was first introduced in Section 1.3 when we "*created* new Y scores from original X scores," and then reviewing Davis (1985), James, Mulaik, and Brett (1982), and Heise (1975: Chapter 6). Note also that indirect effects rest on a stronger philosophical foundation than do direct effects because indirect effects act through other modeled/measured variables whereas direct effects act through unmodeled/unmeasured, and hence mysterious, forces.

2.   See Heise (1975:24) for a discussion of why linear structural equations describing the values of variables can be used to represent coordinated changes between those variables.

3.   Beware of the errors in Fox (1980) as reported in the errata appearing in the next journal volume (p. 119). Additional errors appear in the middle portion of the second equation listed under his number 14; and the eighth summary equation under his number 31 should be deleted. Some of these errors are corrected in the reprint of this article as Pp. 178-202 in Peter V. Marsden (ed.). 1981. *Linear Models in Social Research.* Beverly Hills: Sage.

4.   The matrix power series may be considered a series of pre- or postmultiplications because $B^3 = BBB = (BB)B = B(BB)$ (cf. Greene, 1977).

5.   Note that the diagonal elements of $I_{\eta\eta}$ in the powers of B are unitless, whereas the off-diagonal elements are associated with the specific units of the variables specifying the rows and columns of the matrix. Specifically, the implicit units are the number of *units* of change in the dependent (row) variable *per unit* change in the independent (column)

variable. For the diagonal elements the units of the dependent and independent variables are identical because they are for the same variable. This implies that the diagonals of these matrices would remain the same no matter what scale of measurement we select, whereas the off-diagonal elements would change if the units of measurement were changed. The diagonals can be thought of as the proportion of a change that returns to modify the changed variable. In electrical engineering this concept is called a return effect (Huggins and Entwisle, 1968), and a value greater than 1.0 indicates an unstable system, because this implies the subsequent change is greater than the perturbation inserted into the variable.

6.   Sobel (1982) gives a procedure for obtaining the standard errors of indirect effects. His procedure works as long as the symbolic representations of the indirect effects are differentiable (which seems attainable for *stable* models containing loops). He illustrates his procedure with a recursive model. Sobel (1986) extends the procedure to indirect effects in nonrecursive models.

7.   The formal condition for the stability of the model and hence for the veracity of replacing the infinite sum with an inverse is that the modulus (or length) of all the eigenvalues of B (which are potentially complex numbers containing real and imaginary components) lies within a unit circle in a plane defined by the real and imaginary axes (Arminger, 1987; Protter and Morrey, 1964:371; Varga, 1962:9,82; Wolkowicz and Styan, 1980:472). This matrix algebra is equivalent to the scalar restriction that $|L| < 1.0$ if the geometric series in L is to reduce to $1/(1 - L)$. A sufficient (but not necessary) condition for satisfying the "unit circle" criterion and, hence, the stability of the model is that the largest eigenvalue of $BB' < 1$ (Bentler and Freeman, 1983). The LISREL V manual (p. III.93) erroneously gives BB' as "an equivalent" condition, but the LISREL VI manual (p. III.93) corrects this to read "a sufficient" condition.

The condition that the eigenvalues are of modulus less than 1.0 also satisfies Fisher's (1970) condition for using simultaneous equation systems as approximations of nonsimultaneous systems with time lags that approach zero—that is, where simultaneous equations are used to model causal forces that are not instantaneous but merely act "relatively quickly."

8.   The remaining two formulas in the table for effects on Y (LISREL V and VI, p. III-39) are mislabeled. $\Lambda_y \Gamma$ is the "indirect effect of any given ξ through a single η" (i.e., paths of length 2), and $\Lambda_y (I - B)^{-1} \Gamma - \Lambda_y \Gamma$ furnishes the indirect effects through paths involving two or more η's (paths of length at least 3).

9.   Flow graph theory began with Samuel Mason, an electrical engineer, who developed a general procedure for calculating total effects in models allowing feedback of the earsplitting variety. Mason (1953, 1956) developed a gain formula for total effects that was imbedded in flow graph theory, the electrical engineering equivalent of path models (see also Mason and Zimmerman, 1960; Lorens, 1964). The theory and the formula were introduced into the social sciences by Doris Entwisle (a sociologist/mathematician) after she coauthored a text on electrical engineering (Huggins and Entwisle, 1968). David Heise (1975) popularized flow graph theory and Mason's formula (renaming it a "total effect" formula), and the tradition lives on (cf. Chen, 1983). Mason's formula provides a procedure for hand calculating the total effect between variables in complex models with or without loops, and it duplicates the intent of the matrix procedures we have discussed. We recommend Huggins and Entwisle (1968: Chap. 2) and Heise (1975), but beware of an important error Heise (1975) makes in defining when paths and loops touch. He claimed that paths and loops touch only if they have any variable in common "except the first" variable in a path (Heise, 1975:61). He thus claims that the loop in Figure 8.3B does not touch the path, and he would express the total effect of X_1 on X_2 as b_{21} rather than $b_{21}/(1 - L)$. A clear conception of Heise's error and its perpetuation by Chen (1983:11) is obtained by contrasting Heise (1975:61) with Huggins and Entwisle (1968:87, 96, 132, 136).

10. Though the numerical aspects of calculating the loop effects are straightforward, the artificiality of the loop in Figure 8.6 remains conceptually unwieldy. It is easy to see that personal views about smoking can influence smoking behavior and that smoking behavior can influence views about smoking, but the incremental effects implied by repeated cyclings through this particular loop seem less intuitive because of the 0–1 scaling of the smoking variable. The repeated cyclings involve scale changes of less than 1 full unit and hence scores indicating something other than smokers or nonsmokers. (A similar concern arises for η_2, which was created to have only unit incremental values.) The shift to providing average group descriptions should be seen as a direct attempt to salvage some meaning for these "less than unit changes."

11. For example, the supposedly acceptable "final" standardized model in Lomax (1982) seems considerably less acceptable when we note that it implies a total effect of η_1 on η_2 that exceeds 1.0 when the loop enhancement is considered.

Chapter 9

More and Better

This chapter extends the basic LISREL model in two ways. First, it presents the minor changes required to model two or more groups simultaneously (Section 9.1). Second it introduces models containing means and intercepts, including estimating mean differences between groups and predicting an endogenous concept's value for particular values of the predictor concepts (Section 9.2). Models with means and intercepts require fundamental changes in how we think about the logical implications of structural equation models, the data used to test these models, and even in the formal specification of the general LISREL model itself.

9.1 Stacked Models for Multiple Groups

Researchers often confront data that is most conveniently discussed by splitting it into groups. The grouping may reflect different data sources (different cities, countries, or organizations), different time periods, different experimental conditions, or groupings created from the variables available within a data set (e.g., grouping on sex, religion, or age). As long as group membership is clearly defined (i.e., the groups are mutually exclusive) and as long as the data for each group constitute a random sample from its respective population, a stacked LISREL analysis is appropriate. It is permissible to model husbands and *their* wives, or parents and *their* children. All one needs is an input covariance matrix for the two groups.

Though each such group might be modeled with a separate LISREL run, two or more groups may be stacked together for simultaneous estimation. Stacking groups together *allows some of the effect coefficients to be constrained to be equal between the groups* while other coefficients vary between the groups. Since many variables may display differential effects in the different groups, stacking permits estimation of models containing multiple interactions between the variable providing the grouping and the other variables in the model (cf. Marsden, 1983).

Entering constraints between stacked groups (reflecting the ways the groups are thought to behave similarly) provides fewer coefficients to estimate and hence extra degrees of freedom, which may be used for other purposes. Lomax (1983), for example, uses between-group constraints to provide sufficient degrees of freedom to estimate the mean differences between groups. Entwisle and Hayduk (1982) use stacked models to test whether the estimates from models *developed* on one random half of a data set differ significantly from the estimates obtained when these same models are *tested* on the other "virgin" half of the data—that is, as a test of whether there was substantial "capitalizing on chance" during model development.

Further examples of stacked models are in Alwin and Jackson (1981), Hoelter (1983), Joreskog and Sorbom (1980), Sorbom and Joreskog (1981, 1982), and Werts et al. (1976, 1977). The implications of stacking are clearest in the context of real models, so we introduce the few necessary pieces of statistical theory in the context of a stacked version of the smoking model.

9.1.1 A Stacked Smoking Model

One disturbing aspect of the smoking model introduced in Chapter 4 (Figures 4.1 and 4.3) is the failure of education to influence either AS views or AS acts. One explanation for this failure is that age interacts with education to provide differential patterns of effects for older and younger persons. Differential educational effects might arise because the older respondents gained their education at a time when there were no antismoking health classes in the schools and when a high school education was considered above average. Or differences in effect might be due to the differental lags from the receipt of an education to the present. The younger respondents might recall the content of health or other class materials, and older respondents should be experiencing the long-term status discrepancies created by educational differences. Substantial effects of education on AS views and AS acts for one group (the young) might combine with null effects for the other (the old) to provide the observed minimal effects of education.

Further consideration suggests that age might also interact in several other modeled relationships. Specifically, the failure of sex to influence AS acts might be age dependent. Of the four age/sex combinations, young females should be the most likely to take action against smokers. They have the youthfulness to act, an assertiveness nurtured by the feminist movement, and the protection against retaliation offered by their female status.

The effects of smoking on AS views and acts might also be age dependent. Having developed their habits of smoking/nonsmoking long before the recent wave of AS activism, the older respondents should feel little pressure to make their views conform to their entrenched habits. Younger persons, on the other hand, should be more likely to adopt views consistent with their smoking/nonsmoking status because they cannot plead that it is an ingrained habit that has long since gone beyond their control.

Although one or two interactions with age might be modeled with the strategies discussed in Section 7.5, the possibility of age conditioning several effects suggests that we create separate models for the younger and older cases in our sample. But building two completely separate models would not allow for the fact that there are other effects that we expect should be the same in the two groups. For example, we would expect the same effect of sex on smoking and on AS views for the younger and older groups and simultaneously allow differential effects of sex on AS acts. The solution is to *stack the younger and older models together in a single LISREL run* so that we can enter the desired constraints between the groups and then estimate the models simultaneously with those constraints operative (see Figure 9.1).

To prepare the program input, we first split the overall sample into a group of respondents 33 years old or younger (N_1 = 213) and a group of respondents 34 years old or older (N_2 = 219). Next, we created for each group a covariance matrix containing all the original indicators. These covariance matrices and the means and variances of the indicators are given in Figure 9.2.

Scanning these means, we see that, in comparison to the older group, the younger group has a few more smokers, agrees with fewer AS bylaw provisions, engages in slightly fewer AS acts, attends church less often, and has minimally higher education. Both groups are nearly evenly split on sex; the younger group averages about 26 years of age; the older, 51 years of age. The younger group is much more homogeneous in terms of age than is the older group: the variance of ages for the younger group is only about one tenth the variance for the older group. The variance in AS views for the younger group is about double that for the older group, a clear sign that different causal structures might be operative in the two groups.

The same researcher-determined proportions of error variance presented in Figure 4.5 were then used to calculate and fix the corresponding Θ_ϵ and Θ_δ values for each group, just as if we were preparing

Figure 9.1   The Stacked Smoking Model.[1]

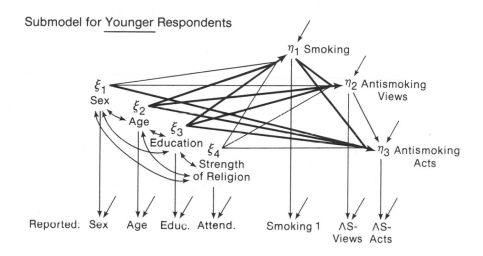

Submodel for Younger Respondents

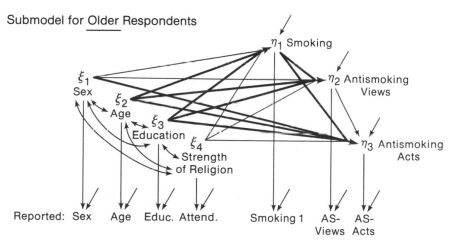

Submodel for Older Respondents

[1]Coefficients allowed to differ between the younger and older groups are indicated by heavy arrows. The effect of age (ξ_2) on antismoking acts (η_3) was excluded from the original smoking model in Chapter 4, but is included here since splitting the respondents into two age groups invalidates the original justification for omitting this coefficient.

Figure 9.2   Means, Variances, and Covariances for the Younger and Older Groups.

Variable	Mean Younger	Mean Older	Variance Younger	Variance Older
Smoking-1	0.479	0.411	0.251	0.243
AS-views	6.225	7.009	4.088	2.273
AS-acts	0.662	0.550	1.359	1.245
Sex	1.469	1.511	0.250	0.251
Age	25.99	50.56	15.68	173.51
Education	8.437	7.932	5.030	8.642
R-Attendance	2.272	3.196	4.171	4.727

Variables required for a later stacked run

Smoking-4	1.854	1.648	1.974	2.046
R-Importance	3.944	4.822	4.723	3.853

Correlation/Covariance

	Smoking 1	Smoking 4	AS-Views	AS-acts	Sex	Age	Educ.	Religion Attend	Religion Import
Smoking-1	1.0	.904	-.328	-.271	-.036	-.076	-.271	-.160	-.023
Current	—	.636	-.332	-.158	-.009	-.151	-.304	-.164	-.025
Smoking-4	.895	1.0	-.394	-.245	-.050	-.105	-.312	-.215	-.058
Physio-strain	.631	—	-1.118	-.401	-.035	-.586	-.983	-.616	-.178
AS-views	-.153	-.142	1.0	.224	.191	.014	.150	.082	.042
	-.114	-.306	—	.527	.193	.113	.682	.339	.183
AS-acts	-.029	-.015	.063	1.0	.156	.059	.217	-.075	-.101
	-.016	-.024	.106	—	.091	.270	.568	-.179	-.255
Sex	-.038	-.062	.130	.003	1.0	.021	.048	.105	.050
	-.009	-.044	.099	.002	—	.042	.053	.107	.055
Age	-.126	-.109	-.003	-.109	-.034	1.0	.260	.067	.099
	-.820	-2.058	-.065	-1.602	-.222	—	2.311	.540	.849
Education	-.098	-.098	.062	.038	-.007	-.269	1.0	.096	.033
	-.141	-.414	.274	.125	-.011	-10.433	—	.442	.162
Religion	-.170	-.162	.123	-.032	.139	.088	.041	1.0	.606
Attendance	-.182	-.504	.404	-.079	.151	2.533	.261	—	2.690
Religion	-.099	-.096	.170	.073	.228	.235	-.069	.647	1.0
Importance	.096	-.269	.502	.160	.225	6.089	-.398	2.760	—

[1]Younger respondents (N=213) provided the upper-right correlation/covariance matrix while the older respondents (N=219) provided the lower-left matrix.

to do a separate LISREL run for each group.

Next, we determined which of the **B** and **Γ** coefficients were to be constrained to be equal in the two groups and which were to be free to differ between the groups, as determined by the possible interactions noted earlier. Specifically, the younger and older groups were allowed to have different estimates for the effects of age on the three endogenous concepts $(\gamma_{12}, \gamma_{22}, \gamma_{32})$, the effects of education on the three endogenous concepts $(\gamma_{13}, \gamma_{23}, \gamma_{33})$, the effects of sex on AS acts (γ_{31}), and the effects of smoking on both AS views and AS acts (β_{21}, β_{31}). The remaining six coefficients in **B** and **Γ** were constrained to be equal across groups, because we found no reason to believe that these effects should differ, depending on the respondent's age grouping.

The program for a stacked model is basically composed of two separate single-group programs stacked one behind the other, but LISREL is cued to treat these as a single model by a statement that these groups are to be estimated simultaneously. Coefficients specified as fixed or free, with no requirement of constraints between groups, are simply entered for each group using the procedures discussed previously (see Figure 9.3). Coefficients that are free, but constrained to receive the same estimate in the two groups, are specified as free for each groups and then recorded on an EQuality statement, such as those appearing near the end of the program in Figure 9.3. The constrained coefficients are designated by their usual row/column indexing, but an additional indexing value precedes the usual designation if a coefficient in another group is being referred to. Thus the statement EQ BE(1,3,2) BE(3,2) constrains the first (younger) group's estimate of BE(3,2) to equal the estimate of BE(3,2) in the current (older or second) group, within whose confines the EQ statement appears in the program. (See Chapter 5 in the LISREL V or VI manuals.)

When the program in Figure 9.3 is run, the younger and older models are simultaneously estimated with some of the previously specified **B** and **Γ** coefficients constrained to take on the same values in the two groups while other **B** and **Γ** coefficients are allowed different estimates in the two groups. In arriving at the maximum likelihood estimates, LISREL attempts to maximize the fit between two **S** matrices and two **Σ** matrices: one observed covariance matrix and one model-implied covariance matrix for each group. To obtain maximum likelihood estimates, LISREL begins with an initial set of estimates for all the free coefficients in the whole model (both groups) that are created in accordance with the constraint that some of the **B** and **Γ** coefficients are equal in the two groups. LISREL then calculates the model-implied covariance matrix (**Σ**) for the first group and compares this **Σ** to the **S** for that group using the maximum likelihood fit function developed as Eq. 5.9. LISREL then calculates the model-implied **Σ** for the second group and compares this to the **S** for that

Figure 9.3   LISREL Program for the Stacked Model with Some $\mathbf{B}$ and Γ Constraints between the Groups.

```
title 'SMOKSTACK.1 a STACKED smoking model in SPSSX'
file handle 8/name='smokcov.youn'
file handle 9/name='smokcov.old'
INPUT PROGRAM
NUMERIC A
END FILE
END INPUT PROGRAM
USERPROC NAME=LISREL
SMOKING-SEX AGE ED RELIG,SMOKSTACK.1 YOUNG1st OLDER2nd THEORY ADDED
DA NI=13 NO=213 MA=CM   NG=2
CM UN=8 FU FO
(8F10.7)
LA
'smoking1Y' 'smoking2Y' 'smoking3Y' 'smoking4Y' 'asviewsY' 'asactsY'
'sexY' 'ageY' 'educationY' 'GodexistsY'
'strongXY' 'attendY' 'importantY'
SE
'smoking1Y' 'asviewsY' 'asactsY'
'sexY' 'ageY' 'educationY' 'attendY'/
MO NY=3 NX=4 NE=3 NK=4 LY=ID LX=ID BE=FU,FI GA=FU,FR C
 PH=FU,FR PS=DI,FR TE=DI,FI TD=DI,FI
FR BE(2,1) BE(3,1) BE(3,2)
VA .0251  TE(1,1)
VA .2044 TE(2,2)
VA .0678 TE(3,3)
VA .0025 TD(1,1)
VA .7840 TD(2,2)
VA .5030 TD(3,3)
VA .4171 TD(4,4)
OU ML AL TM=20
SMOKING - SEX AGE ED RELIG,SMOKSTACK.1 OLDER "half" of sample
DA NI=13 NO=219 MA=CM
CM UN=9 FU FO
(8F10.7)
LA
'smoking1O' 'smoking2O' 'smoking3O' 'smoking4O' 'asviewsO' 'asactsO'
'sexO' 'ageO' 'educationO' 'GodexistsO'
'strongXO' 'attendO' 'importantO'
SE
'smoking1O' 'asviewsO' 'asactsO'
'sexO' 'ageO' 'educationO' 'attendO'/
MO NY=3 NX=4 NE=3 NK=4 LY=ID LX=ID BE=FU,FI GA=FU,FR C
 PH=FU,FR PS=DI,FR TE=DI,FI TD=DI,FI
FR BE(2,1) BE(3,1) BE(3,2)
EQ BE(1,3,2) BE(3,2)
EQ GA(1,1,1) GA(1,1)
EQ GA(1,1,4) GA(1,4)
EQ GA(1,2,1) GA(2,1)
EQ GA(1,2,4) GA(2,4)
EQ GA(1,3,4) GA(3,4)
VA  0243  TE(1,1)
VA .1137 TE(2,2)
VA  0623 TE(3,3)
VA .00251 TD(1,1)
VA 8.676 TD(2,2)
VA .8642 TD(3,3)
VA .4727 TD(4,4)
OU ML AL TM=20
end user
```

group, again using the fit function from Eq. 5.9.

The fit of the stacked model is calculated as a weighted average of the fits attained for each of the groups individually. Parallel to Eq. 5.9, we can represent the fit for any one group (the gth group) as

$$F_g = \text{tr}(S_g \Sigma_g^{-1}) + \log|\Sigma_g| - \log|S_g| - (p + q) \qquad 9.1$$

The fit of the overall stacked model is the weighted average of the fits achieved for each group separately, namely

$$F = \sum_{\substack{\text{over all} \\ \text{groups}}} \left(\frac{N_g}{N}\right) F_g \qquad 9.2$$

(cf. LISREL V or VI, Eq. V-1).

LISREL then tries to improve the initial estimates by raising some estimates and lowering others [as dictated by the partial derivatives of the fit function with respect to the free parameters (Section 6.4)]. *All the free parameters are provided new values, but LISREL must enter exactly the same new B and Γ estimates for the coefficients constrained to be equal in the two groups.* LISREL then recalculates the Σ for each group separately, compares the Σ and S for each group, and develops a revised estimate of the overall fit on the basis of the weighted average of the fits in Eq. 9.2.

This process is repeated (with the equality constraints respected at each iteration) until LISREL locates the set of estimates providing the best fit between the Σ's and S's the estimates maximizing the (weighted) likelihood that the S's could have appeared as mere sampling fluctuations if the Σ's were the population covariance matrices.

The closeness of the match between the stacked model-implied Σ's and the observed S's can be judged using the likelihood ratio χ^2 in Eq. 6.2. Once the maximum of the fit function (Eqs. 9.1, 9.2) for the stacked model is known, a stacked model χ^2 can be obtained by multiplying this maximum by the overall number of cases. Multiplying by N cancels with the N in the denominator of Eq. 9.2 and leaves a series of terms to be summed across the groups. Each term is of the form $N_g F_g$, which is one of the forms for a χ^2 variate in Eq. 6.2. Summing these "independent" χ^2's (i.e., $N_g F_g$ terms) gives a quantity that is also a χ^2 variate, because the sum of independent χ^2's is also a χ^2. Note the requirement for independent random samples to provide the independence between the χ^2's, and recall the discussion in Section 6.1.2 and its footnote on why sums of χ^2's are also χ^2's (or see McGaw and Joreskog, 1971:161).

The degrees of freedom for the overall stacked model χ^2 are, as before, the total number of pieces of known input information (variances and covariances) minus the number of unknown parameters being estimated (t). Since the total number of input variances/covariances is the number of variances/covariances in any group ($\frac{1}{2}(p + q)(p + q + 1)$)) times the number of groups, the general formula for the degrees of freedom in a stacked model is

$$d.f. = \text{(number of groups)}(\tfrac{1}{2})(p + q)(p + q + 1) - t \qquad 9.3$$

For our smoking model,

$$d.f. = 2(\tfrac{1}{2})(7)(8) - 50 = 6 \qquad 9.4$$

The 50 coefficients to be estimated arise from the 6 coefficients constrained to be equal in the two groups, plus 22 unconstrained coefficients in each of the younger and older submodels (2 β's, 7 γ's, 3 Ψ's, and 10 Φ's in each). The χ^2 for the stacked smoking model is 3.35 with the anticipated 6 $d.f.$, and this has a very acceptable probability of .76.

We highlight only a few implications of the maximum likelihood estimates given in Figure 9.4. First, there is confirmation that higher education decreases smoking and increases AS acts among the younger respondents, whereas education level is much less strongly related to the smoking and AS acts of the older respondents. This is precisely the pattern we postulated as accounting for the unexpectedly modest effects of education in the original model. There is no obvious reason why AS views do not respond to education among the younger respondents.

The prediction that the younger respondents should display a greater consistency between their smoking behavior and AS views and AS acts is also strongly confirmed. The effects (β_{21} and β_{31}) for the young are about three times as strong as for the older respondents. The only significant effect of sex on AS actions (γ_{31}) is for the young, and the positive sign of the relationship indicates it is the young females who are likely to engage in AS acts, as predicted.

Though no specific predictions were made about the basis for potential differences in age effects between the submodels, the absence of effects for the younger group may reflect the restriction on the range of variation in age for the younger group. Recall that the variance in age for the younger group was only about one tenth the variance in age among the older group. Indeed, you might question whether age should even be entered in the stacked model, given that we created the two submodels by dichotomizing the age variable. The reasonableness of including age is demonstrated by the significant negative effect of age on smoking for the older group. This informs us that the respondents in their late thirties and early forties are more likely to smoke than the oldest respondents in the survey. Clearly, *dichotomizing a variable to create two stacked groups does not guarantee that all the effects of that variable have been accounted for*. Indeed, the

Figure 9.4 Estimates for the Stacked Model Allowing Differences between the Younger and Older Groups.

YOUNG

LISREL ESTIMATES (MAXIMUM LIKELIHOOD)

BETA

	smoking1	asviewsY	asactsY
smoking1	0.0	0.0	0.0
asviewsY	-1.356 *	0.0	0.0
asactsY	-0.527 *	0.061 †	0.0

GAMMA

	sexY	ageY	educatio	attendY
smoking1	-0.015	0.001	-0.064 *	-0.036 *
asviewsY	0.485 *	-0.018	0.059	0.042
asactsY	0.307 *	0.001	0.082 *	-0.058 *

PHI

	sexY	ageY	educatio	attendY
sexY	0.243			
ageY	0.042	14.895		
educatio	0.054	2.311	4.527	
attendY	0.107	0.541	0.435	3.758

PSI

smoking1	asviewsY	asactsY
0.200	3.265	1.083

THETA EPS

smoking1	asviewsY	asactsY
0.025	0.204	0.068

THETA DELTA

sexY	ageY	educatio	attendY
0.002	0.784	0.503	0.417

SQUARED MULTIPLE CORRELATIONS FOR STRUCTURAL EQUATIONS

smoking1	asviewsY	asactsY
0.113	0.151	0.149

OLDER

LISREL ESTIMATES (MAXIMUM LIKELIHOOD)

BETA

	smoking1	asviewsO	asactsO
smoking1	0.0	0.0	0.0
asviewsO	-0.453 †	0.0	0.0
asactsO	-0.126	0.061 †	0.0

GAMMA

	sexO	ageO	educatio	attendO
smoking1	-0.015	-0.006 *	-0.025 †	-0.036 *
asviewsO	0.485 *	-0.001	0.025	0.042
asactsO	0.005	-0.009	0.001	-0.058 *

PHI

	sexO	ageO	educatio	attendO
sexO	0.248			
ageO	-0.222	164.836		
educatio	-0.011	-10.433	7.778	
attendO	0.151	2.525	0.262	4.250

PSI

smoking1	asviewsO	asactsO
0.203	2.049	1.162

THETA EPS

smoking1	asviewsO	asactsO
0.024	0.114	0.062

THETA DELTA

sexO	ageO	educatio	attendO
0.003	8.676	0.864	0.473

SQUARED MULTIPLE CORRELATIONS FOR STRUCTURAL EQUATIONS

smoking1	asviewsO	asactsO
0.069	0.062	0.032

CHI-SQUARE WITH   6 DEGREES OF FREEDOM IS   3.35 (PROB. LEVEL = 0.764)

† T>1.5   * T>2.0

zero slope for the young and the negative slope for the older respondents imply that there is a curvilinear relationship between age and smoking.

To test the significance of the interaction implied by a difference in the estimates of any particular coefficient (or set of coefficients) in the younger and older portions of the model, we can use the difference χ^2 procedure (Section 6.1.2). We first estimate a stacked model with the coefficients constrained to be equal between the groups, and we then reestimate the model with the coefficients unconstrained. The difference between the χ^2's and $d.f.$'s is a test of whether the freeing of the coefficients gave a significant improvement in fit. For example, we can test for the combined need for all the nine postulated interactions by forcing all the $\mathbf{B}$ and $\mathbf{\Gamma}$ coefficients to be equal between the groups. This gives a model with $\chi^2 = 23.88$ with 15 $d.f.$[1] Since the original model allowing the nine interactions (nine $\mathbf{B}$ and $\mathbf{\Gamma}$ coefficients differing between the groups) had a χ^2 of 3.35 with 6 $d.f.$, then the difference χ^2 is 20.53 with 9 $d.f.$ and a probability .02. Thus we can claim that the data would have been significantly less well fit had we not allowed the nine coefficient estimates to differ between the groups (the nine interactions).

We end our discussion of the stacked model by noting that in comparison to the basic smoking model in which the young and old respondents were bunched together, we have minimally increased the amount of explained variance for the younger group and decreased the explained variance for the older group. The original model explained 8.2%, 13.1%, and 6.5% of the variance in the three endogenous concepts, respectively, and the corresponding values for the young group are 11.3%, 15.1%, and 14.9%; for the old group, 6.9%, 6.2%, and 3.2%. These values reflect the fact that the younger submodel displays more and stronger effects than the older submodel. These proportions of explained variance come from the standardized $\mathbf{\Psi}$ for the stacked smoking model, and you should remind yourself of the potentiality for incorrect standardizing in stacked models as discussed in Section 6.6.1.2.

9.2 Modeling Means

The LISREL model discussed in the preceding chapters assumes that the model's indicators and concepts have zero means. Analyzing the covariance matrix and its implicit deviations from the mean guarantees that the input data comply with this assumption, and the zero indicator means in combination with the structural equations guarantee the concepts also comply with this assumption, which will become clear later. Hence, though these assumptions are implicit in the formal model, they are never problematic because they are never violated.

The difficulty with the assumption of zero indicator and concept means resides in the constraints this places on the research questions that can be addressed by LISREL. We simply are unable to estimate any means, be it the mean of an experimental or control group, or a group receiving a special social program, or even the mean of some conceptual "true" scores. For instance, entering a dummy exogenous variable to capture the effect of an experimental manipulation provides a Γ "slope" that can be interpreted as a mean difference between the experimental and control groups (cf. Section 2.3.1), but the actual levels of performance of both groups remain unknown.

Since the intercepts in structural equations are required in order to derive the mean (expected values) of dependent variables (recall Section 2.3), *intercepts will have to be introduced into LISREL if we are to "model means"* and thereby gain the ability to describe the portions of the measurement scales occupied by the observed cases.

Attempts to model means using LISREL (cf. Cole and Maxwell, 1985; Faulbaum, 1987; Hoelter, 1983b; Lomax, 1983; McArdle and Epstein, 1987) have not been widespread, despite several discussions of the topic by Joreskog and Sorbom (Joreskog, 1979c, 1981a; Joreskog and Sorbom, 1978a: Appendix B, 1980, 1981, 1984; Schoenberg, 1982; Sorbom, 1974, 1982; Sorbom and Joreskog, 1982). Some factors contributing to this lack of use are the late appearance of procedures for handling means (means were not discussed in the early editions of the LISREL manual), the relegation of means to appendixes and parts of final chapters even within the later manuals, and inconsistencies between the programs and results in these later discussions. But there are more fundamental reasons for the avoidance of means, such as the required changes in the basic model and input data, the need for a few more pieces of statistical jargon, and the fact that social scientists trained in the tradition of path analysis (Duncan, 1975) have simply become accustomed to seeing equations without intercepts. LISREL with means will ultimately replace the basic LISREL model, but this may take years. "Bringing the means back in" will constitute a revolution in structural equation modeling because the classic sources (such as Duncan, 1975) are mute on the subject and simply cannot be superficially modified to incorporate models with means and intercepts.

The rest of this chapter tries to spell out more clearly the nature of the conceptual revolution, to present the basic relevant mathematics in a more accessible form, and to illustrate the use of means in models with and without stacked groups.

9.2.1 Returning to the Basics

In Chapter 4 we showed that a specific structural equation model implies a specific variance/covariance matrix among the variables in that model. *A structural equation model also implies specific relations among the means of the variables in the model.* Recall from Chapter 1 that if a variable Y is created from a variable X by $Y = a + bX$, then the *means* and variances of X and Y are linked by the following equations: $E(Y) = a + bE(X)$ and $Var(Y) = b^2 Var(X)$. The preceding chapters use the second equation [and its generalizations involving multiple variables (Section 4.4)] to estimate and test models. Specifically, the match between the model-implied covariance matrix Σ and the observed covariance matrix S provided the fit function underlying both MLE and the χ^2 test.

The equation linking the means of the variables can be put to the same uses. Specifically, *information on means should contribute to the estimation of model coefficients, and a model should be rejected if it implies patterns of means that are inconsistent with the observed means.* Consider the model in Figure 9.5. The three equations constituting this model are

$$X = \lambda_x \xi + \delta \tag{9.5}$$

$$Y = \lambda_y \eta + \epsilon \tag{9.6}$$

$$\eta = \gamma \xi + \zeta \tag{9.7}$$

Taking expectations of both sides of these equations (Section 1.3.3) gives

$$E(X) = \lambda_x E(\xi) + E(\delta) \tag{9.8}$$

$$E(Y) = \lambda_y E(\eta) + E(\epsilon) \tag{9.9}$$

$$E(\eta) = \gamma E(\xi) + E(\zeta) \tag{9.10}$$

We previously assumed that the means of all the error variables were zero [$E(\epsilon) = E(\delta) = E(\zeta) = 0$], and we guaranteed that the means of the observed variables $E(X)$, $E(Y)$ were zero by expressing the input data as covariances in which the variables are always expressed as deviations from their means (Section 1.1 and LISREL VI:I-5). Since the λ values are assumed to be nonzero, these assumptions force the means of the concepts to also be zero. For example, inserting the zero indicator mean and the zero error variable mean into Eq. 9.8 gives

$$0 = \lambda_x E(\xi) + 0 \tag{9.11}$$

Figure 9.5   A Model Introducing Means.

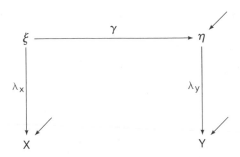

The only way this equation can hold for any nonzero value of λ is if $E(\xi) = 0$. Similarly, $E(\eta)$ must equal zero because of Eq. 9.9 and the assumptions about the zero means for Y and ϵ. In short, Eqs 9.8–9.10 all reduce to the trivial statement that $0 = 0$, given the assumptions about the zero means of the observed and error variables.

From now on, we will maintain the assumption that the error variables have mean zero $[E(\epsilon) = E(\delta) = E(\zeta) = 0]$, *but we will allow the means of the observed variables to be nonzero.* Equations 9.8–9.10 now demand a specific structure among the means if the model in Figure 9.5 is to be accepted. Specifically,

$$E(X) = \lambda_x E(\xi) \qquad\qquad 9.12$$

$$E(Y) = \lambda_y E(\eta) \qquad\qquad 9.13$$

$$E(\eta) = \gamma E(\xi) \qquad\qquad 9.14$$

For example, if λ_x and λ_y are set at 1.0 to provide identical measurement scales for the concepts and indicators, the mean of the concept ξ must equal the mean of the observed variable X, and the mean of η must equal the mean of Y (from Eqs. 9.12 and 9.13). That is, the concepts and indicators are now forced to have scales that are identical in terms of the size of their scale units *and their implicit origins.*

Neither of these statements is testable, because the means of the concepts are not directly available to us. A testable statement about the means arises, however, as soon as γ is given a value. For example, if $\gamma =$ .5, Eq. 9.14 implies that the mean of η must equal half the mean of ξ, and because the concepts and their corresponding indicators have identical means (Eqs. 9.12, 9.13 with λ's = 1.0), the mean of the observed Y values must equal half the mean of the observed X values, if this structural equation model is to hold. Providing the structural coefficients λ_x, λ_y, and γ any specific numerical values will imply a specific relation between the means of X and Y, and checking this implied relation between the means against the actual observed means can support or contradict the model in Figure 9.5. Models containing equations with several independent variables also imply specific patterns among the means of the variables (Eq. 1.25), and these predicted patterns may be used to test the models.

Observed means may also help us estimate model coefficients. To continue the preceding example, suppose the observed mean of Y was .6 of the observed mean of X, and we continue to specify λ_x and λ_y = 1. It should be clear that estimating γ as .6 would make the model fit the data, and that the further the value of γ is from .6 the worse the fit between the model's predictions and the data. That is, the observed information on the means can help us estimate model coefficients as well as test the model.

Though information on either means or variances/covariances might be used to estimate γ, it would be preferable to have an estimate of γ that is sensitive to the fit of both the observed means and variances/covariances. That is, *we would prefer an estimate of γ that takes into account all the kinds of data for which γ has implications.* This requires, however, that the importance of discrepancies in fitted means be compared against the importance of discrepancies in fitted covariances. The likelihood that any given discrepancy is a mere sampling fluctuation provides a common "scale" on which the seriousness of discrepancies in both means and covariances can be rated (recall Sections 5.1 and 5.2). That is, we should choose as estimates of the model's coefficients those values that maximize the likelihood that all the observed data (means and covariances simultaneously) could arise as mere sampling fluctuations around the model-implied means and covariances. We pursue the details of the maximum likelihood fit function in Section 9.2.3.2.

9.2.1.1 On intercepts and means

Consider what Eq. 9.12 looks like if an intercept ν (nu) is introduced so that this equation parallels Eq. 1.21.

$$E(X) = \nu + \lambda_x E(\xi) \qquad 9.15$$

Imagine further that λ_x continues to be specified as 1.0, even though the following argument would apply to any specific value for λ_x, such as that determined by the covariances among the variables. Under these circumstances, does the observed mean of X uniquely determine the mean of the concept ξ? The answer is no; the mean of ξ is not uniquely determined if the corresponding intercept is free. If X has mean 100, the mean of ξ might be 50 if ν is 50, or 60 if ν is 40. In general, *the observed indicator mean is insufficient to uniquely identify both the intercept and the concept mean* (even with a specified λ_x value) because once an acceptable pair of estimates is obtained, other pairs of equally acceptable estimates can be obtained by increasing the mean of ξ by any number of units (say "c" units) and decreasing the intercept estimate by $\lambda_x c$ units.

If a value for the intercept ν can be specified, it determines what the mean of ξ must be if the equation's implications are to agree with the observed data (mean); or if the mean of the concept is specified, it determines the value the intercept must have if the equation's implications are to agree with the observed data. Thus the estimations of intercepts and means become intertwined in that estimating (or specifying) an intercept determines the corresponding concept mean, or estimating (specifying) the concept mean determines the corresponding intercept. The observed indicator mean, by itself, provides insufficient new information to uniquely identify both the intercept and concept means, even if the corresponding λ is known.

We are now ready to specify the general form of a LISREL model with means.

9.2.2 LISREL with Means and Intercepts

The general model representing LISREL with means and intercepts is patterned on Eqs. 4.1, 4.4, and 4.6. When vectors of intercepts are added into these equations, they become

$$\eta = \alpha + B\eta + \Gamma\xi + \zeta \qquad\qquad 9.16$$

$$y = \nu_y + \Lambda_y \eta + \epsilon \qquad\qquad 9.17$$

$$x = \nu_x + \Lambda_x \xi + \delta \qquad\qquad 9.18$$

The intercepts in the equations for the endogenous concepts appear in the $m \times 1$ column vector α (alpha), and the intercepts in the equations for the y and x indicators appear in the $p \times 1$ and $q \times 1$ vectors ν_y (nu-y) and ν_x (nu-x), respectively. The matrices B, Γ, Λ_y, Λ_x, Θ_ϵ, Θ_δ, Φ, and Ψ retain their original meanings.

In addition to the three vectors of intercepts, the *means of the exogenous concepts* $[E(\xi)]$ *are also estimated*, and they are recorded in the $n \times 1$ column vector κ (kappa). *The means of the endogenous concepts*

[E(η)] *do not have to be estimated* because they are determined by the structural coefficients in the model and the estimates of the exogenous concept means, as shown by the following. First, rearrange Eq. 9.16 by moving $\mathbf{B}\eta$ to the left side of the equation, factoring out η, and then premultiplying both sides of the equation by the inverse of $(\mathbf{I} - \mathbf{B})$ to obtain

$$\eta = (\mathbf{I} - \mathbf{B})^{-1}(\alpha + \Gamma\xi + \zeta) \qquad 9.19$$

Then take expectations of both sides of the equation, realizing that E(ζ) = $\mathbf{0}$ and that all the other matrices except η and ξ contain constants and can be factored out of the expectation operator.

$$E(\eta) = (\mathbf{I} - \mathbf{B})^{-1}(\alpha + \Gamma E(\xi)) \qquad 9.20$$

Thus, any model implies specific means of the η concepts as soon as we have a specific set of estimates for the $\mathbf{B}$ and Γ slopes, the α intercepts, and the means of the exogenous concepts. Since we denote the means of the exogenous concepts [E(ξ)] by the vector κ, the preceding equation is identical to

$$E(\eta) = (\mathbf{I} - \mathbf{B})^{-1}(\alpha + \Gamma\kappa) \qquad 9.21$$

Though it is not immediately obvious, a little thought should convince you that Eq. 9.19 directly parallels Eq. 1.24 (it expresses a set of dependent variables as a linear function of several independent variables), and Eq. 9.20 parallels Eq. 1.25 (because we have merely taken expectations in both cases). It is like repeatedly calculating Eq. 1.25 with the successive η's serving as dependent variables. You might find it easier to see these connections if we carry out the implicit multiplication in Eq. 9.19:

$$\eta = (\mathbf{I} - \mathbf{B})^{-1}\alpha + (\mathbf{I} - \mathbf{B})^{-1}\Gamma\xi + (\mathbf{I} - \mathbf{B})^{-1}\zeta \qquad 9.22$$

Note that the dimensioning of the matrices implies that $(\mathbf{I} - \mathbf{B})^{-1}\alpha$ is a column vector of constants and that $(\mathbf{I} - \mathbf{B})^{-1}\Gamma$ is a single matrix of constants such that each row of the matrix links the exogenous variables to a particular endogenous concept. (The term involving ζ disappears as soon as expectations are taken, because E(ζ) = $\mathbf{0}$.) The parallel of Eqs. 1.24 and 1.25 to 9.19 and 9.20 is important because it demonstrates that the way a structural model implies means for the endogenous concepts is precisely the same "way" an equation for a dependent variable (1.24) implies that the dependent variable's mean can be calculated by inserting the means of the independent variables into equation (1.25). We draw on this parallel repeatedly in the next few paragraphs.

The model-implied means of the observed variables are obtained in a similar fashion. We take expectations of Eqs. 9.17 and 9.18 and then insert equivalent representations for the means of the exogenous concepts [$E(\xi) = \kappa$] and endogenous concepts (Eq. 9.21) into the appropriate portions of these equations.

$$E(\mathbf{x}) = \nu_x + \Lambda_x E(\xi) \qquad\qquad 9.23$$

$$E(\mathbf{x}) = \nu_x + \Lambda_x \kappa \qquad\qquad 9.24$$

and

$$E(\mathbf{y}) = \nu_y + \Lambda_y E(\eta) \qquad\qquad 9.25$$

$$E(\mathbf{y}) = \nu_y + \Lambda_y (\mathbf{I} - \mathbf{B})^{-1}(\alpha + \Gamma\kappa) \qquad\qquad 9.26$$

These equations tell us how to calculate the model-implied means for the observed indicators once any set of estimates are placed in the matrices on the right of the equations.

9.2.3 Recasting the Mathematics

Two essentially mathematical topics and one procedural topic remain to be discussed. First, what changes in the input data are required to allow estimation of the newly inserted intercepts and means? Second, how, in detail, can we justify using an estimation strategy that was developed for estimating coefficients in covariance models (MLE) to estimate models containing both means and covariances? And third, how can LISREL with means be represented in an acceptable form for input into the basic LISREL program, which has no means? These issues are addressed in the next three sections, respectively.

9.2.3.1 The moment matrix

The presence of additional coefficients in LISREL models with means (the intercepts and exogenous concept means) implies additional data will typically be required before all the coefficients can be estimated. *The input data for LISREL with means is the usual covariance matrix among the observed indicators (*$\mathbf{S}$*) plus a vector of observed indicator means* (denoted $\bar{\mathbf{z}}$). The means in $\bar{\mathbf{z}}$ are ordered in the same sequence as the variables in $\mathbf{S}$, so $\bar{\mathbf{z}}$ contains the means of the Y variables followed by the means of the X variables. This section describes how the information on the covariances and means is preserved in the moment matrix (**M**). The moment matrix is central to the discussion of the fit function that follows.

The **first moment** of variable X is another name for $E(X)$, the expected value, mean, or average of that variable. The **second moment** of X is $E(X^2)$, or the average of the squared values of variable X. The **joint moment** of variables X and Y is $E(XY)$, or the average of the product of the two variables, without the deviations from the means that would make this a covariance. The moment matrix **M** contains the first, second, and joint moments for all the observed variables: it contains the average of the variables, the average of the squared variables, and the average of the cross products of the variables.

The link between variance/covariances and first, second, and joint moments becomes clear if we examine the equations

$$\text{Var}(X) = E[(X - E(X))^2] = E(X^2) - (E(X))^2 \qquad 9.27$$

$$\text{Cov}(XY) = E[(X - E(X))(Y - E(Y))] = E(XY) - E(X)E(Y) \qquad 9.28$$

The first equal sign in each of these equations constitutes a definition (of variance and covariance, respectively) as discussed in Section 1.1. The second equal sign in each equation links these definitions to "moments" as just defined. The proofs justifying the second equal signs are given in note 2.

In words, these equations tell us that the variance of a variable equals the difference between its second moment and the square of its first moment, and the covariance between two variables equals the joint moment minus the product of the first moments of the variables. (Try reexpressing these equations using the terminology of mean or averages.)

Rearranging these equations to

$$E(X^2) = \text{Var}(X) + (E(X))^2 \qquad 9.29$$

and

$$E(XY) = \text{Cov}(XY) + E(X)E(Y) \qquad 9.30$$

shows how the second and joint moments of variables can be calculated from the variance/covariances and means (first moments) of those variables.

Thus, from an input covariance matrix and vector of means, we can calculate a matrix containing the joint and second moments of the included variables by adding the product of the corresponding variables' means to each covariance and the square of the variable's mean to each variance. This is what is accomplished if the column vector of means is postmultiplied by its transpose ($\bar{z}\,\bar{z}'$) and added to the covariance matrix for the variables. That is, a matrix of *second and joint* moments for all variables in **S** can be obtained as

$$\text{matrix of joint and second moments} = \mathbf{S} + \bar{z}\bar{z}' \qquad 9.31$$

The matrix variously referred to in LISREL as the moment matrix, the augmented moment matrix, or **M** is this matrix of joint moments with an extra row and column of means (first moments) attached to its bottom and right side.

$$\mathbf{M} = \begin{bmatrix} \mathbf{S} + \bar{z}\bar{z}' & \bar{z} \\ \bar{z}' & 1 \end{bmatrix} \qquad 9.32$$

The meaning of the "1" in the lower right of this matrix becomes clear if we consider **M** as the joint moment matrix created from a set of variables containing all the observed variables plus (augmented by) a final variable having value 1.0 for all the cases. We name this variable ONE to reflect its construction. Note that the value of 1 in the lower right of the moment matrix can be thought of as either E(ONE), thereby emphasizing that this is the last element in the last row of the matrix (the row of means), or as the E(ONE²), thereby emphasizing that this is also the second moment (the average squared value) of the variable ONE, appearing as the last diagonal element of the matrix.

The means appearing along the last row (and last column) of the matrix can be thought of as the joint moment of ONE with each of the other variables. For example, the joint moment of ONE with a variable X, or E(X, ONE), is the average of the product of the X and ONE values, but since ONE always takes on value 1, this is the same as the average of the X values [E(X)].

Thus the matrix of first, second, and joint moments (**M**) is a matrix containing information on the expected values (means) of all the variables as well as on their variances/covariances, now combined with the information on means and represented in the upper left portion of the matrix.

The foregoing describes how a sample moment matrix **M** can be constructed. A population moment matrix Ω (omega) can be defined by applying a corresponding set of procedures to the population means, variances, and covariances.

$$\Omega = \begin{bmatrix} \Sigma + \mu\mu' & \mu \\ \mu' & 1 \end{bmatrix} \qquad 9.33$$

9.2.3.2 The fit function for the moment matrix

The previous section develops a notation for matrices containing information on the means and covariances of variables simultaneously—the sample moment matrix $\mathbf{M}$ and the population moment matrix Ω. This section shows how these matrices can be used to create maximum likelihood estimates of the coefficients in the model, including the newly inserted intercepts and means.

Chapter 5 discussed the idea that if specific numerical values are provided for a model's coefficients, then that model implies (predicts) a specific covariance matrix among the observed indicators. The discussion in Section 9.2.2 shows that if specific numerical values are provided for the newly inserted intercepts and the means of the exogenous concepts (as well as all the previous model coefficients), then the model implies (predicts) a vector of observed indicator means. Since a moment matrix can be created from knowledge of the covariances and means (Eq. 9.32, 9.33), these statements together show that a model containing numerical estimates implies a specific moment matrix.

We denote this model-implied moment matrix by Ω to emphasize that we will be considering this matrix as a possible population matrix. We inquire about the likelihood that the observed moment matrix $\mathbf{M}$ could arise as mere sampling fluctuations around Ω. Specifically, we will select as estimates of the model coefficients those values that imply an Ω that maximizes the likelihood that the observed data (now $\mathbf{M}$) could arise as mere sampling fluctuations around Ω (cf. Sections 5.1 and 5.2). Recall that when the model-implied covariance matrix Σ was viewed as the possible population matrix from which the observed covariance matrix $\mathbf{S}$ arose, selecting the coefficient estimates that maximized the likelihood that the remaining differences were mere sampling fluctuations was equivalent to minimizing the fit function (Eq. 5.9)

$$\text{fit} = \log|\Sigma| + \text{tr}(\mathbf{S}\Sigma^{-1}) - \log|\mathbf{S}| - (p + q) \qquad 9.34$$

A reasonable guess at a procedure for maximizing the likelihood of $\mathbf{M}$ arising as mere sampling fluctuations around Ω is to select as coefficient estimates those values that minimize a function that parallels Eq. 9.34 but where Σ is replaced by Ω and $\mathbf{S}$ is replaced by $\mathbf{M}$.

$$\text{fit-}\mathbf{M} = \log|\Omega| + \text{tr}(\mathbf{M}\Omega^{-1}) - \log|\mathbf{M}| - (p + q + 1) \qquad 9.35$$

The 1 at the end of the new fit function compensates for the fact that the augmented moment matrix has one more row and column than the $\mathbf{S}$ and Σ matrices (the row and column of means). This extra row implies that the matrix product $\mathbf{M}\Omega^{-1}$, which would be an identity matrix if the model's implications (Ω) perfectly matched the observed $\mathbf{M}$, will have one more diagonal 1 than it had previously. Thus the trace (the sum of the diagonal elements) of $\mathbf{M}\Omega^{-1}$ would be 1 unit larger than before, and this should be reflected in the corresponding $p + q + 1$ part of the implicit perfect model fit contained in the two right-hand terms (recall Section 5.2).

Despite the ease with which one attains an intuitive grasp that means are amenable to maximum likelihood estimation (recall Section 5.1), a rather lengthy series of mathematical steps is required to prove that minimizing the fit function in which the covariance matrices have been replaced with moment matrices results in maximum likelihood estimates of the coefficients in models containing means and intercepts. Appendix D contains most of these details and corresponds to a retracing of the developmental steps in Joreskog (1971a), Sorbom (1974), and Joreskog and Sorbom (1980, 1981: Appendix B). None of these steps are provided in the LISREL V or VI manuals.

The implication of this proof is that we now have in hand all the necessary components for maximum likelihood[3] estimation of the coefficients in models with means and intercepts. That is, we enter some initial estimates and then iteratively improve those estimates until the maximum likelihood (minimum of the fit function) is attained, as discussed in Chapter 5. We are now ready to exemplify the setup required for some real data.

9.2.4 Two Smoking Models with Means

Our first example illustrates the insertion of means and intercepts into the basic nonstacked smoking model developed in Chapter 4. The literature has preferred to identify the extra coefficients in models with means and intercepts by stacking groups together and then using equality constraints between the groups to provide the necessary identifying constraints (cf. the examples in the LISREL manuals). But this is not the only way to obtain identified models. Researchers knowledgable about their data collection procedures may feel comfortable fixing enough measurement coefficients to leave only as many (or fewer) new coefficients to estimate as there are new pieces of data (the indicator means). This, in effect, adjusts for the measurement biases specified by the researcher. Naturally, as with all cases of researcher-specified coefficients (be they null effects or specific researcher estimates), one should always examine the output for any signs of unreasonable researcher decisions. ("Signs" of unacceptable model specifications are discussed in Chapter 6.)

The second model with means is a young/old stacked model in which multiple indicators of two of the concepts have been inserted. This model demonstrates the fixing, freeing, and constraining of intercepts between groups, as well as demonstrating simultaneous estimation of measurement bias and measurement error variance.

It is possible to estimate only a vector of means for each group, and then to test for the equality of one or more of the means between groups, without postulating any causal structure among the concepts [much like Hotelling's T^2 (Morrison, 1976)]. We do not illustrate this because the requisite procedures become obvious after you understand the ideas grounding the more general case in which influences among the concepts are allowed.[4]

9.2.4.1 Means in a single group

This section inserts means and intercepts into the basic smoking model discussed in Chapter 4 and diagrammed in Figure 4.1. The structural equations for the model with means and intercepts are given in Figure 9.6. The LISREL program used for input appears as Figure 9.7, and the LISREL maximum likelihood estimates are reported in Figure 9.8.

The model specification in Figure 9.6 allows the model with means to be represented in the original LISREL notation. All the concepts are treated as endogenous η's, and hence all the indicators are formally Y variables (cf. Section 7.2), though the labeling of some of the concepts as ξ's and others as η's, and some indicators as X's and others as Y's is retained to assist comparison to Chapter 4. The use of a partitioned η vector implies that B and Λ_y are partitioned. The program designations (in Figure 9.7) for all the coefficients appearing in the partitioned B are merely B subscripted with the appropriate row and column subscripts.

The concept labeled ONE and appearing at the top of the list of concepts is identical to the indicator Y-ONE, the variable with value 1.0 for all the cases. The fixed value of 1.0 in the lower left of the Λ_y matrix and the fixed 0.0 at the bottom of the vector of measurement errors guarantee the equivalence of the concept ONE and the indicator Y-ONE.

The first column of both the B and Λ_y matrices contain the "effects" of ONE on the other concepts and the indicators, and hence it is these columns that ultimately contain the estimates of the means and intercepts. [Think of ONE (or 1.0) times the value of these coefficients as being added into the appropriate dependent variables.] Thus the partitioned B matrix includes the original B and Γ matrices, the new intercepts in the equations for the endogenous concepts (α's), and the means of the

Figure 9.6   The Smoking Model with Means and Intercepts.

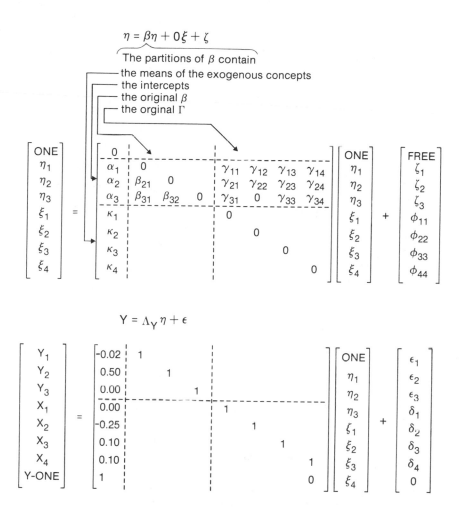

[1]Note the variable names and coefficients are labeled to correspond to their original designations in Figures 4.1–4.3, not as they are designated in the program and output Figures 9.7 and 9.8. The blank portions of the β and Λ_y matrices represent "zero" elements.

Figure 9.7   LISREL Program for the Model with Means and Intercepts.

```
title 'SMOKING  LISREL with means IN SPSSX'
file handle 8/name='smokcov1.M'
INPUT PROGRAM
NUMERIC A
END FILE
END INPUT PROGRAM
USERPROC NAME=LISREL
SMOKING -- SEX AGE EDUCATION RELIGION  with MEANS***      smoklis22
DA NI=14 NO=432 MA=MM
ME UN=8 FO
(8F10.7)
CM UN=8 FU FO
(8F10.7)
LA
'smoking1' 'smoking2' 'smoking3' 'smoking4' 'asviews' 'asacts' 'sex'
'age' 'education' 'Godexists' 'strongX' 'attend' 'important' 'ONE'
SE
'smoking1' 'asviews' 'asacts'
'sex' 'age' 'education' 'attend' 'ONE'/
MO  NY=8   NE=8  LY=FU,FI   BE=FU,FI   PS=FU,FI   TE=DI,FI
FR BE(3,2) BE(4,2) BE(4,3)
FR BE(2,5) BE(2,6) BE(2,7) BE(2,8)
FR BE(3,5) BE(3,6) BE(3,7) BE(3,8)
FR BE(4,5)         BE(4,7) BE(4,8)
FR BE(2,1) BE(3,1) BE(4,1) BE(5,1) BE(6,1) BE(7,1) BE(8,1)
FR PS(1,1)
FR PS(2,2) PS(3,3) PS(4,4)
FR PS(5,5)
FR PS(6,5) PS(6,6)
FR PS(7,5) PS(7,6) PS(7,7)
FR PS(8,5) PS(8,6) PS(8,7) PS(8,8)
VA 1.0 LY(1,2) LY(2,3) LY(3,4) LY(4,5) LY(5,6) LY(6,7) LY(7,8)
VA -.02 LY(1,1)
VA .5   LY(2,1)
VA 0.0  LY(3,1)
VA 0.0  LY(4,1)
VA -.25 LY(5,1)
VA .10  LY(6,1)
VA .10  LY(7,1)
VA 1.0  LY(8,1)
VA .05 BE(2,1) BE(3,1) BE(4,1)
VA .5  BE(5,1)
VA 20. BE(6,1)
VA 6.0 BE(7,1)
VA 1.0 BE(8,1)
VA 0.5 PS(1,1)
VA  .0247 TE(1,1)
VA  .1657 TE(2,2)
VA  .0651 TE(3,3)
VA  .0025 TE(4,4)
VA 12.3449 TE(5,5)
VA  .6909 TE(6,6)
VA  .4657 TE(7,7)
ST -.05 BE(3,2) BE(4,2) BE(4,3)
ST .05 BE(2,5) BE(2,6) BE(2,7) BE(2,8)
ST .05 BE(3,5) BE(3,6) BE(3,7) BE(3,8)
ST .05 BE(4,5)         BE(4,7) BE(4,8)
ST .25 PS(2,2) PS(3,3) PS(4,4) PS(5,5)
ST 200. PS(6,6)
ST 5.0 PS(7,7) PS(8,8)
ST .05 PS(6,5) PS(7,5) PS(7,6) PS(8,5) PS(8,6) PS(8,7)
OU ML AL NS TM=7
end user
```

Figure 9.8   Maximum Likelihood Estimates of the Smoking Model with Means and Intercepts.

```
LISREL ESTIMATES (MAXIMUM LIKELIHOOD)

    LAMBDA Y
```

	ETA 1	ETA 2	ETA 3	ETA 4	ETA 5	ETA 6	ETA 7	ETA 8
smoking1	-0.020	1.000	0.0	0.0	0.0	0.0	0.0	0.0
asviews	0.500	0.0	1.000	0.0	0.0	0.0	0.0	0.0
asacts	0.0	0.0	0.0	1.000	0.0	0.0	0.0	0.0
sex	0.0	0.0	0.0	0.0	1.000	0.0	0.0	0.0
age	-0.250	0.0	0.0	0.0	0.0	1.000	0.0	0.0
educatio	0.100	0.0	0.0	0.0	0.0	0.0	1.000	0.0
attend	0.100	0.0	0.0	0.0	0.0	0.0	0.0	1.000
ONE	1.000	0.0	0.0	0.0	0.0	0.0	0.0	0.0

```
    BETA
```

	ETA 1	ETA 2	ETA 3	ETA 4	ETA 5	ETA 6	ETA 7	ETA 8
ETA 1	0.0	0.0	0.0	0.0	0.0	0.0	0.0	0.0
ETA 2	1.033	0.0	0.0	0.0	0.0	0.0	0.0	0.0
ETA 3	4.567	-0.879	0.0	0.0	-0.015	-0.004	-0.037	-0.036
ETA 4	-0.092	-0.312	0.068	0.0	0.545	0.017	0.047	0.042
ETA 5	1.491	0.0	0.161	0.0	0.0	0.0	0.046	-0.068
ETA 6	38.704	0.0	0.0	0.0	0.0	0.0	0.0	0.0
ETA 7	8.081	0.0	0.0	0.0	0.0	0.0	0.0	0.0
ETA 8	2.641	0.0	0.0	0.0	0.0	0.0	0.0	0.0

```
    PSI
```

	ETA 1	ETA 2	ETA 3	ETA 4	ETA 5	ETA 6	ETA 7	ETA 8
ETA 1	1.000							
ETA 2	0.0	0.204						
ETA 3	0.0	0.0	2.729					
ETA 4	0.0	0.0	0.0	1.153				
ETA 5	0.0	0.0	0.0	0.0	0.247			
ETA 6	0.0	0.0	0.0	0.0	0.166	233.982		
ETA 7	0.0	0.0	0.0	0.0	0.016	-7.265	6.202	
ETA 8	0.0	0.0	0.0	0.0	0.139	7.252	0.232	4.180

```
    THETA EPS
```

smoking1	asviews	asacts	sex	age	educatio	attend	ONE
0.025	0.166	0.065	0.002	12.345	0.891	0.466	0.0

```
CHI-SQUARE WITH  1 DEGREES OF FREEDOM IS    2.28 (PROB LEVEL = 0.131)
```

exogenous concepts (κ's). The partitioned Λ_y contains the original Λ_x and Λ_y matrices and a vector of intercepts in the equations for the observed indicators (ν's).

The partitioned ζ vector results in a Ψ matrix including the original ζ error variances and the Φ's corresponding to the exogenous concepts. The partitioned ϵ vector includes the original ϵ and δ measurement errors, and implies a corresponding partitioning of the vector of measurement error variances Θ_ϵ.

The fixed 1.0 values running diagonally down the Λ_y matrix fix the scale of each concept to equal the scale of the corresponding observed indicator. That is, a 1 unit change in the score for the concept results in a 1 unit change in the corresponding observed indicator. The first column of Λ_y is composed of three ν_y intercepts and four ν_x intercepts (cf. Eqs. 9.17

and 9.18) and the 1.0 equating the concept ONE to Y-ONE in the data matrix. In accordance with the philosophy of incorporating as much as possible about known aspects of the measurement process into the model before seeking estimates of the remaining coefficients, the ν values are fixed at researcher-specified values.[5]

The nature of the considerations required in determining these values becomes clear if we consider the equation for X_2 in the model. The equation is

$$X_2 = -.25(ONE) + 1.0\xi_2 + \delta_2 = -.25 + \xi_2 + \delta_2 \qquad 9.36$$

Taking expectations gives

$$E(X_2) = -.25 + E(\xi_2) \qquad 9.37$$

From this, it is clear that the observed scores on variable age (X_2) are thought to be .25 units lower, on average, than the true ages of the respondents. This reflects our judgment about the magnitude of the tendency of people to underestimate their age. Specifically, it implies that we expect that on average about one person in four will underestimate his or her age by one year, or that one in eight will underestimate by two years.

Fixing the intercept for age at $-.25$ builds in a suspected age bias of .25 years, but it also implicitly fixes the mean of the concept ξ_2. Returning to Eq. 9.37, we see that if the observed mean age $E(X_2)$ is 38.45, then the mean of the concept of true age $E(\xi_2)$ must be $38.45 + .25 = 38.70$. Similarly, fixing the intercept for any given indicator simultaneously specifies the anticipated bias in measurement and determines the mean of the corresponding underlying concept. This has implications for what is "tested" by the overall model χ^2 as discussed later.

You should be able to see that the column of fixed values in Λ_y embodies the following researcher judgments. The $-.02$ reflects a suspicion that smoking will be slightly underreported and, in particular, that 1 in 50 respondents (about 1 in 25 smokers) will fail to report their occasional smoking. The .5 arises because respondents might have found it difficult to report their disagreement with current city legislation, and hence received overly high scores on the AS views scale. Social desirability might also have prompted respondents to report taking more AS actions than they did, but the possibility of respondents having forgotten actions they took, and the very small mean for the AS acts variable suggest there is no overall bias in this indicator (hence the first 0.0). Similarly, there was no reason to expect a bias in reported sex, so the ν_x linking X_1 to ξ_1 is also fixed at 0.0. The $-.25$ for age has already been discussed. The .10 for both education and religious service attendance reflect a suspected slight exaggeration on both these scales. Although our justifications for all these fixed coefficients are reasonable, we caution that they are given in less detail and with less supporting documentation than should accompany "real" LISREL models.

We have now included the researcher's judgment about both bias and error variance in the model (bias by fixing the ν values, and error variance by adopting the fixed Θ variances justified in Chapter 4). Note that if a single respondent reports a value lower than his or her true score, it introduces a downward bias, but it may either increase or decrease the variance of the variable (i.e., if the case was initially below or above the mean, respectively). If all the cases in a distribution report values that are a fixed number of units below their true scores, a bias, but no error variance, is introduced. If some cases report unjustifiably high scores and others report unjustifiably low scores, error variance may be introduced even though there is no overall bias.

Turning to the first column of the **B** matrix, we note that all the elements are to be estimated except the top one, which must be zero due to its location on the diagonal of **B**. The last four entries in the first column will be the estimates of the means of the exogenous concepts. The fixing of the ν values as discussed in the preceding paragraphs determines what these values should be, so when the actual estimates are obtained we will be able to check whether the estimated means correspond to those implied by our previous fixings. The three remaining elements in the first column are the intercepts in the equations for the endogenous concepts, and they are to be estimated. The means of the endogenous concepts are not estimated directly but can be calculated from Eq. 9.21 after the means of the exogenous concepts, the intercepts, **B**, and Γ are known (estimated).

Note that in comparison to the original smoking model in Chapter 4, we have introduced a total of eight new coefficients to be estimated and eight new pieces of information. The eight new coefficients are the seven coefficients in the first column of **B** and ψ_{11} (corresponding to the first element of the ζ vector), which we discuss in detail later. The eight new pieces of information are the vector of seven means for the observed indicators and the 1.0 mean for the extra variable Y-ONE.

It is easy to demonstrate that all eight coefficients can be estimated with the eight new pieces of information—that is, they are "identified." We begin by noting that since all the other coefficients were identified by the variances and covariances of the observed indicators as discussed in Chapter 4, and since we continue to include the information on variances and covariances (even if in the form of the moment matrix) all the other model coefficients remain identified. The means of the exogenous and endogenous concepts are identified, because fixing the ν values in the first column of Λ_y specifies the values of the concepts' means by repeated application of the logic applied to ξ_2 in Eq. 9.37. Thus the last four entries in the first column of **B** are identified. The remaining three estimated elements in the first column of **B** are the intercepts in the equations for the η variables. If we take the equation for η_3 as an example, we can write the equation and its expectation as

$$\eta_3 = \alpha_3 + \beta_{31}\eta_1 + \beta_{32}\eta_2 + \gamma_{31}\xi_1 + \gamma_{33}\xi_3 + \gamma_{34}\xi_4 + \zeta \qquad 9.38$$

$$E(\eta_3) = \alpha_3 + \beta_{31}E(\eta_1) + \beta_{32}E(\eta_2) + \gamma_{31}E(\xi_1) + \gamma_{33}E(\xi_3) + \gamma_{34}E(\xi_4) \qquad 9.39$$

Since all slopes (B's and Γ's) are identified (by the covariances of the input variables) and the means of all the concepts are identified (as just discussed), α_3 is the only unknown in this equation. Hence its value is also well determined (identified). Parallel arguments apply to the equations containing α_1 and α_2, so these coefficients are also identified.

The only remaining new coefficient to estimate corresponds to the "free" element heading the ζ vector. The as yet unused 1.0 value at the lower right of the input moment matrix ultimately constrains the value of the corresponding estimated coefficient to be 1.0, but we again postpone detailed discussion of this coefficient until we encounter the model degrees of freedom.

9.2.4.2 The estimates and their implications

When we enter the model in Figure 9.6 into LISREL using the commands in Figure 9.7 (which specifies start values for all the Λ_y coefficients in compliance with LISREL VI, p. V-20), we obtain the estimates in Figure 9.8. The estimates of **B** and Γ are identical to those produced previously, and the estimates in Φ and Ψ are practically the same as those obtained previously. (They differ by at most 2 units in the third significant figure.) The standard errors of the **B** and Γ coefficients (and hence the T values for these coefficients) are identical to those in Chapter 4, and the standard errors and T's for Φ and Ψ display the same kind of trivial differences evident in the coefficient estimates themselves.

The means for the exogenous concepts (κ's) agree perfectly with the means implied by fixing the ν values. The intercepts in the equations for the η's also function as expected. For example, inserting the estimate for α_3, the estimated slopes, and the means of all the concepts on the right of Eq. 9.39 (as calculated from their indicator means and biases in the first column of Λ_y) gives a predicted mean for η_3 of .612. Since several of the **B** estimates contain only two significant digits, this estimate agrees well with the mean of .605 obtained by using the Y_3 mean and its 0.0 bias to predict the mean of η_3. The use and interpretation of intercepts in the conceptual level equations are discussed in Section 9.2.4.4.

A slightly different perspective on this model arises if we consider the fit between the input data M and the model-implied Ω. The model has χ^2 = 2.28 with 1 $d.f.$ and p = .131. These values are identical to the χ^2, degrees of freedom, and probability reported for the original model. There has been no change in χ^2, because entries in the new row of the input matrix (the means) are exactly reproducible by the model and hence have zero residuals and make no contribution to any ill fit of the model.

This prompts two observations. First, the model as specified has not really tested anything about the model-implied means and intercepts. We estimated just as many new coefficients as pieces of new data, and there were no leftover degrees of freedom to contribute to any testing. Or, equivalently, fixing the indicator intercepts (the ν's in the first column of Λ_y) implied that all the new coefficients were just identified because there are just as many new pieces of data as new unknown coefficients in the model. The estimates of the κ's and α's demonstrate the implications of the researcher's decisions about the biases implicit in the fixing the indicator intercepts (ν's) but there is insufficient information in the single-group–single-indicator model to provide any test of these specifications. In this sense, the model χ^2 continues to test the fit provided for the variances and covariances, but it does not test the researcher's assumptions about bias. We reconsider the issue of bias tests when we investigate the stacked model.

The second implication of the χ^2 leads us to the promised discussion of the free element at the top of the ζ vector. Actually, the free coefficient estimated is the ψ_{11} "variance" corresponding to this first element (which is merely a variable's name), and its estimated value is 1.0. This coefficient is of interest for two reasons: its inherent meaning may be unclear; and our freeing of this coefficient is inconsistent with what Joreskog and Sorbom have done on several occasions.

What does the coefficient mean? Since the estimated coefficient corresponds to the first element in the ζ vector, its estimate (1.00) appears as the first element in Ψ and hence would routinely be interpreted as an error variance for an endogenous concept. But note that the equation in which this ζ appears has no error variance because the endogenous concept (ONE) and the observed indicator to which it corresponds (Y-ONE) have *no* variance, let alone error variance. Indeed, the empty first row of the B matrix in Figure 9.6 informs us that the "variable" ONE equals this first ζ, and the unity element in the last row of Λ_y informs us that ONE equals Y-ONE. Thus, ONE, Y-ONE, and the first "free" element of ζ all take on the value 1.0 for each case. This makes the first ζ unlike any of the other ζ's in this or previous models, because it implies that its mean is 1.0 rather than the usual 0.0 mean for error variables.[6]

But why is ψ_{11} estimated as 1.0 if the first ζ has no variance? The mystery is resolved by considering what Eqs. 4.48 and 4.49 look like for a ζ composed entirely of 1's. The transition from 4.48 to 4.49 demonstrates that Ψ arises as the $E(\zeta\zeta')$, and, *for variables with zero means* (cf. Section 3.1.1.1), this implies that the variances and covariances of the error variables appear in Ψ. But note that if a ζ happens to have a nonzero mean, as does the ζ of interest here, obtaining $E(\zeta\zeta')$ produces the expected value of the squared error values (cf. ζ in 4.21), which is a second moment, not a variance.

The second moment of a variable composed entirely of 1.0 values is 1.0; therefore when ψ_{11} is estimated, 1.0 appears. Thus it is neither a mean nor a variance, but the second moment of the first ζ variable that is estimated in ψ_{11}. This estimated 1.0 value provides a perfect fit for the 1.0 in the lower right of the input moment matrix through the equivalence of the first ζ, ONE, and Y-ONE variables.

The remaining elements of Ψ continue to be variances or covariances, because the other error variables continue to have zero means. All other estimates in Ψ are identical (within rounding error) to the error variances and Φ variances and covariances in the basic model.

With the meaning of ψ_{11} clear, we can examine whether it should be estimated or be fixed at 1.0 on the basis of the preceding logic. LISREL manuals IV, V and VI and the literature following the manuals (e.g., Lomax, 1983) have routinely treated ψ_{11} as fixed[7] at 1.0, in contrast to our freeing it and observing that its estimated value is 1.0. Our reason for questioning the fixing of ψ_{11} is that, as it stands, the smoking model with means has a χ^2 and a single degree of freedom identical to the single degree of freedom for the original smoking model. Fixing ψ_{11} at 1.0 gives estimates identical to those in Figure 9.8, but the model has 2 *d.f.* (since one less coefficient is estimated), and a higher probability level (.320) appears. Achieving the same χ^2 with an increase in degrees of freedom would claim we have achieved a better fit than previously even though precisely the same estimates were produced! In our view the "improvement" is totally illusory because we do not deserve the extra degree of freedom we could obtain by fixing ψ_{11}. Although we know from our creation of the moment matrix that ψ_{11} will be 1.0, we have no justification for gaining an extra degree of freedom for using this knowledge. Using this information amounts to using the data matrix directly to fix a coefficient, which contrasts with the acceptable procedure of resorting to either theory or a substantive knowledge of the relevant subject matter to fix model coefficients.

There is no substantive knowledge about any real variable contained in the 1.0 value at the lower right of a moment matrix. The 1.0 in the moment matrix is "counted" among the total number of pieces of input data (and hence contributes to the degrees of freedom), but estimating ψ_{11} (i.e., leaving this a free coefficient) subtracts a compensating degree of freedom and returns us to the degrees of freedom we deserve. In sum, whenever a moment matrix is entered as data, the Ψ value corresponding to the variable ONE should be estimated as a way of guaranteeing the proper degrees of freedom for the model.

We end our discussion of the single-group model with means and intercepts with a caution about some of the output of LISREL VI. Inconsistencies between the output for the smoking model with and without means appear in the squared multiple correlations for the Y variables and the standardized solution. It appears the program is using the moment matrix instead of a covariance matrix in doing these calculations. We recommend that if a model with means is estimated, the corresponding model without means be run to obtain the proper values for these portions of the output.

9.2.4.3 Means in a stacked model with multiple indicators

This section extends the younger/older stacked model reported in Figures 9.1 and 9.4 to include means, intercepts, and multiple indicators of both smoking and religion. A diagram of the model would differ from Figure 9.1 only in that smoking and religion would each have two indicators in each group. The model allows some $\mathbf{B}$ and $\mathbf{\Gamma}$ coefficients to differ between the groups while others are constrained to be equal between the groups, as discussed in Section 9.1. The input data are the covariance matrices and vectors of means for the younger and older groups reported in Figure 9.2.

The matrix representation of this model is too lengthy to present in its entirety, though it would look much like Figure 9.6 duplicated once for the younger group and once for the older group, and it would be of the same general form as the output in Figure 9.11. The structual equations among the concepts differ from Figure 9.6 because β_{46} is freed and the six $\mathbf{B}$ and $\mathbf{\Gamma}$ coefficients are constrained to be equal between the groups, as discussed in Section 9.1. The intercepts in the equations for the endogenous concepts (the α's in Figure 9.6) are constrained to be equal between the groups, as is the mean for the concept sex (κ_1 in Figure 9.6), because there seemed to be no reason to expect differences of these types between the groups. Differences in mean age, education, and religious strength were expected, so the corresponding concept means (κ_2, κ_3, and κ_4 in Figure 9.6) were allowed different estimates in the younger and older groups.

If we had a special interest in testing for the significance of the difference between the groups on the exogenous concept means, we could do it by using the difference χ^2 procedure (Section 6.1.2) for comparing a model allowing separate estimates of these means and a model constraining the means (κ's) to be equal between the groups. There is no direct procedure for testing for the significance of group differences in the endogenous concept means because differences in effects (**B** and **Γ**), differences in intercepts (α), and differences in the exogenous concept means all contribute to the model-implied differences in the endogenous concept means (Eqs. 9.20 and 9.21). We return to this issue in Section 9.2.4.5.

More substantial changes between the single-group and two-group models appear in the measurement structure due to the multiple indicators, as detailed in Figure 9.9. Since much of the following discussion focuses on these multiple indicators, we examine them closely. The first smoking indicator (smoking-1) is the one we have been using all along: nonsmokers coded 0.0, smokers coded 1.0, researcher-judged error variance of 10%, and a bias toward slight underreporting (the intercept $-.02$). This indicator continues to be used to scale the concept smoking (η_1) by a unity λ value.

The second smoking indicator (called smoking-4 because it is the fourth smoking indicator in Figure 4.4) is a more detailed description of smoking. The nonsmokers are divided into the "never smokers = 0" and the "former smokers = 1," and the smokers are divided into the currently "light smokers = 2," "moderate smokers = 3," and "heavy smokers = 4." This indicator can be thought of as a more finely scaled indicator of the respondent's contact with smoking or as an indicator of the degree to which the respondent's physiology is being stressed through current or previous smoke consumption. Smoking-4 must have a considerably higher mean and variance than does smoking-1, due to the nature of this scale's construction. The bottom portion of Figure 9.9 indicates that the proportion of error variance (the Θ_ϵ corresponding to ϵ_2), the "bias" (a ν coefficient modeled as λ_{21}), and the extent to which smoking-4 is derived from the underlying concept of smoking (the "loading" λ_{22}) are all free to be estimated but that the loading of smoking-4 on the concept of smoking (λ_{22}) is forced to be equal in the younger and older groups to force the concept to have a similar meaning in the two groups.

The two indicators of strength of religion are handled in a parallel fashion. The original indicator of religion—current religious service attendance (scored from 0 = never to 7 = several times a week, as detailed in Figure 4.4)—continues to set the scale of the underlying religion concept ($\lambda_{88} = 1.0$) and to have a researcher-specified bias of .1 and an error variance of 10%. The new indicator of religion is the degree to which the respondent agreed or disagreed with the statement "My religion is important to me now" (scored from 1 = strongly disagree to 7 = strongly agree). The "bias," error variance, and loading of this indicator on

Figure 9.9   The Smoking Model with Means and Multiple Indicators of Smoking and Religion.

$$\eta = \beta\eta + 0\xi + \zeta$$

The conceptual structural model is similar to the top of Figure 9.6 as is explained in the text.

$$Y = \Lambda_Y\,\eta + \epsilon$$

Smoking 1		-0.02	1							ONE	ϵ_1	
Smoking 4		free	equal							η_1	ϵ_2	← free
AS views		0.50		1						η_2	ϵ_3	
AS acts		0.00			1					η_3	ϵ_4	
Sex		0.00				1				ξ_1	δ_1	
Age	$=$	$-0.02\cdots$ $\cdots 0.03$					1			ξ_2	δ_2	$+$
Education		0.10						1		ξ_3	δ_3	
R-attend		0.10							1	ξ_4	δ_4	
R-important		free							equal		δ_5	← free
Y-ONE		1.00							0		0	

religious strength are all free to be estimated, but the loading is constrained to be equal in the younger and older groups to preserve the identity of the concept religious strength between the groups.

The researcher judgments about the proportion of error variance in all the other indicators continue to be those specified in Figure 4.5, and the biases continue to be those justified in conjunction with the model for means in a single group (Section 9.2.4.1), except that slightly more bias is expected in the age reports of the older respondents than in the younger (hence the intercept of $-.03$ as opposed to $-.02$ for the younger group). The details of the program specification are in Figure 9.10, and the maximum likelihood estimates are in Figure 9.11.

The model does not fit all that well ($\chi^2 = 54.97$ with 38 $d.f.$ and $p = .037$). The estimates corresponding to the **B**, **Γ**, **Ψ**, and **Φ** coefficients in the stacked model (Figure 9.4) are so similar in magnitude and significance to the current estimates that they require no further discussion.

The means estimated in the first column of **B** indicate that besides being older, the older group has a lower average education and holds more

Figure 9.10 LISREL Program for a Stacked Model with Means.

```
title 'SMOKE stacked with theory,means,EQintercepts,MULTIPLE IND***'
file handle 8/name='smokcov.youM'
file handle 9/name='smokcov.oldM'
INPUT PROGRAM
NUMERIC A
END FILE
END INPUT PROGRAM
USERPROC NAME=LISREL
SMOKING young 1/2 STACKED MEANS EQintercepts+ MULTI-IND smoklis30
DA NI=14 NO=213 MA=MM  NG=2
ME UN=8 FO
(8F10.7)
CM UN=8 FU FO
(8F10.7)
LA
'smoking1Y' 'smoking2Y' 'smoking3Y' 'smoking4Y' 'asviewsY' 'asactsY'
'sexY' 'ageY' 'educationY'
'GodexistsY' 'strongXY' 'attendY' 'importantY' 'ONEY'
SE
'smoking1Y' 'smoking4Y' 'asviewsY' 'asactsY'
'sexY' 'ageY' 'educationY' 'attendY' 'importantY' 'ONEY'/
MO NY=10  NE=8  LY=FU,FI    BE=FU,FI   PS=FU,FI    TE=DI,FI
FR BE(3,2) BE(4,2) BE(4,3)
FR BE(2,5) BE(2,6) BE(2,7) BE(2,8)
FR BE(3,5) BE(3,6) BE(3,7) BE(3,8)
FR BE(4,5) BE(4,6) BE(4,7) BE(4,8)
FR BE(2,1) BE(3,1) BE(4,1) BE(5,1) BE(6,1) BE(7,1) BE(8,1)
FR PS(1,1)
FR PS(2,2) PS(3,3) PS(4,4)
FR PS(5,5)
FR PS(6,5) PS(6,6)
FR PS(7,5) PS(7,6) PS(7,7)
FR PS(8,5) PS(8,6) PS(8,7) PS(8,8)
VA 1.0 LY(1,2) LY(3,3) LY(4,4) LY(5,5) LY(6,6) LY(7,7) LY(8,8)
FR LY(2,2) LY(9,8)
ST 3.0 LY(2,2)
ST 1.1 LY(9,8)
VA -.02 LY(1,1)
FR LY(2,1)
VA  .50 LY(2,1)
VA .5   LY(3,1)
VA 0.0  LY(4,1)
VA 0.0  LY(5,1)
VA -.20 LY(6,1)
VA .10  LY(7,1)
VA .10  LY(8,1)
FR LY(9,1)
VA 1.1  LY(9,1)
VA 1.0  LY(10,1)
ST 1.0 BE(2,1)
ST 6.0 BE(3,1)
ST .4  BE(4,1)
ST 1.5 BE(5,1)
ST 26. BE(6,1)
ST 8.0 BE(7,1)
ST 2.0 BE(8,1)
ST .95 PS(1,1)
VA  .0251 TE(1,1)
FR  TE(2,2)
VA  .1974 TE(2,2)
VA  .2044 TE(3,3)
VA  .0679 TE(4,4)
VA  .0025 TE(5,5)
VA  .7840 TE(6,6)
VA  .5030 TE(7,7)
VA  .4171 TE(8,8)
FR TE(9,9)
VA  .2362 TE(9,9)
ST -.05 BE(3,2) BE(4,2) BE(4,3)
ST .05 BE(2,5) BE(2,6) BE(2,7) BE(2,8)
ST .05 BE(3,5) BE(3,6) BE(3,7) BE(3,8)
ST .05 BE(4,5) BE(4,6) BE(4,7) BE(4,8)
ST .25 PS(2,2) PS(5,5)
ST 3.0 PS(3,3)
ST 1.0 PS(4,4)
ST 15. PS(6,6)
ST 4.0 PS(7,7) PS(8,8)
ST .05 PS(6,5) PS(7,5) PS(7,6) PS(8,5) PS(8,6) PS(8,7)
OU ML AL NS TM=30
SMOKING older 1/2 STACKED MEANS EQintercepts+ MULTI-IND smoklis30
DA NI=14 NO=219 MA=MM
ME UN=9 FO
(8F10.7)
```

Figure 9.10 Continued.

```
CM UN=9 FU FO
(8F10.7)
LA
'smoking10' 'smoking20' 'smoking30' 'smoking40' 'asviews0' 'asacts0'
'sex0' 'age0' 'education0'
'Godexists0' 'strongX0' 'attend0' 'important0' 'ONE0'
SE
'smoking10' 'smoking40' 'asviews0' 'asacts0'
'sex0' 'age0' 'education0' 'attend0' 'important0' 'ONE0'/
MO  NY=10    NE=8   LY=FU,FI    BE=FU,FI    PS=FU,FI    TE=DI,FI
FR BE(3,2) BE(4,2) BE(4,3)
FR BE(2,5) BE(2,6) BE(2,7) BE(2,8)
FR BE(3,5) BE(3,6) BE(3,7) BE(3,8)
FR BE(4,5) BE(4,6) BE(4,7) BE(4,8)
FR BE(2,1) BE(3,1) BE(4,1) BE(5,1) BE(6,1) BE(7,1) BE(8,1)
FR PS(1,1)
FR PS(2,2) PS(3,3) PS(4,4)
FR PS(5,5)
FR PS(6,5) PS(6,6)
FR PS(7,5) PS(7,6) PS(7,7)
FR PS(8,5) PS(8,6) PS(8,7) PS(8,8)
EQ BE(1,2,1) BE(2,1)
EQ BE(1,3,1) BE(3,1)
EQ BE(1,4,1) BE(4,1)
EQ BE(1,5,1) BE(5,1)
EQ BE(1,4,3) BE(4,3)
EQ BE(1,2,5) BE(2,5)
EQ BE(1,2,8) BE(2,8)
EQ BE(1,3,5) BE(3,5)
EQ BE(1,3,8) BE(3,8)
EQ BE(1,4,8) BE(4,8)
VA 1.0 LY(1,2) LY(3,3) LY(4,4) LY(5,5) LY(6,6) LY(7,7) LY(8,8)
FR LY(2,2) LY(9,8)
EQ LY(1,2,2) LY(2,2)
EQ LY(1,9,8) LY(9,8)
ST 3.0 LY(2,2)
ST 1.1 LY(9,8)
VA -.02 LY(1,1)
FR LY(2,1)
VA  .50 LY(2,1)
VA  .50 LY(3,1)
VA 0.0  LY(4,1)
VA 0.0  LY(5,1)
VA -.30 LY(6,1)
VA .10  LY(7,1)
VA .10  LY(8,1)
FR LY(9,1)
VA 1.1  LY(9,1)
VA 1.0  LY(10,1)
ST 1.0 BE(2,1)
ST 6.0 BE(3,1)
ST .4  BE(4,1)
ST 1.5 BE(5,1)
ST 50  BE(6,1)
ST 8.0 BE(7,1)
ST 3.0 BE(8,1)
ST 0.5 PS(1,1)
VA  .0743 TE(1,1)
FR TE(2,2)
VA  .2046 TE(2,2)
VA  .1137 TE(3,3)
VA  .0623 TE(4,4)
VA  .00251 TE(5,5)
VA  8.676 TE(6,6)
VA  .8642 TE(7,7)
VA  .4727 TE(8,8)
FR TE(9,9)
VA  .1927 TE(9,9)
ST -.05 BE(3,2) BE(4,2) BE(4,3)
ST .05 BE(2,5) BE(2,6) BE(2,7) BE(2,8)
ST .05 BE(3,5) BE(3,6) BE(3,7) BE(3,8)
ST .05 BE(4,5) BE(4,6) BE(4,7) BE(4,8)
ST .25 PS(2,2) PS(5,5)
ST 2.0 PS(3,3)
ST 1.0 PS(4,4)
ST  160. PS(6,6)
ST 6.0 PS(7,7) PS(8,8)
ST .05 PS(6,5) PS(7,5) PS(7,6) PS(8,5) PS(8,6) PS(8,7)
OU ML AL NS TM=30
end user
```

Figure 9.11   Estimates for the Stacked Model with Means.

young

LISREL ESTIMATES (MAXIMUM LIKELIHOOD)

LAMBDA Y

	ETA 1	ETA 2	ETA 3	ETA 4	ETA 5	ETA 6	ETA 7	ETA 8
smoking1	-0.020	1.000	0.0	0.0	0.0	0.0	0.0	0.0
smoking4	0.428	2.860	0.0	0.0	0.0	0.0	0.0	0.0
asviewsY	0.500	0.0	1.000	0.0	0.0	0.0	0.0	0.0
asactsY	0.0	0.0	0.0	1.000	0.0	0.0	0.0	0.0
sexY	0.0	0.0	0.0	0.0	1.000	0.0	0.0	0.0
ageY	-0.200	0.0	0.0	0.0	0.0	1.000	0.0	0.0
educatio	0.100	0.0	0.0	0.0	0.0	0.0	1.000	0.0
attendY	0.100	0.0	0.0	0.0	0.0	0.0	0.0	1.000
importan	2.473	0.0	0.0	0.0	0.0	0.0	0.0	0.677
ONEY	1.000	0.0	0.0	0.0	0.0	0.0	0.0	0.0

BETA

	ETA 1	ETA 2	ETA 3	ETA 4	ETA 5	ETA 6	ETA 7	ETA 8
ETA 1	0.0	0.0	0.0	0.0	0.0	0.0	0.0	0.0
ETA 2	1.113	0.0	0.0	0.0	-0.022	0.002	-0.067	-0.038
ETA 3	5.751	-1.524	0.0	0.0	0.477	-0.016	0.045	0.042
ETA 4	0.217	-0.535	0.063	0.0	0.245	-0.017	0.068	-0.055
ETA 5	1.491	0.0	0.0	0.0	0.0	0.0	0.0	0.0
ETA 6	26.194	0.0	0.0	0.0	0.0	0.0	0.0	0.0
ETA 7	8.341	0.0	0.0	0.0	0.0	0.0	0.0	0.0
ETA 8	2.181	0.0	0.0	0.0	0.0	0.0	0.0	0.0

PSI

	ETA 1	ETA 2	ETA 3	ETA 4	ETA 5	ETA 6	ETA 7	ETA 8
ETA 1	1.000							
ETA 2	0.0	0.192						
ETA 3	0.0	0.0	3.157					
ETA 4	0.0	0.0	0.0	1.098				
ETA 5	0.0	0.0	0.0	0.0	0.247			
ETA 6	0.0	0.0	0.0	0.0	0.042	14.815		
ETA 7	0.0	0.0	0.0	0.0	0.054	2.313	4.513	
ETA 8	0.0	0.0	0.0	0.0	0.105	0.582	0.422	3.767

THETA EPS

smoking1	smoking4	asviewsY	asactsY	sexY	ageY	educatio	attendY	importan	ON
0.025	0.161	0.204	0.068	0.002	0.784	0.503	0.417	2.814	

older

LISREL ESTIMATES (MAXIMUM LIKELIHOOD)

LAMBDA Y

	ETA 1	ETA 2	ETA 3	ETA 4	ETA 5	ETA 6	ETA 7	ETA 8
smoking1	-0.020	1.000	0.0	0.0	0.0	0.0	0.0	0.0
smoking4	0.415	2.860	0.0	0.0	0.0	0.0	0.0	0.0
asviewsO	0.500	0.0	1.000	0.0	0.0	0.0	0.0	0.0
asactsO	0.0	0.0	0.0	1.000	0.0	0.0	0.0	0.0
sexO	0.0	0.0	0.0	0.0	1.000	0.0	0.0	0.0
ageO	-0.300	0.0	0.0	0.0	0.0	1.000	0.0	0.0
educatio	0.100	0.0	0.0	0.0	0.0	0.0	1.000	0.0
attendO	0.100	0.0	0.0	0.0	0.0	0.0	0.0	1.000
importan	2.725	0.0	0.0	0.0	0.0	0.0	0.0	0.677
ONEO	1.000	0.0	0.0	0.0	0.0	0.0	0.0	0.0

BETA

	ETA 1	ETA 2	ETA 3	ETA 4	ETA 5	ETA 6	ETA 7	ETA 8
ETA 1	0.0	0.0	0.0	0.0	0.0	0.0	0.0	0.0
ETA 2	1.113	0.0	0.0	0.0	-0.022	-0.006	-0.027	-0.038
ETA 3	5.751	-0.437	0.0	0.0	0.477	-0.002	0.023	0.042
ETA 4	0.217	-0.041	0.063	0.0	0.079	-0.004	0.022	-0.055
ETA 5	1.491	0.0	0.0	0.0	0.0	0.0	0.0	0.0
ETA 6	50.894	0.0	0.0	0.0	0.0	0.0	0.0	0.0
ETA 7	7.833	0.0	0.0	0.0	0.0	0.0	0.0	0.0
ETA 8	3.082	0.0	0.0	0.0	0.0	0.0	0.0	0.0

PSI

	ETA 1	ETA 2	ETA 3	ETA 4	ETA 5	ETA 6	ETA 7	ETA 8
ETA 1	1.000							
ETA 2	0.0	0.205						
ETA 3	0.0	0.0	2.041					
ETA 4	0.0	0.0	0.0	1.167				
ETA 5	0.0	0.0	0.0	0.0	0.248			
ETA 6	0.0	0.0	0.0	0.0	-0.221	164.037		
ETA 7	0.0	0.0	0.0	0.0	-0.011	-10.402	7.733	
ETA 8	0.0	0.0	0.0	0.0	0.169	3.142	0.181	4.194

THETA EPS

smoking1	smoking4	asviewsO	asactsO	sexO	ageO	educatio	attendO	importan	ON
0.024	0.227	0.114	0.062	0.003	8.676	0.864	0.473	2.025	

CHI-SQUARE WITH 38 DEGREES OF FREEDOM IS   54.97 (PROB. LEVEL = 0.037)

strongly to religion than the younger group. The mean for sex (which was constrained to be equal in the two groups) demonstrates a nearly even split between males and females (1.491).

9.2.4.4 The intercepts

The three intercepts in the first column of **B** (the first three nonzero entries) provide the values required to make the means of the corresponding endogenous concepts mesh with the means of the predictor concepts and their effect coefficients. The concept means have been determined by the specified indicator intercepts (in Λ_y), and the slopes have been estimated, so this parallels solving for a in Eq. 1.16 if b and the means are known.

Once the intercepts are known, we are prepared to reap one of the major benefits of having included the means and intercepts in the model. We can use the structural equations *to predict the expected values of the dependent concepts for any desired combination of values on the independent concepts* (cf. Section 2.3). For example, we can predict the number of AS bylaw provisions agreed to by a smoking male who is 50 years old with the highest level of education and lowest church attendance. We merely obtain the equation for the relevant concept (AS views) from the matrix equations (Figure 9.11, older group)

$$\text{AS views} = 5.751 - .437(\text{smoking}) + .477(\text{sex}) - .002(\text{age})$$
$$+ .023(\text{ed}) + .042(\text{religion/attend}) \qquad 9.40$$

and insert the researcher-specified values for the concepts on the right to obtain

$$\text{AS views} = 5.751 - .437(1) + .477(1) - .002(50) + .023(15) + .042(0)$$
$$= 6.0 \qquad 9.41$$

We can now predict that a person with these "true" characteristics is likely to agree with six of the eight AS bylaw provisions.

Similarly, a young smoking, male with the lowest reported education and church attendance provides (from the younger stacked group)

$$\text{AS views} = 5.751 - 1.524(1) + .477(1) - .016(18) + .045(1) + .042(0)$$
$$= 5.0 \qquad 9.42$$

This tells us that even the most recalcitrant individuals in the survey agree with a majority (five out of eight) of the nonsmoking bylaw provisions. Statements like this make the results accessible to nonacademics, who may not comprehend the details of the implicit adjustments for bias and measurement error.

The advance provided by the preceding interpretation is that it involves the "real" scales of measurement of the various variables. The standardized solution (in models without means or intercepts) can provide the predicted "relative" score on the dependent variable for individuals posessing a specific combination of relative values (in standard deviations) on the predictor variables, but the required use of standard deviation units remains a considerable hindrance to nontechnical presentations.

9.2.4.5 Intercepts and experiments

Another major use of intercepts in the conceptual level equations arises in the context of comparing groups after adjusting for measurement bias and unreliability. Though our stacked example does not exactly fit the paradigm of comparing experimental and control groups, it should not be too difficult to imagine an experimental group and a control group stacked together and for which a separate intercept estimate has been obtained in each group for some endogenous concept. In parallel to Eqs. 9.39 or 9.20, the intercept in each of these equations represents the difference between the mean of the endogenous concept and the mean that would be predicted on the basis of the means and slopes for the predictor variables in the equation. Thus *each intercept reflects the net upward/downward effect exerted by all the variables excluded from the equation*, and the difference between the intercepts for the two groups reflects the difference in the net effects of the excluded variables.

If an experimental group and a control group are stacked together, no variable corresponding to the experimental/control status appears among the list of predictors, so any upward or downward effect provided by being in the experimental group contributes to the α for that group. This should result in a corresponding elevation/decline in the α for the experimental group relative to the α for the control group; hence testing for the significance of the difference between the α's for the groups provides a method of testing for the significance of the impact of the experimental treatment.[8]

That the α's for each group and the differences in α's between groups contain the combined effects of all the excluded variables meshes with methodological admonitions to reduce the influences of all extraneous factors within experiments and to make the experimental and control groups as similar as possible in all nontreatment respects. The more numerous and effective the variables on the list of predictor variables, and the greater the group differences in means for these variables, the more strongly we have supplemented actual laboratory controls with statistical controls that reduce the possibility of systematic extraneous effects being confused with the experimental treatment.

The nature of the supplementary controls or adjusting provided by concepts on the list of predictors is most easily visualized in the context of equations of the form of Eq. 9.39, where the mean of an endogenous concept is expressed as a function of the structural coefficients (α and slopes) and the means of the predictor concepts. If a new and effective predictor concept is added to the list of predictors in both groups, and if the mean of this new variable differs between the groups, a difference between the groups that used to be modeled by (appear as a difference between) the α's would now be modeled by the new concept's structural coefficient ("slope") and its difference in means. Similarly, when there are multiple predictors, we have controlled for the differences between the means of the groups on any of the predictor variables. The α's reflect any differences in the dependent concept's mean *not accounted for by group differences in the means of the predictor concepts.*

Do the **B** slopes have to be the same in the groups before α can be used to test for experimental effects?[9] This does *not* seem to be necessary. Slopes differing between the experimental and control groups reflect the differential effectiveness of the variables in the groups. Since the extent to which a predictor concept's mean contributes to the dependent concept's mean is determined by the effectiveness with which that predictor concept influences the dependent concept, differences in effectiveness between groups should be allowed. Differences in α that remain in models allowing slope differences between groups can be interpreted as *differences in the mean of the dependent concept that remain after we have adjusted for both the mean differences in the predictor concepts and the differential effectiveness of those concepts in the two groups.* All of this adjusting is above and beyond adjustments for measurement bias and unreliability, because it is carried out at the conceptual level.

Returning to the smoking model, we might consider the implications of constraining the endogenous concept intercepts to be equal between the groups in light of the foregoing considerations. Our specification of equal α's amounts to a claim that the variables excluded from these equations provide no net differential elevation or reduction in the scores of the older and younger persons. That is, it embodies our belief that there is nothing akin to an experimental treatment that *systematically* boosts or reduces the scores of the younger respondents on smoking, AS views, or AS acts.

A further perspective on the possible uses of intercepts arises from the special way we defined the older and younger groups. Note that the model's predictions for a smoker of age 33.5 might be obtained by an equation from either the younger or the older model, because this individual is exactly on the age cutpoint defining the two groups. The rather substantial difference in the effect of smoking on AS views in the younger and older groups (contrast -1.524 with $-.437$) implies that a different prediction regarding "views" might appear for this smoker of age 33.5, depending on whether the equation for the older group or the younger group is used. Allowing separate estimates of the intercepts in the two equations might compensate for such problems.

9.4.4.6 The indicators

We turn now to what is by far the most informative part of this model regarding its ill fit—the multiple indicators of religion and smoking. We focus first on religion. The loading of the "importance of religion" indicator on the concept religion strength is estimated as .677, its intercepts are estimated as 2.473 and 2.725 in the young and old groups, respectively, and its measurement error variances (Θ_ϵ's) are 2.814 and 2.025, respectively. The story told by the loading and error variances is very clear. The importance of religion indicator does not match well with the concept of religious strength as defined by its fixed relations to the indicator religious attendance. Specifically, of the total variance in importance ratings for the young respondents (4.723), 60% is error variance, whereas 53% of the variance in the older respondents' importance ratings (3.853) is error variance. The low loading of importance on the concept strength of religion (.677) reflects the fact that less than half the variance in importance scores is acting in concert with the respondents' actual religious service attendance. Thus, the behavior of attending services differs substantially from cognitive ratings of the importance of religion. Although the two indicators might be substantially linked to one another (r's over .6 are reported in Figure 9.2), they *do not behave in a sufficiently coordinated fashion toward the other variables in the model* to provide covariances that would justify a claim that a single concept, such as religious strength, could be the fount of both religious service attendance and self-ratings of religion's importance.

Concerning bias in the reports of the importance of religion, it might be surprising that we can say nothing about bias in the importance measurements, even though the intercepts in the equations for the importance indicators have been estimated. The reason is that the two religion indicators have substantively different scales. To see the connection between these issues, we begin by collecting the relevant information, focusing on only the younger group (though parallel comments apply to the older group). The estimated importance intercept (λ_{91}) is 2.473, the mean of the concept religious strength (β_{81}) is estimated as 2.181, and the observed means for attendance and importance are 2.272 and 3.944, respectively. The estimated concept mean is close to being .1 unit lower than the mean of the attendance indicator, as it should be if the predetermined bias in this indicator $(\lambda_{81} = .1)$ was being respected. Using Eqs. 1.16, 1.25, or their generalizations in Eqs. 9.23–9.26, we see that the intercept estimate for importance is merely the value that makes the observed indicator mean consistent with both the concept mean and the loading .677:

$$\text{indicator mean} = \text{intercept} + .677(\text{concept mean})$$

$$3.944 = \text{intercept} + .677(2.181)$$

9.43

so the intercept must equal 2.467. This agrees with the estimated intercept to as many significant digits as the loading is accurately reported. (The same is true for the older group.) *Thus, the reported intercepts are merely those values that make the estimated concept means and loadings consistent with the observed means.*

But what does this tell us about bias in the reporting of importance? Clearly, if the respondents had consistently over- or underestimated the importance of religion to them, it could have changed the value of the intercept by changing the numerical value on the left of the preceding equation. But can we recover the amount of bias from the actual magnitude of the intercept estimate? No, we cannot. The value of the intercept is really determined by two factors, and we cannot recover unique assessments of the impact of both these factors from the single intercept estimate. The two factors are the amount of bias and the differences in the substantive meanings of the measurement scales for importance and attending. The scale for attending is at issue because it was used to fix the scale for the underlying concept. The fundamental issue is not whether the scales merely contain the same number of points between the lowest and highest possible responses (both are seven-point scales) but whether *the psychological meanings of the scales to the respondents are different.* If different portions of the scales are used in providing responses to the two indicators, it might mean that one set of responses was biased, or it might mean that the respondents were just thinking of different things in determining what the various points on the scale mean when responding to the two questions.

If someone was challenged to justify his or her extreme response (say seven) on both scales, the nature of the things that could be pointed to as justifying an extreme attendance score would be very different from the things that justify an extreme score on the importance of religion to them. The substantive difference between the *meanings* of the scales created by the context of answering different questions makes it impossible to claim that a score of seven, or any other scale point, on the two scales should be treated as objectively identical.

Nor is this even a function of the fact that the scale points refer to different kinds of entities—*frequency* of attendance versus *degree* of agreement with a statement of importance. Even if both indicators had been measured on seven-point "strongly disagree to strongly agree" scales, there is no guarantee that the meaning of those scales would be precisely the same during its two uses. To make matters worse, there are further problems with determining whether the different groups place the same meaning on the scales,[10] or indeed even if different individuals place the same meaning on even the same scale.

We are touching fundamental issues in psychology and philosophy, and they must be addressed in some way before we can claim to have tested or estimated the amount of bias in a scale. Although we can supply grounds for suspected sizes of biases (as was done in determining the fixed values for the intercepts for all single indicators in this model), we are on a very different scientific foundation when testing or estimating bias. To claim to have included, and hence adjusted for, a certain amount of expected bias is a far weaker claim and is easier to justify than a claim that we have located a procedure for estimating and testing biases that is somehow independent of the researcher's judgments.

The net result of all this is that *although the estimated intercepts are informative because they tell us about the additive constant required to provide consistency between any particular indicator and the other indicators in the model, they cannot stand as estimates of bias per se.* This would have required that we provide additional evidence (which is lacking) that the two indicators were expected to have been responded to in the same way by the respondents (i.e., by their using the same portions of the available response scales and not merely that any individual's responses should be in the same *relative* positions on the two scales). Extreme intercepts might provide grounds for suspicion but not proof of bias.

Returning to the observation that the two religion indicators do not seem to be compatible when they are forced to respond to the single underlying religious strength concept, we might consider ways of improving the model. One option is to modify the meaning of the concept of religious strength by freeing the error variance associated with religious service attendance. We prefer not to do this because it changes the meaning of the concept sufficiently to force us to reexamine all our decisions about religious strength. A second strategy would be to model attendance and importance as single indicators of two separate religion concepts. This strategy is preferable from the perspective of substantive theory, but we do not pursue it because it implies building a substantially altered model.

Another, weaker, strategy is to eliminate the importance indicator from the model. Though we will not pursue the details of this approach either, we note that a stacked model with means, two smoking indicators, and a single religious-strength indicator (attendance) fits the data very well (χ^2 = 20.55 with 23 $d.f.$ and p = .608) and gives estimates differing only minimally from those in the stacked model with two religion indicators. This confirms our hunch that the inconsistencies between the two religion indicators are providing the borderline unacceptable χ^2 reported earlier.

Turning to the multiple smoking indicators, we observe a very different pattern than was seen for the multiple religion indicators. For smoking, the estimates of the error variance in the second smoking indicators are surprisingly close to the 10% error variance we postulated for all the smoking indicators. For the young group, the estimated error variance (.161) is 8.7% of the total variance, whereas for the older group the estimated error variance (.227) is 13.8% of the total variance. The complement of these observations is that the vast majority of the variance in these indicators is being modeled as arising from the smoking concept common to the two smoking indicators. The relatively large λ_{22} (2.86) expressing the dependence of the second smoking indicator on the smoking concept reflects the larger variance in the second indicator compared to the variance in the smoking concept (which should be 10% less than the variance of the first smoking indicator).

The intercepts (.428 and .415 for the younger and older groups, respectively) may surprise you. The second indicator was supposed to provide merely a more finely graduated smoking measure, so why are the intercepts not *negative* (to reflect the underreporting of smoking) as was the fixed intercept for the first smoking indicator?

There seems no way the questionnaire items or routing could account for underreporting of smoking and overreporting of amount, so we must look elsewhere for an explanation. In desperation we might consider the explanation that our estimate of the underreporting of smoking is much too small. For example, if the intercept λ_{11} were fixed at $-.2$ (instead of $-.02$), it would raise the mean of the concept smoking and thus require a small (and possibly negative) intercept to coordinate this mean with the observed mean of the second smoking indicator. Inserting a $-.2$ value in the model with two smoking indicators but a single religion indicator (attendance) does indeed make the intercept estimates negative (though small). But to assume a bias of $-.2$ in the smoking-1 indicator implies, in the context of the 0–1 coding for this indicator, that 20% of the respondents, or about 40% of those claiming to be nonsmokers, are providing invalid responses. Clearly, trying to force a negative intercept by this strategy leads to a cure that is less acceptable than the disease.

Like all good mysteries, the mystery of the positive intercepts is resolved by pointing to a tacit though untenable assumption. The incorrect assumption is precisely the assumption we tried to dissuade you of just a few paragraphs ago! There is no sound reason to expect that these intercepts must be negative. Although the $-.02$ intercept for the first indicator of smoking can be viewed as a *bias specification because this value determines the relative standing of the means of the first indicator and the underlying concept, the estimated intercept for the second smoking indicator does not play the same kind of role.* The intercept for the second indicator cannot determine the relative standing of the means of the second indicator and the concept. This difference has already been determined by the fact that both these means are specified—one by data observations and the other by the previously specified relation to the mean of the first indicator. *An intercept for the second indicator can only make the mean of the second indicator "consistent" with the mean of the concept.* Thus, the estimated intercept is not an estimated bias, and hence there is no reason to expect a negative value as if it were a bias assessment. In short, the mystery of why the intercepts are not negative is solved by noting that we had no justifiable reason to expect them to be negative!

Does this mean we are unable to estimate (or test for) the amount of bias in the multiple indicators in this, or any, model? It seems there is no way around the preceding argument that this particular multiple-indicator model has not estimated or tested the bias of the "second" indicators, but the possibilities for testing bias in models in general are more promising. Testing for bias is, in principle, possible whenever the theory of the indicators' measurement scales or the methodology providing the measurement scales justifies hypothesizing equality constraints between two or more intercepts. This requirement seems attainable in the context of psychological profiles, educational achievement tests, and numerous other research contexts that include truly "equivalent" indicator items. In the context of "different yet related" indicators (as in our smoking model), we may have to settle for the intercepts as informing us about the requirements for consistency between the different scales rather than about bias per se.

Matters are made worse because there is some fundamental unclearness in the notion of bias itself. Traditional statistical discussions of bias (as the mean of a sampling distribution failing to correspond to the population mean) do not help much because they always assume that the scores (values) in the population and sample are recorded on the same scales and because they are aimed at discovering the implications of some sampling procedure or estimation strategy. This contrasts sharply with our current problem of demonstrating the equivalence (or nonequivalence) of the scales of unobserved concepts and observed indicators. This is a classic problem of any science, and it persists even if we have all the data from the relevant population.

Here are two examples of the types of as yet unresolved complexities that arise in the context of LISREL applications. Suppose two concepts have three indicators: each concept has one indicator solely responsive to it, and the third indicator is shared between ("loads" on) both concepts. How can we talk about the bias of this shared indicator, given that it originates in two substantively different concepts that may have radically different measurement scales specified by the nonshared indicators?

For a second disconcerting example, consider a model that has a single indicator of a concept and where the intercept linking the indicator and concept has been estimated in conjunction with all the other coefficients in some model. Now suppose that a new concept is added to the model that only influences the indicator of the previous concept. That is, we included one of the potentially many previously excluded causes of the single indicator. Assuming that the new concept has a nonzero mean and a nonzero loading on the indicator, the concept would contribute to the equation of form 9.23 for the indicator, so the estimate of the intercept in the new model would differ from the original intercept estimate. The dilemma this poses is that although the bias of the indicator at reflecting the original concept seems to have changed (the intercept changed), it was done by merely naming and including in the model one of the possibly many other variables contributing to the intercept in the original model. Can we claim to have changed the bias of an indicator if we merely name and estimate the size of the effect of the factor providing some of the bias? That is, does bias cease to be bias the instant we remove it from the intercept by estimating it in conjunction with some additional variable (in the same way that error variance is transformed into explained variance with the addition of a new variable)? There is currently no consensus on such questions because they have received insufficient discussion to clarify the logical foundations necessary for forging a consensus. It may be more than trivial to note that both error variables and intercepts encapsulate the net effects of excluded variables.

We conclude our discussion of models containing means and intercepts with a reminder of the fundamental advances these models provide. First, with the inclusion of means, we have shown that a model is consistent (hopefully) with a wider range of evidence than if mere covariances were used. Second, we will, in general, obtain better estimates of the structural coefficients (B and Γ, etc.), because these estimates must now be consistent with a wider range of evidence (the indicator means). Third, the actual estimates of the means and intercepts may be of inherent interest if we are attempting to compare groups or to assess the effectiveness of interventions. Lastly, once we know the *slopes and intercepts* in a structural equation we can use that equation *to predict the values of the dependent concepts for any desired combination of the real values on the predictor concepts.* Recall our use of the equation for AS views to predict the number of AS bylaw provisions we would expect to be agreed to by a 50-year old, smoking male who has a high level of education and low church attendance. Would you like to know the views that are likely to be expressed by a young nonsmoking female with high education and average church attendance?

Notes

1.   This model does not test whether there is any difference at all between the groups, because we continue to allow the groups to have different Φ, Ψ, and Θ values. To test for the presence of any difference between groups, we can use the procedures described by Joreskog and Sorbom (1981:V-2; 1984:V-2) for testing the equality of covariance matrices. Essentially, we create a nonmodel in which all the indicators directly and perfectly reflect a set of ξ concepts (i.e., Λ's equal 1, and Θ's $= 0$) for each group. This stacked model allows perfect reproduction of the observed covariances if the Φ coefficients are free and allowed to differ between groups (i.e., a χ^2 of 0.0 with 0.0 *d.f.*). The χ^2 for a stacked model forcing the Φ matrices to be equal in the two groups provides a test for the significance of the differences between the covariance matrices. If the covariance matrices do not differ significantly, there is no reason to expect the coefficients that we are attempting to estimate from those covariances to differ between the groups.

    The equality of correlation matrices can be tested by fixing the Λ values at the standard deviations of the indicators within each group (which standardizes the ξ's) and testing for the equality of Φ as before.

2.   By definition

$$\text{Cov}(XY) = E[(X - E(X))(Y - E(Y))] \qquad 9.44$$

Expanding the product gives

$$\text{Cov}(XY) = E[XY - YE(X) - XE(Y) + E(X)E(Y)] \qquad 9.45$$

Using the rules for the algebra of expectations

$$\text{Cov}(XY) = E(XY) - E(Y)E(X) - E(X)E(Y) + E(X)E(Y) \qquad 9.46$$

$$\text{Cov}(XY) = E(XY) - E(X)E(Y) \qquad 9.47$$

as required.

    Because the variance of a variable is the covariance of that variable with itself (Section 1.2), the preceding equation also shows that

$$\text{Var}(X) = \text{Cov}(XX) = E(XX) - E(X)E(X) = E(X^2) - (E(X))^2 \qquad 9.48$$

as required.

3.   Generalized least square and unweighted least squares can also be used to provide estimates of models with means and intercepts (Joreskog, 1981a:80). The proofs of this would parallel Appendix D and use the fit functions appearing as Eqs. 12 and 13, as opposed to 14 (the maximum likelihood fit function), from Joreskog (1981a:74).

4.   This can be done by letting only the constant variable ONE and an error variable influence each variable. The "effects" of the constant variable are the estimated means, and the free variances and covariances among the error variables model the variances and covariances among the input variables. The input data continues to be the moment matrix. The procedure is illustrated in the LISREL VI manual beginning on p. V-30.

5.   It is impossible to estimate all the intercepts and means in models for single groups without a substantial number of fixed coefficients. The reason for this should be clear. We will be including only seven new pieces of information (the seven indicator means), and hence we can expect to estimate at most seven new coefficients. (The mean of the indicator Y-ONE might be considered as an eighth piece of new information, but we postpone discussion of this until later.) If all the intercepts and means were free, there would be 14 new coefficients [seven intercepts in the equations for the indicators (ν's), four exogenous concept means (κ's), and three intercepts in the equations for the endogenous concepts (α's)]. Clearly not all the desired coefficients can be estimated. Since knowledge of the identity of the indicator variables and the data collection strategy provide reasonable estimates of the intercepts in the indicator equations, these seven values are fixed on the basis of this extra information.

6.   Another way of seeing why the expected value of the first ζ cannot be zero arises in the context of Eqs. 9.16 and 9.19. Since our model (Figure 9.6) has no Γ or α coefficients (they are formally just some of the columns within B), Eq. 9.16 can be written as

$$\eta = \mathbf{B}\eta + \zeta \qquad 9.49$$

Doing the rearrangements discussed just prior to Eq. 9.19 gives

$$\eta = (\mathbf{I} - \mathbf{B})^{-1}\zeta \qquad 9.50$$

Taking expectations gives

$$E(\eta) = (\mathbf{I} - \mathbf{B})^{-1}E(\zeta) \qquad 9.51$$

If the $E(\zeta)$ vector was composed entirely of variables with zero means, then $E(\eta)$ would have to be zero. But this is impossible (at least the "old ξ" variables in the partitioned η vector should have nonzero means), so not all the elements of $E(\zeta)$ can be zero.

The equality of η_1 and the first ζ implies that the expected value of the first ζ is 1.0, and all the other ζ means are 0.0. Combining this information with Eq. 9.51 shows that the means of all the concepts (the η's) will appear as the first column of $(\mathbf{I} - \mathbf{B})^{-1}$.

7.   This has not been obvious because they typically model the variable ONE as a ξ and then use the "FixedX" option within LISREL. Using this option implies the fixing of the corresponding Λ to 1.0 and Θ to 0.0 and *the fixing of the corresponding Φ variance*.

8.   This can be done by the difference in χ^2 procedure if the intercepts are first provided separate estimates and later constrained to be equal.

In the LISREL V and VI manuals, Joreskog and Sorbom fix α at zero for one of the groups to help identify the model parameters. Hence the magnitude of the difference they observe is merely the value of the α for the "other" group. Thus, testing to see if this other α is significantly different from zero (the usual T test) amounts to testing for the difference in α's between the groups.

9. The consistent testing for the equivalence of B's between groups immediately before testing for intercept differences between the groups and the repeated use of the statement "Thus we can treat the B's as equal. Then it is meaningful to talk about α as a measure of effect" (LISREL V, p. V-29; LISREL VI, p. V-28, Sorbom and Joreskog, 1982:404; Sorbom, 1982:189) leads us (erroneously) to suspect that differences between α's are only interpretable as effects if the same B's are used in the two groups.

10. The issue of the meaning of test items is as appropriate between groups as it is between items. For example, Joreskog and Sorbom have used the mere application of the same test to two different groups as prima facie evidence that the same λ's and ν's apply to the different groups (LISREL V p. V-20, LISREL VI p. V-19; Sorbom, 1974, 1982). The questionableness of this procedure becomes obvious if we imagine the test as an IQ test, and the groups as blacks and whites. The issues of "culture fairness" and the "meaning of the test" to the members of the different groups are missed if we merely adopt application of the same test instrument as ample evidence to guarantee measurement equivalence.

Chapter 10

Odds and Endings

This chapter covers a series of disjoint topics important enough to deserve comment but which do not fit the structures of the previous chapters.

10.1 Old Beta

When beginning our discussion of LISREL in Chapter 4, we presented Eq. 4.1 as embodying the relationships among the modeled concepts.

$$\eta = \mathbf{B}\eta + \Gamma\xi + \zeta \qquad 10.1$$

This basic equation is used in LISREL V and VI, but all the earlier versions of LISREL used the equation

$$(\text{old}\mathbf{B})\eta = \Gamma\xi + \zeta \qquad 10.2$$

To see that Eq. 10.2 is merely a rearrangement of Eq. 10.1, subtract $\mathbf{B}\eta$ from both sides of 10.1 and factor out η from the two terms now appearing on the left of the equation. This shows that the matrix called $\mathbf{B}$ in the earlier versions of LISREL is merely $\mathbf{I}$ minus the $\mathbf{B}$ we are accustomed to using. That is,

$$\text{old}\mathbf{B} = (\mathbf{I} - \mathbf{B}) \qquad 10.3$$

Thus, the old$\mathbf{B}$ appearing in the earlier versions of LISREL has 1's on the diagonal, instead of the current 0's, and the off-digaonal elements are the negative of the current **B** values.

Equation 10.2 places the endogenous variables on the left of the equation and the exogenous variables on the right, but having to continually reverse the signs of the old**B** coefficients before entering them on diagrams or discussing them as effect coefficients eventually forced the notation change. The placement of **B** makes no difference to the magnitudes of the coefficient estimates, χ^2, standard errors, or any other aspect of LISREL's output—it just reverses the signs of the **B** coefficients. Caution is therefore warranted when reading works based on LISREL III or IV. The signs of the **B** coefficients should, but may not, have been reversed before the **B** estimates were entered on path diagrams or discussed as effect coefficients.

10.2 LISREL in SPSSX and Pairwise Matrices

The distribution of LISREL VI as a supplement to SPSSX is most welcome, but it needs two cautions because the LISREL user can now create the input covariance matrix S using SPSSX conventions as an antecedent portion of a LISREL program. The first caution may save you money. Creating a covariance matrix is an expensive task, especially if both N and the number of indicators are large. By storing the covariance matrix and then recalling it for subsequent LISREL runs, you can avoid the cost of repeatedly recalculating the input covariance matrix.

The second caution concerns the uses of pairwise versus listwise covariance matrices. The mathematics grounding the calculation of maximum likelihood estimates (Chapter 5) assumes we have a covariance matrix created by recording the value of each individual (case) on all the variables included in the input data matrix—a listwise matrix. In pairwise calculations, each covariance is based on all the cases having information available for only the relevant pair of variables (not all the variables on the list); therefore different covariances can be based on different sets of cases.

Pairwise matrices are easily obtained from SPSSX, and their use in LISREL is likely to increase. In general, use of pairwise matrices in LISREL should be avoided but not blindly condemned. We must weigh the relative costs of violating the assumption of a listwise matrix against the cost of using an unrealistic listwise matrix. The costs involved in violating the assumption of a listwise matrix are unknown, but they are likely to be proportional to both the degree of nonoverlap between the cases providing the various covariances and the size of the smallest N's on which any of the covariances are based.[1] Both of these factors influence the degree to which the entered covariances are likely to depart from their true population values.

The costs involved in using an unrealistic listwise matrix arise because a terribly unusual type of individual may be needed to provide a complete set of information on all the variables of interest. In the usual social science survey, for example, these respondents are "available," willing to talk to an interviewer about some topic in which they have little interest, and sufficiently polite or compliant to answer all the "difficult" questions. Clearly only a special type of respondent is likely to survive this implict screening, and, equally clearly, the extent to which issues like this are problematic will depend on the details of the data collection process. Thus, listwise selection also involves somewhat of an "unknown cost" in general, even if it might be well determined for any given data set.

We have encountered situations in which a longitudinal data collection procedure was combined with shifts in data collection strategies in ways such that few cases had full information on all the variables of interest but where we nonetheless felt comfortable depending on pairwise matrices (Entwisle and Hayduk, 1982). We have also seen models based on matrices containing thousands of cases that we view as relatively worthless because many more thousands of cases were excluded during calculation of a listwise matrix.[2]

In models using pairwise matrices, the reasonableness of ultimate estimates should be assessed carefully, and the overall output should be viewed tentatively because the response of χ^2, standard errors, and other program output to pairwise matrices is unknown and may well be unknowable given the range of "reasons" prompting the use of pairwise matrices. Whether using a listwise or pairwise matrix, researchers should rerun the final model using the "other" matrix and report any differences in results. This procedure will help us when accumulating evidence on this as yet minimally investigated topic, and it may clarify the portions of the sampled population to which the results can or cannot be generalized. Whenever use of a pairwise matrix is contemplated, the various alternatives to outright elimination of cases (mean substitutions, regression based estimates, etc.) should be carefully considered (cf. Hertel, 1976). New procedures for handling missing data are also emerging (cf. Lee, 1986; Skinner, 1986).

10.3 Some Data-related Issues

10.3.1 Ordinal Measures

Versions V and VI of the LISREL manual include a brief chapter instructing LISREL users on how to obtain polychoric and polyserial correlations for input as S. The manual recommends replacing the usual Pearson, or product moment, correlation coefficient with these other measures of association whenever there are endogenous or nonfixed exogenous variables measured as ordinal variables having few categories or strongly skewed distributions. The general strategy is to replace the corresponding element(s) of S with a polychoric correlation (if both variables are discrete ordinal) or a polyserial correlation (if one variable is continuous and the other is discrete ordinal) and proceed as before.

The suggested procedure is only appropriate if we intend to analyze a correlation matrix and not a covariance matrix. The procedure does not eliminate the normality assumption. Indeed, it explicitly assumes that the observed nonnormal variables correspond to true underlying variables that are multivariate normally distributed, and that the observed data appear as nonnormal merely because poor cutpoints were used to specify the observation categories. The recommended procedure amounts to searching for a better estimate of the correlations in the "true" underlying multivariate normal distribution, based on the assumption that all the skewness and clumping of cases within categories is an artifact of the arbitrary specification of category boundaries. The procedure is designed to "improve" data that missed being interval and multivariate normal because of poor category specification. The process is not designed to handle truly categoric variables, such as religious affiliation or preferred political party. For such variables we are forced to use multiple dummy variables as in multiple regression, or to create stacked models.

The replacement of product moment correlations seems most beneficial (given a truly multivariate normal population distribution) when the arbitrariness in the categorization process has produced oppositely skewed categoric distributions for multiple indicators of a single concept. Under these circumstances, using the ordinary correlation coefficient as input can result in the appearance of "more" common factors in exploratory factor analyses than are justified (Olsson, 1979b). Using the ordinary correlation coefficient with similarly skewed multiple indicators or nonskewed indicators seems much less problematic.

Johnson and Creech (1983) investigate the impact of categorization error on a "causal chain" model. They come to the relatively optimistic conclusion that "while categorization errors do produce distortions in multiple indicator models, under most of the conditions explored these distortions were not of sufficient magnitude to strongly bias the estimates of the important parameters" (Johnson and Creech, 1983:406). They do warn that variables having four or fewer categories can be troublesome if they appear in models with small sample sizes. The reactions of models other than simple causal chains and exploratory factor models to the various categorization errors remain unknown.[3]

Polychoric and polyserial correlations are undoubtedly preferable if the variables do originate from poor classification of truly multivariate normal variables, but several issues remain. First, there has been no mathematical demonstration that polychoric and polyserial correlations can be justifiably entered into the fit function for maximum likelihood estimation (Eq. 5.9) or its relatives for other fitting procedures. Sometimes this is clearly unjustified because the S containing the replaced correlations may not be positive definite, so S has zero determinant and the maximum likelihood procedure breaks down. Matrices with a few entries replaced by polychoric or polyserial correlations are much like pairwise matrices in that the fundamentals of the estimation process are violated in unanticipated ways and the implications of the changes may be "data set" dependent.

Second, if the underlying variables are nonnormally distributed, it remains unknown whether more harm is done by living with the ordinary correlation coefficient and the skewed distribution or correcting the skew by even further emphasizing the untenable assumption of multivariate normality through calculation of the polychoric correlation. That is, if the problematic skewness really does originate from a skewed or otherwise nonmultivariate normal population distribution, we might be doing more harm than good by "rectifying" the problem of categoric data at the expense of stressing the unfounded assumption of multinormality. Though Joreskog and Sorbom (1985a) give dramatic evidence on how poorly the usual Pearson product moment matrix might perform, it seems much more must be learned about these particular procedures before they can be unconditionally recommended.

Those wishing to pursue the replacement of Pearson correlations with polychoric or polyserial correlations can read Joreskog (1986), Muthen (1983) and the numerous references therein, Olsson (1979a,b), Olsson, Drasgow, and Dorans (1982), and Winship and Mare (1983). Less mathematically demanding background readings are Guilford and Fruchter (1978:304ff) and Borgatta and Bohrnstedt (1981). Muthen (1984) suggests a general strategy for dealing with mixed levels of measurement, but this work is too new to have had any impact.

The section on categoric data transformations will be removed from the LISREL VII manual and replaced with a stand-alone PRELIS program for the preprocessing of the raw data (Joreskog and Sorbom, 1985a, 1987). In addition to dealing with ordinal data, PRELIS will have an enhanced ability to provide descriptive univariate and bivariate statistics, to transform variables, to adjust for "censored" variables (variables whose recording provides an artificial ceiling or floor value), and to provide an "asymptotic covariance matrix" that will be required for a new estimation technique (diagonally weighted least squares) that will appear in LISREL VII.

Warning: If a data matrix is preprocessed with PRELIS and entered as input into LISREL, the same conceptual meanings should be placed on the variables in both PRELIS and LISREL. The preprocessing demands that the conceptualization of the adjusted variables be normal and hence demands a corresponding commitment to a precise theoretical (conceptual) meaning for the adjusted variable. When the adjusted matrix is subsequently inserted into LISREL as input data, the researcher must be careful to maintain precisely the same conceptual meanings on the LISREL concepts as was inherent in the adjustments made at the data preprocessing stage. Note specifically that no discussions have appeared about the specification of measurement reliabilities for adjusted variables.

10.3.2 Robust Covariance Matrices

Ever since LISREL III, Joreskog and Sorbom have appended a cautionary note to their discussion of the maximum likelihood fit function. The note suggests it is wise to "robustify" the covariance matrix if the observed variables have a distribution that is far from multivariate normal. The references they provide in this context include Andrews, Gnanadesikan, and Warner (1973), Devlin, Gnanadesikan, and Kettenring (1975), Gnanadesikan and Kettenring (1972), and Gnanadesikan (1977).

The consistency of these admonitions stands in stark contrast to the infrequent reports of corrective actions. Some data sets may contain no outliers or influential data points and therefore may need no adjustments. Other researchers may have simply eliminated a small percentage of the extreme cases or "retrieved" the outliers by coding the highest categories with "or more" type codings. Still others may actually have done various forms of outlier adjustment but failed to report it, lest they be accused of "altering the sacred data" with a not-so-veiled hint that the alteration in some way biased the results in support of the researcher's views. Others were undoubtedly intimidated by the complexity of the cited sources.

We cannot urge all LISREL users to robustify their covariance matrices as a routine part of their LISREL modeling habits. The decision to engage in data adjustments involves evaluating the relative seriousness of risking poorer estimates caused by more or less serious violation of the assumptions of the estimation procedure against the seriousness of having good estimates but for data that no longer completely and accurately reflect the population of interest. We can encourage all researchers to review the implications of outliers and nonnormality in their data and to report any action or inaction in this regard. The availability of relatively simple introductions to these topics (cf. Bollen and Jackman, 1985; Belsley, Kuh, and Welsch, 1980; Comrey, 1985) should help propagate these procedures. Muthen and Kaplan (1985) report that the χ^2 test is "quite robust" with respect to skewed and kurtotic variables. Such variables tend to make the observed χ^2 larger than expected, so researchers are being conservative if they use the traditional .05 level with such variables.

The PRELIS program, which will be a companion to LISREL VII, will provide an alternative procedure for dealing with nonnormal ordinal variables (see Section 10.3.1).

10.3.3 Sample Selection Bias

Though we assume we have a random sample, we rarely achieve this ideal. The implications of selection bias for LISREL models are likely to be similar to the implications for ordinary regression. Berk (1983) summarizes several important implications of selection bias and provides an entree into the extensive literature on this topic. Muthen and Joreskog (1983) integrate the issue of selection bias with structural equation modeling more generally but not to LISREL specifically. Bielby (1981) discusses how cluster samples can be used to estimate "neighborhood" or clustering effects in LISREL.

10.4 Locating What Is Wrong in a Program

Though Joreskog and Sorbom have made LISREL as "user friendly" as possible, the sheer variety of models estimated and the many things researchers can bungle mitigate against a complete set of diagnostic error messages. Messages regarding too short a time limit or insufficient computer memory are usually clear, but there are numerous other fatal programming mistakes that are not as obvious. For example, failure to use uppercase letters in portions of the program can lead to a message that an unidentified character has been encountered or that the specified option is not available. Forgetting the slash (/) at the end of the variable selection statement can lead to a complaint ostensibly concerning the "model" card that usually follows variable selection.

You should routinely check the input covariance matrix when the LISREL program is first written. If problems appear, begin by rechecking the covariance matrix number by number. Wildly incorrect entries may give the message that the data matrix is "not positive definite" or result in nonconvergence of the iterative fit procedure (Section 5.2) due to inconsistencies between the "erroneous" data and a perfectly good model. Simple recopying errors (e.g., misplacement of a decimal), using a completely wrong matrix, inadvertently including missing value codes as real values, including rows/columns referring to incorrect variables, and incorrect sequencing of the variables selected from the input matrix (endogenous should precede exogenous) are not uncommon and can only be detected by scrutinizing each element of S.

If the number of input variables (NI) is erroneously specified as too small, the computer will not have finished reading all the lines of the input covariance matrix before trying to read the remaining lines of the covariance matrix as model definition cards, resulting in the error message that the program has encountered an "unidentified line name." If you try to have too many data lines read (i.e., if NI on the data card is greater than the number of variables that actually appear, or if a covariance matrix created from too few variables is entered), LISREL will read all the data and try to read the next program line as further data. In doing so, it reads a line containing letters where it expects numbers and therefore gives the error message that there is an "illegal character in a real field."

It is sometimes difficult to determine precisely which line of a LISREL program is creating problems. One useful trick is to insert a line containing unexpected characters (such as %%%%) early in the program and to progressively move this strange line further down in the program on successive runs until LISREL stops complaining about it and starts complaining about the real error. This locates the line in which LISREL encountered a problem, but the actual programming problem could be in any of the preceding lines. An earlier incorrect model specification may have been "accepted" until the current "correct" line highlighted an inconsistency.

If LISREL begins the iterative estimation process but fails to converge, we can often locate the problem by investigating which of the estimated coefficients are being driven to extreme or unreasonable values. (Specifiying an extremely short time limit can interrupt the process after only a few iterations.) Nonconvergent models occasionally result from sampling fluctuations in covariance matrices or from inconsistencies among multiple indicators (cf. Boomsma, 1982, 1985; Anderson and Gerbing, 1984; Driel, 1978). Starting with a few well-established indicators can solve the indicator problem, but we usually have no way of checking for or rectifying the contribution of sampling fluctuations.

10.4.1 Start Values

Usually LISREL's automatic start values are adequate to begin convergence to an acceptable solution. Boomsma (1985) compares the final estimates that result from starting with LISREL's automatic start values, the proper values as start values, or real but wild start values. He concludes that all three start values converged to the same final estimates for his models. Naturally, the wild estimates took more iterations, so these runs cost more, but the ultimately reported estimates were comparable.

Although we can often calculate reasonable start values by doing multiple regressions (without considering measurement error), I prefer using the start values to test my hunches about the ultimate model's operation. Specifically, I often enter my own best guesses for effect sizes by simply guessing how many units of change will appear in the dependent variable if the independent variable is changed 1 unit and all the other effect routings are rendered inoperative.[4] This guessing procedure often produces initial estimates that are worse than the estimates that might be obtained with regression, but they are substantially better at assisting me with an even more important task. The size of the differences between my initial estimates (start values) and the final estimates tell me where my mental bets are most inconsistent with the data. I will gladly pay an extra computing dollar to gain a clear picture of my mental biases, especially if this knowledge saves me the embarrassment of being publicly censured for my biases.

If you do not care to make specific estimates, zero start values (especially for $\mathbf{B}$, $\mathbf{\Gamma}$, $\mathbf{\Psi}$, $\mathbf{\Theta}_e$, $\mathbf{\Theta}_\delta$ and off-diagonal elements of $\mathbf{\Phi}$) are often reasonable and simple to program.[5] But beware that LISREL is occasionally sensitive to zero start values. For example, if the model in Figure 7.8 is given a start value of 0.0 for β_{34}, no acceptable solution is found, but if a start value of 0.2 is given the model converges to the expected solution. I have also encountered models that converge to the same acceptable solution with start values slightly to either side of zero, though not from zero itself, so I suspect this is a problem with LISREL's programming and not a true discontinuity in the fitting function. So, avoid zero start values if you encounter convergence problems.

10.5 Alternative Estimation Strategies

The preceding chapters focused on maximum likelihood estimation of the model coefficients. LISREL offers the user a choice of several different estimation strategies: maximum likelihood (ML), generalized least squares (GL), unweighted least squares (UL), two-stage least squares (TS), and instrumental variables (IV). Choosing among these amounts to choosing the quantity to be minimized or maximized in arriving at the ultimate estimates (cf. Joreskog, 1978a); the assumptions you are willing to make to obtain the estimates; and the additional quantities (such as χ^2) required. For example, no overall goodness of fit measures are available for the IV, TS, or UL estimation procedures. Naturally, we chose to discuss MLE because it seems to be the most widely applicable procedure. Generalized least squares estimation is probably the second most widely applicable procedure, because it does not assume multivariate normality, and it still provides a χ^2 test of the model fit. But the other strategies also have traditional applications and dedicated supporters.

We recommend that novice LISREL users rerun their "final" model with the two or three estimation strategies most common to the appropriate literature. My experience has been that the choice of estimation strategies usually makes little difference, but it has the side benefit of creating a footnote that quiets much potential opposition. Naturally, if any substantial differences are found, they should be investigated in detail. Brown (1986a,b), for example, found that a program error in LISREL VI led to incorrect estimates whenever automatic start values were used with GLS. Useful discussions of alternative estimation strategies appear in Huba and Harlow (1987), Joreskog (1978a, 1981a), and Joreskog and Goldberger (1972). A wider variety of statistical alternatives are cited in the following section. LISREL VII will include two additional estimation procedures: weighted least squares (WLS), which is an asymptotically distribution-free procedure, and diagonally weighted least squares (DWLS) (Joreskog and Sorbom, 1985a, 1987).

10.6 A Guide to the Literature

You should have no pressing need to read other introductions to LISREL, but I recommend that you skim whichever of the following sources are most likely to be read by researchers in your particular substantive area: Anderson (1987), Bollen (1987), Bynner and Romney (1985), Carmines (1986), Carmines and McIver (1981), Cuttance (1985), Fox (1984:282ff.), Herting (1985), Joreskog (1982a), Joreskog and Sorbom

(1982), Long (1976, 1981, 1983a,b), Lomax (1982), Saris and Stronkhorst (1984), and Wolfle (1982). It should make you sensitive to the different styles of LISREL modeling, and it may even provide sufficient contrast for you to appreciate the perspectives embodied in this book. Everitt (1984) has written a brief introduction designed for applied statisticians.

Cliff (1985) clearly discusses how some classic problems in correlational analysis extend to structural equation modeling. Jackson and Chan (1980) and Steiger (1979) give cautionary notes focusing on factor analysis.

For the usefulness of various goodness of fit tests, see Anderson and Gerbing (1984), Boomsma (1982), Hoelter (1983a), Matsueda and Bielby (1986), and Saris, Pijper, and Zegwaart (1979). Allison (1977) and Busemeyer and Jones (1983) raise several fundamental issues about scaling and the assumption that one has the correct scales with the correct zero points.

For handling growth curves, developmental models, and longitudinal data, see Gollob and Reichardt (1987), Hertzog and Nesselroade (1987), Joreskog (1979c), Joreskog and Sorbom (1980, 1985b), Kessler and Greenberg (1981), McArdle and Epstein (1987), and Sorbom and Joreskog (1981b). Molenarr (1985) links LISREL to multivariate time series anlaysis. Readers fluent in German can read Jagodzinski (1984). Arminger (1986) and Otter (1986) discuss dynamic LISREL models.

Matsueda and Bielby (1986) and Satorra and Saris (1985) discuss the power of the χ^2 test to discriminate between competing models and parameter specifications.

Herting and Costner (1985) use LISREL's ability to obtain a model-implied Σ for models composed entirely of fixed coefficients to evaluate the ability of various diagnostic indicators to guide researchers in recovering the "true" model. They also illustrate how separate estimation of a complex original model and a supposedly equivalent simplified model can be an additional check on correct model specification.

More sophisticated discussions of theoretical statistics are Bartholomew (1985), Bentler (1983a), Browne (1974, 1982, 1984), Dijkstra (1983, 1985), McDonald (1978, 1979, 1980, 1986), and McDonald and Krane (1978).

For further discussion of moment structures, see Bentler (1983a, 1986), Bentler and Weeks (1985), and McArdle and McDonald (1984).

Other estimation strategies or programs are discussed in Bentler (1983b,c, 1986, 1987), Bentler and Lee (1983), Bentler and Newcomb (1986), Bentler and Weeks (1979, 1980, 1982, 1985), Brown (1986a) Fraser (1980), Horn and McArdle (1980), Joreskog (1979c), Joreskog and Sorbom (1985a, 1986), Lee and Jennrich (1979, 1984), Lee and Tsui (1982), Long (1983b:85), McArdle (1980), McArdle and Epstein (1987),

McArdle and McDonald (1984), McDonald (1978, 1979, 1980, 1985), Schoenberg (1982), and Sorbom (1978). On the EM (Expectation-step, Maximization-step) method in particular see Bentler and Tanaka (1983), Bye, Gallicchio, and Dykacz (1985), Rubin and Thayer (1982, 1983), and Schoenberg and Richtand (1984).

The COSAN program of McDonald (1985) may be computationally more efficient than LISREL in the special case of numerous time points in a block-recursive panel model. COSAN's ability to declare any parameter to be a "possibly nonlinear" function of other model parameters (cf. McDonald, 1985:148) makes this attractive to those attempting to fit the types of models discussed near the end of Chapter 7. LISREL VII will also allow arbitrary constraints between coefficients, so COSAN's advantage is only temporary. The distinction between LISREL and the RAM model of McArdle and McDonald is largely a question of notation. McDonald (1985:154) says that "LISREL is a special case of RAM. Conversely, RAM from another point of view is a special case of LISREL." Although the RAM conceptualization of error variables is equivalent to LISREL's error variables (cf. McDonald, 1985:153), RAM diagrams are unconventional in their presentation of error covariances (cf. McDonald, 1985:134, 132n; Nesselroade and McArdle, 1985). COSAN is available from C. Fraser, Department of Measurement, Evaluation and Computer Applications, The Ontario Insititute for Studies in Education, 252 Bloor Street West, Toronto, Ontario, Canada M5S 1V6.

The EQS program developed by Bentler (1983b) also allows inequality constraints, and it does not require matrix algebra, but it cannot do multiple-group analyses. EQS is available from BMDP Statistical Software, 1964 Westwood Blvd., Suite 202, Los Angeles, California 90025. Dillon, Kumar, and Mulani (1987:130) describe how to represent Bentler's structural factor analysis model within LISREL.

The SIMPLIS program (a fast and SIMPle version of LISREL) (Joreskog and Sorbom, 1986, 1987) will introduce LISREL to a wider audience by avoiding matrix algebra and Greek characters, but its output is very limited, and the broadening of the user base will undoubtedly increase the incidents of published rubbish.

Works of Joreskog or Sorbom are also of interest for historical and contextual reasons. Joreskog (1963, 1966, 1967, 1968, 1969, 1970a,b, 1971a,b, 1973a,b, 1977a,b,c, 1978a,b, 1979a,b, 1981a,b, 1982a,b, 1983, 1986), Joreskog and Goldberger (1972, 1975), Joreskog, Gruvaeus, and van Thillo (1970), Joreskog and Lawley (1968), Joreskog and Sorbom (1976a,b, 1977, 1978a,b, 1979, 1980, 1981, 1982, 1983, 1984, 1987), Joreskog and Wold (1982a,b,c), McGaw and Joreskog (1971), Rock et al. (1977), Sorbom (1974, 1975, 1976, 1978, 1986), Sorbom and Joreskog (1981a,b, 1982), Werts, Joreskog, and Linn (1973, 1985), Werts, Linn, and Joreskog (1971, 1974, 1977, 1978), and Werts et al. (1976, 1977). Many of their earlier works are reprinted in Joreskog and Sorbom (1979).

10.7 The End

Do not be discouraged if you feel you do not know every detail of LISREL despite having persisted to the end. Structural equation modeling is advancing, and there are many unresolved issues and loose ends. Do resolve to read at least a few of the cited references and do glance at even the "tough" new works that will undoubtedly appear. Do not let a concern for the technical side of LISREL blind you to the greatest hindrance to LISREL modeling—namely a paucity of theory and substantive knowledge. Ill-formed and imprecise theories ruin many models, and flagging expertise dooms many more. If you know your subject area, proceed with caution and attention and build models with only the best indicators, you will learn a lot from LISREL—and maybe even chance upon a theory implying a Σ that is acceptably close to S.

Notes

1.   If a pairwise matrix is used, the N entered into LISREL should usually be the minimum N for any covariance in the pairwise matrix. This may be overly conservative if almost all of the covariances in the matrix are calculated on larger N's.

2.   The listwise/pairwise dichotomy is not as hard and fast as one might assume. Creative researchers will undoubtedly construct covariance matrices that are part listwise and part pairwise. For example, one might create a listwise matrix for the variables where listwise deletion results in the loss of only a few cases, and then supplement this with the best available pairwise covariances for the problematic variables.

3.   In Chapter 9 we encountered some indirect data suggesting that even dichotomous data can be sufficiently well behaved to remove the need for polychoric correlational adjustments. Recall the dichotomous smoker/nonsmoker variable used as an indicator of the concept smoking. In Chapter 9 we discussed a model allowing multiple indicators of the concept smoking, where one indicator was this dichotomous smoker/nonsmoker indicator and the second indicator was a five-point scale reflecting the physiological strain created by varying amounts of current and past smoking. Though a five-point scale is far from interval, we certainly would have expected major modeling problems to appear with the simultaneous modeling of the two-point and five-point scales if

the dichotomous indicator was poorly behaved. The model incorporating these two smoking indicators was well behaved and displayed no problems with the dichotomous indicator.

This absence of problems cannot be attributed to an inability to distinguish a two-point scale from a five-point scale. Creating a two-indicator model using the dichotomous smoker/nonsmoker indicator with either the three-point "smoking history" scale or the four-point "amount smoked" scale (Figure 4.4) instead of the five-point physiology scale produces ill-fitting models. The reason these models fail is also clear. These particular indicators amount to expanding one or the other of the two categories in the original dichotomous scale and hence force some nonlinearity between the original dichotomy and these other indicators. Thus the indicators are nonlinearily related while the model formally maintains that they are multiple and linear indicators of a single concept. The implicit nonlinearity between the dummy and those other scales can be seen by plotting the original 0–1 coding against the new coding provided by the amount and history scales in Figure 4.4.

4.   One of the most difficult aspects of making mental estimates is keeping track of the real metrics of the variables, especially when trying to estimate free variances/covariances (Θ_ϵ, Θ_δ, Ψ) in the context of multiple indicators. Equation 1.26 on the partitioning of the variance of a linear composite, even if only roughly calculated, is most useful because it highlights variances and the squaring of structural coefficients implicit in the calculation of the variance of a "new" dependent variable.

5.   Zero is the default start value if the creation of automatic start values is suppressed.

Appendix A

Summation Notation

By definition

$$\sum_{i=1}^{n} x_i = x_1 + x_2 + x_3 + x_4 + \cdots + x_n \qquad \text{A.1}$$

This implies

$$\sum_{i=2}^{5} x_i = x_2 + x_3 + x_4 + x_5 \qquad \text{A.2}$$

If the index i begins with the value 1 and continues until all possible values of i have been considered, the beginning and ending designations on the sum are often omitted. Thus,

$$\sum_i x_i = x_1 + x_2 + x_3 + x_4 + \cdots + x_n \qquad \text{A.3}$$

Because

$$a(x_1 + x_2 + x_3 + \cdots + x_n) = ax_1 + ax_2 + ax_3 + \cdots + ax_n \qquad \text{A.4}$$

$$a\sum_i x_i = \sum_i ax_i \qquad \text{A.5}$$

Because

$$(ax_1 + by_1) + (ax_2 + by_2) + (ax_3 + by_3) + \cdots + (ax_n + by_n)$$
$$= (ax_1 + ax_2 + ax_3 + \cdots + ax_n) \qquad \text{A.6}$$
$$+ (by_1 + by_2 + by_3 + \cdots + by_n)$$

$$\sum_i (ax_i + by_i) = \sum_i ax_i + \sum_i by_i \qquad \text{A.7}$$

Summation may cumulate complex entities.

$$\sum_i a(x_iy_i)^2 = a(x_1y_1)^2 + a(x_2y_2)^2 + a(x_3y_3)^2 + \cdots + a(x_ny_n)^2 \qquad \text{A.8}$$

The X scores for a set of cases may be written in the cells of a table having r rows and c columns instead of a single long row (as in Eq. A.1). If i indexes the row in which a case appears and j indexes the column in which it appears, the overall sum of the X values for these cases (the sum of the cells) is

$$\sum_i^r \sum_j^c x_{ij} = \sum_j^c x_{1j} + \sum_j^c x_{2j} + \sum_j^c x_{3j} + \cdots + \sum_j^c x_{rj} \qquad \text{A.9}$$

That is, the overall sum is the sum of the row totals. Equivalently, the overall sum also equals the sum of the column totals.

$$\sum_j^c \sum_i^r x_{ij} = \sum_i^r x_{i1} + \sum_i^r x_{i2} + \sum_i^r x_{i3} + \cdots + \sum_i^r x_{ic} \qquad \text{A.10}$$

Appendix B

LISREL Output for the Smoking Model

```
17 JAN 86    SPSS-X RELEASE 2.1  FOR IBM VM/MTS                          PAGE    1
10:51:24     University of Alberta


For                        University of Alberta             License Number 30

    1    O          title 'SMOKING  LISREL IN SPSSX'
    2    O          file handle 8/name='smokcov1'
    3    O          INPUT PROGRAM
    4    O          NUMERIC A
    5    O          END FILE
    6    O          END INPUT PROGRAM
    7    O          USERPROC NAME=LISREL
    8    O          SMOKING -- SEX AGE EDUCATION RELIGION(attendance)        smoklist
    9    O          DA NI=13 NO=432 MA=CM
   10    O          CM UN=8 FU FO
   11    O          (8F10.7)
   12    O          LA
   13    O          'smoking1' 'smoking2' 'smoking3' 'smoking4' 'asviews' 'asacts'
   14    O          'sex' 'age' 'education' 'Godexists' 'strongX' 'R-attend' 'important'
   15    O          SE
   16    O          'smoking1' 'asviews' 'asacts'
   17    O          'sex' 'age' 'education' 'R-attend'/
   18    O          MO NY=3 NX=4 NE=3 NK=4 LY=ID LX=ID BE=FU,FI GA=FU,FR C
   19    O           PH=FU,FR PS=DI,FR TE=DI,FI TD=DI,FI
   20    O          FR BE(2,1) BE(3,1) BE(3,2)
   21    O          FI GA(3,2)
   22    O          PL GA(3,2)
   23    O          VA   .0247   TE(1,1)
   24    O          VA   .1657  TE(2,2)
   25    O          VA   .0651  TE(3,3)
   26    O          VA   .0025  TD(1,1)
   27    O          VA  12.3449  TD(2,2)
   28    O          VA   .6909  TD(3,3)
   29    O          VA   .4657  TD(4,4)
   30    O          OU ML AL TM=15
   31    O          end user

THERE ARE    64984 BYTES OF SPSS MEMORY AVAILABLE.
THE LARGEST CONTIGUOUS AREA HAS    64984 BYTES.
SPSS IS USING   65536 BYTES OF SYSTEM MEMORY.
THERE ARE AT LEAST  2031616 BYTES OF SYSTEM MEMORY AVAILABLE
```

```
17 JAN 86    SMOKING  LISREL IN SPSSX                                    PAGE    2
10:52:39     University of Alberta


                      L I S R E L   VI - VERSION 6.6

                                  BY

               KARL G JORESKOG AND DAG SORBOM


SMOKING -- SEX AGE EDUCATION RELIGION(attendance)        smoklist

THE FOLLOWING LISREL CONTROL LINES HAVE BEEN READ :


DA NI=13 NO=432 MA=CM
CM UN=8 FU FO
(8F10.7)
LA
'smoking1' 'smoking2' 'smoking3' 'smoking4' 'asviews' 'asacts'
'sex' 'age' 'education' 'Godexists' 'strongX' 'R-attend' 'important'
SE
'smoking1' 'asviews' 'asacts'
'sex' 'age' 'education' 'R-attend'/
MO NY=3 NX=4 NE=3 NK=4 LY=ID LX=ID BE=FU,FI GA=FU,FR C
 PH=FU,FR PS=DI,FR TE=DI,FI TD=DI,FI
FR BE(2,1) BE(3,1) BE(3,2)
FI GA(3,2)
PL GA(3,2)
VA   .0247   TE(1,1)
VA   .1657  TE(2,2)
VA   .0651  TE(3,3)
VA   .0025  TD(1,1)
VA  12.3449  TD(2,2)
VA   .6909  TD(3,3)
VA   .4657  TD(4,4)
OU ML AL TM=15
```

```
17 JAN 86    SMOKING  LISREL IN SPSSX                                      PAGE    3
10:52:43     University of Alberta


                        L I S R E L  VI - VERSION 6.6


SMOKING -- SEX AGE EDUCATION RELIGION(attendance)         smoklis1


                    NUMBER OF INPUT VARIABLES  13

                    NUMBER OF Y - VARIABLES     3

                    NUMBER OF X - VARIABLES     4

                    NUMBER OF ETA - VARIABLES   3

                    NUMBER OF KSI - VARIABLES   4

                    NUMBER OF OBSERVATIONS    432


                        OUTPUT REQUESTED

                    TECHNICAL OUTPUT          YES

                    STANDARD ERRORS           YES

                    T - VALUES                YES

                    CORRELATIONS OF ESTIMATES YES

                    FITTED MOMENTS            YES

                    TOTAL EFFECTS             YES

                    VARIANCES AND COVARIANCES YES

                    MODIFICATION INDICES      YES

                    FACTOR SCORES REGRESSIONS YES

                    FIRST ORDER DERIVATIVES   YES

                    STANDARDIZED SOLUTION     YES

                    PARAMETER PLOTS           YES

                    AUTOMATIC MODIFICATION     NO
```

```
17 JAN 86    SMOKING  LISREL IN SPSSX                                      PAGE    4
10:52:43     University of Alberta

SMOKING -- SEX AGE EDUCATION RELIGION(attendance)         smoklis1

            COVARIANCE MATRIX TO BE ANALYZED

          smoking1    asviews     asacts     sex       age       educatio  R-attend
smoking1    0.247
asviews    -0.234     3.314
asacts     -0.084     0.291      1.302
sex        -0.010     0.153      0.044     0.250
age        -0.907     4.850     -1.366     0.167    246.898
educatio   -0.213     0.375      0.357     0.016     -7.251    6.909
R-attend   -0.188     0.553     -0.154     0 139      7.239    0.233     4.657


            DETERMINANT = 0.148151E+04
```

```
17 JAN 86   SMOKING  LISREL IN SPSSX                                    PAGE   5
10:52:43      University of Alberta


SMOKING -- SEX AGE EDUCATION RELIGION(attendance)         smoklis1


PARAMETER SPECIFICATIONS

        BETA

              smoking1     asviews      asacts
smoking1          0            0            0
asviews           1            0            0
asacts            2            3            0

        GAMMA

               sex          age        educatio     R-attend
smoking1          4            5            6            7
asviews           8            9           10           11
asacts           12            0           13           14

        PHI

               sex          age        educatio     R-attend
sex              15
age              16           17
educatio         18           19           20
R-attend         21           22           23           24

        PSI

              smoking1     asviews      asacts
                 25           26           27

        THETA EPS

              smoking1     asviews      asacts
                  0            0            0

        THETA DELTA

               sex          age        educatio     R-attend
                  0            0            0            0
```

```
17 JAN 86   SMOKING  LISREL IN SPSSX                                    PAGE   6
10:52:44      University of Alberta


SMOKING -- SEX AGE EDUCATION RELIGION(attendance)         smoklis1


INITIAL ESTIMATES (TSLS)

        BETA

              smoking1     asviews      asacts
smoking1         0.0          0.0          0.0
asviews        -0.878        0.0          0.0
asacts         -0.313       0.068         0.0

        GAMMA

               sex          age        educatio     R-attend
smoking1       -0.015       -0.004       -0.037       -0.036
asviews         0.545        0.017        0.047        0.042
asacts          0.160        0.0          0.045       -0.067

        PHI

               sex          age        educatio     R-attend
sex             0.248
age             0.187      234.553
educatio        0.016       -7.251        5.218
R-attend        0.139        7.239        0.233        4.191

        PSI

              smoking1     asviews      asacts
                0.204        2.736        1.157

        THETA EPS

              smoking1     asviews      asacts
                0.025        0.166        0.065

        THETA DELTA

               sex          age        educatio     R-attend
                0.002       12.345        0.691        0.466

SQUARED MULTIPLE CORRELATIONS FOR Y - VARIABLES

              smoking1     asviews      asacts
                0.900        0.950        0.950

TOTAL COEFFICIENT OF DETERMINATION FOR Y - VARIABLES IS  1.000

SQUARED MULTIPLE CORRELATIONS FOR X - VARIABLES

               sex          age        educatio     R-attend
                0.890        0.950        0.900        0.900
```

TOTAL COEFFICIENT OF DETERMINATION FOR X - VARIABLES IS   1.000

SQUARED MULTIPLE CORRELATIONS FOR STRUCTURAL EQUATIONS

smoking1   asviews   asacts
0.082   0.131   0.064

TOTAL COEFFICIENT OF DETERMINATION FOR STRUCTURAL EQUATIONS   IS   0.159

BEHAVIOR UNDER MINIMIZATION ITERATIONS

ITER	TRY	ABSCISSA	SLOPE	FUNCTION
1	0	0.0	-0.98424071E-05	0.26460064E-02
	1	0.10000000E+01	-0.51260675E-07	0.26410584E-02
2	0	0.0	-0.95650992E-09	0.26410584E-02
	1	0.10000000E+01	0.48129816E-11	0.26410580E-02

```
17 JAN 86    SMOKING  LISREL IN SPSSX                                      PAGE    9
10:52:46     University of Alberta


SMOKING -- SEX AGE EDUCATION RELIGION(attendance)        smoklis1


LISREL ESTIMATES (MAXIMUM LIKELIHOOD)

        BETA

            smoking1    asviews     asacts
smoking1      0.0         0.0         0.0
asviews      -0.879       0.0         0.0
asacts       -0.312       0.068       0.0

        GAMMA

            sex         age         educatio    R-attend
smoking1     -0.015      -0.004      -0.037      -0.036
asviews       0.545       0.017       0.047       0.042
asacts        0.161       0.0         0.046      -0.068

        PHI

            sex         age         educatio    R-attend
sex          0.248
age          0.166     234.553
educatio     0.016      -7.282       6.218
R-attend     0.139       7.269       0.233       4.191

        PSI

            smoking1    asviews     asacts
              0.204       2.736       1.156

        THETA EPS

            smoking1    asviews     asacts
              0.025       0.166       0.065

        THETA DELTA

            sex         age         educatio    R-attend
              0.002      12.345       0.691       0.466

        SQUARED MULTIPLE CORRELATIONS FOR Y - VARIABLES

            smoking1    asviews     asacts
              0.900       0.950       0.950

        TOTAL COEFFICIENT OF DETERMINATION FOR Y - VARIABLES IS   1.000

        SQUARED MULTIPLE CORRELATIONS FOR X - VARIABLES

            sex         age         educatio    R-attend
              0.990       0.950       0.900       0.900
```

```
17 JAN 86    SMOKING  LISREL IN SPSSX                                      PAGE   10
10:52:47     University of Alberta


        TOTAL COEFFICIENT OF DETERMINATION FOR X - VARIABLES IS   1.000

        SQUARED MULTIPLE CORRELATIONS FOR STRUCTURAL EQUATIONS

            smoking1    asviews     asacts
              0.082       0.131       0.065

        TOTAL COEFFICIENT OF DETERMINATION FOR STRUCTURAL EQUATIONS   IS   0.100



            MEASURES OF GOODNESS OF FIT FOR THE WHOLE MODEL :

   CHI-SQUARE WITH   1 DEGREES OF FREEDOM IS      2.28 (PROB. LEVEL = 0.131)

                    GOODNESS OF FIT INDEX IS 0.998

                ADJUSTED GOODNESS OF FIT INDEX IS 0.958

                ROOT MEAN SQUARE RESIDUAL IS      0.221
```

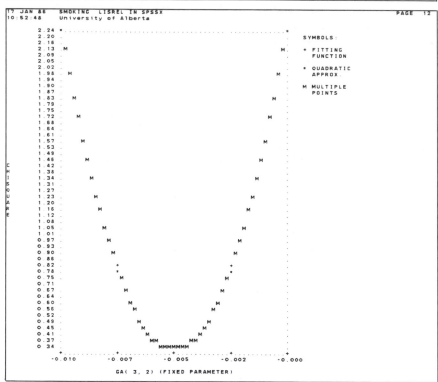

```
17 JAN 86    SMOKING  LISREL IN SPSSX                                          PAGE  11
10:52:47     University of Alberta


SMOKING -- SEX AGE EDUCATION RELIGION(attendance)        smoklis1


MODIFICATION INDICES

        BETA

             smoking1     asviews      asacts
smoking1        0.0          0.0        2.272
asviews         0.0          0.0        2.272
asacts          0.0          0.0         0.0

        GAMMA

             sex          age          educatio    R-attend
smoking1        0.0          0.0          0.0         0.0
asviews         0.0          0.0          0.0         0.0
asacts          0.0          2.272        0.0         0.0

        PHI

             sex          age          educatio    R-attend
sex             0.0
age             0.0          0.0
educatio        0.0          0.0          0.0
R-attend        0.0          0.0          0.0         0.0

        PSI

             smoking1     asviews      asacts
                0.0          0.0         0.0

        THETA EPS

             smoking1     asviews      asacts
                2.272        2.272       0.0

        THETA DELTA

             sex          age          educatio    R-attend
                2.272        0.0          2.272       2.272

        MAXIMUM MODIFICATION INDEX IS    2.27 FOR ELEMENT ( 3, 3) OF THETA DELTA
```

```
17 JAN 86    SMOKING  LISREL IN SPSSX                                          PAGE  12
10:52:48     University of Alberta

      2.24  * ............................................ . *
      2.20  .                                                 .      SYMBOLS:
      2.16  .                                                 .
      2.13  M                                                 M.     + FITTING
      2.09  .                                                 .        FUNCTION
      2.05  .                                                 .
      2.02  .                                                 .      * QUADRATIC
      1.98  . M                                             M .        APPROX.
      1.94  .                                                 .
      1.90  .                                                 .      M MULTIPLE
      1.87  .                                                 .        POINTS
      1.83  . M                                           M   .
      1.79  .                                                 .
      1.75  .                                                 .
      1.72  .   M                                       M     .
      1.68  .                                                 .
      1.64  .                                                 .
      1.61  .                                                 .
      1.57  .     M                                   M       .
      1.53  .                                                 .
      1.49  .                                                 .
      1.46  .       M                               M         .
    C 1.42  .                                                 .
    H 1.38  .                                                 .
    I 1.34  .       M                             M           .
    S 1.31  .                                                 .
    Q 1.27  .                                                 .
    U 1.23  .       M                           M             .
    A 1.20  .                                                 .
    R 1.16  .      M                             M            .
    E 1.12  .                                                 .
      1.08  .                                                 .
      1.05  .        M                         M              .
      1.01  .                                                 .
      0.97  .        M                       M                .
      0.93  .                                                 .
      0.90  .         M                     M                 .
      0.86  .                                                 .
      0.82  .           +                 +                   .
      0.78  .           *               *                     .
      0.75  .           M               M                     .
      0.71  .                                                 .
      0.67  .          M                 M                    .
      0.64  .                                                 .
      0.60  .           M               M                     .
      0.56  .                                                 .
      0.52  .                                                 .
      0.49  .            M             M                      .
      0.45  .            M             M                      .
      0.41  .             M           M                       .
      0.37  .             MM         M                        .
      0 34  .               MMMMMMM                           .
            + ............. + ........... + ........... + .... +
          -0.010      -0.007       -0.005      -0.002    -0.000

                     GA( 3, 2) (FIXED PARAMETER)
```

```
17 JAN 86    SMOKING  LISREL IN SPSSX                                          PAGE   13
10:52:51     University of Alberta


SMOKING -- SEX AGE EDUCATION RELIGION(attendance)          smoklis1


STANDARD ERRORS

      BETA
               smoking1      asviews       asacts
smoking1         0.0          0.0           0.0
asviews          0.194        0.0           0.0
asacts           0.129        0.033         0.0

      GAMMA
               sex           age           educatio      R-attend
smoking1         0.047        0.002         0.010         0.012
asviews          0.168        0.006         0.037         0.045
asacts           0.110        0.0           0.023         0.029

      PHI
               sex           age           educatio      R-attend
sex              0.017
age              0.379        16.819
educatio         0.063        2.020         0.471
R-attend         0.052        1.670         0.273         0.317

      PSI
               smoking1      asviews       asacts
                 0.016        0.199         0.084

      THETA EPS
               smoking1      asviews       asacts
                 0.0          0.0           0.0

      THETA DELTA
               sex           age           educatio      R-attend
                 0.0          0.0           0.0           0.0
```

```
17 JAN 86    SMOKING  LISREL IN SPSSX                                          PAGE   14
10:52:51     University of Alberta


SMOKING -- SEX AGE EDUCATION RELIGION(attendance)          smoklis1


T-VALUES

      BETA
               smoking1      asviews       asacts
smoking1         0.0          0.0           0.0
asviews         -4.538        0.0           0.0
asacts          -2.416        2.049         0.0

      GAMMA
               sex           age           educatio      R-attend
smoking1        -0.308       -2.362        -3.717        -2.855
asviews          3.240        2.885         1.287         0.934
asacts           1.458        0.0           1.982        -2.399

      PHI
               sex           age           educatio      R-attend
sex             14.533
age              0.439        13.946
educatio         0.246       -3.605         13.212
R-attend         2.652        4.352         0.851         13.212

      PSI
               smoking1      asviews       asacts
                12.999        13.720        13.820

      THETA EPS
               smoking1      asviews       asacts
                 0.0          0.0           0.0

      THETA DELTA
               sex           age           educatio      R-attend
                 0.0          0.0           0.0           0.0
```

SMOKING -- SEX AGE EDUCATION RELIGION(attendance)   smoklis1

CORRELATIONS OF ESTIMATES

	BE 2 1	BE 3 1	BE 3 2	GA 1 1	GA 1 2	GA 1 3	GA 1 4	GA 2 1	GA 2 2	GA 2 3
BE 2 1	1.000									
BE 3 1	0.002	1.000								
BE 3 2	0.014	0.267	1.000							
GA 1 1	-0.001	-0.003	-0.000	1.000						
GA 1 2	-0.008	-0.007	0.002	0.013	1.000					
GA 1 3	-0.008	-0.012	0.000	-0.003	0.225	1.000				
GA 1 4	-0.002	0.015	0.000	-0.142	-0.264	-0.104	1.000			
GA 2 1	0.014	-0.001	-0.002	0.023	0.000	-0.000	-0.003	1.000		
GA 2 2	0.134	-0.001	-0.004	0.000	0.022	0.004	-0.006	0.015	1.000	
GA 2 3	0.217	-0.003	-0.009	-0.000	0.003	0.021	-0.003	0.000	0.246	1.000
GA 2 4	0.166	0.004	0.017	-0.003	-0.007	-0.004	0.022	-0.138	-0.236	-0.085
GA 3 1	-0.007	-0.031	-0.161	0.014	-0.000	-0.000	-0.002	-0.002	0.001	0.001
GA 3 3	-0.001	0.176	-0.033	-0.001	-0.002	0.011	0.002	0.000	0.000	-0.002
GA 3 4	-0.002	0.175	-0.085	-0.003	-0.002	-0.003	0.017	0.000	0.000	0.001
PH 1 1	-0.000	0.000	0.000	0.000	0.000	-0.000	-0.000	-0.002	-0.000	0.000
PH 2 1	-0.000	0.000	-0.000	0.006	0.000	0.000	-0.000	-0.008	-0.001	-0.000
PH 2 2	-0.000	0.000	-0.000	0.000	0.008	0.000	-0.000	-0.000	-0.010	-0.000
PH 3 1	-0.000	-0.000	-0.000	0.020	0.000	0.000	-0.000	-0.007	-0.000	-0.002
PH 3 2	-0.000	0.001	-0.001	0.000	0.019	0.006	-0.000	-0.000	-0.007	-0.007
PH 3 3	-0.000	-0.000	0.000	-0.000	0.001	0.028	-0.000	0.000	-0.000	-0.010
PH 4 1	-0.000	-0.000	-0.000	0.016	-0.000	-0.000	0.000	-0.005	0.000	0.000
PH 4 2	-0.000	-0.001	0.002	0.000	0.015	0.000	0.000	0.006	-0.005	-0.000
PH 4 3	-0.000	0.000	0.000	-0.000	-0.000	0.015	0.019	0.000	0.000	-0.005
PH 4 4	-0.000	0.000	-0.000	-0.000	-0.001	-0.000	0.022	0.000	0.000	0.000
PS 1 1	0.033	0.020	0.001	-0.003	0.010	0.028	0.018	0.000	0.004	0.007
PS 2 2	0.033	-0.001	-0.003	0.000	-0.000	-0.000	-0.000	-0.001	-0.008	-0.005
PS 3 3	-0.000	0.020	-0.004	-0.000	-0.000	-0.000	0.000	0.000	0.000	0.000

CORRELATIONS OF ESTIMATES

	GA 2 4	GA 3 1	GA 3 3	GA 3 4	PH 1 1	PH 2 1	PH 2 2	PH 3 1	PH 3 2	PH 3 3
GA 2 4	1.000									
GA 3 1	-0.002	1.000								
GA 3 3	-0.001	0.002	1.000							
GA 3 4	-0.003	-0.122	-0.004	1.000						
PH 1 1	0.000	-0.001	-0.000	0.000	1.000					
PH 2 1	0.000	0.000	0.000	-0.000	0.030	1.000				
PH 2 2	0.000	0.000	0.000	-0.000	0.000	0.030	1.000			
PH 3 1	0.000	-0.011	-0.001	0.000	0.017	-0.176	-0.005	1.000		
PH 3 2	0.000	0.000	0.002	-0.002	0.000	0.008	-0.246	0.019	1.000	
PH 3 3	0.000	-0.000	-0.015	0.000	0.000	-0.003	0.031	0.017	-0.246	1.000
PH 4 1	-0.001	0.013	0.000	-0.001	0.181	0.215	0.006	0.042	0.003	0.001
PH 4 2	-0.007	-0.000	-0.003	0.003	0.004	0.130	0.296	-0.021	0.003	-0.010
PH 4 3	-0.007	0.000	0.013	-0.010	0.002	-0.020	-0.053	0.129	0.204	0.058
PH 4 4	-0.007	-0.000	-0.000	0.018	0.017	0.039	0.046	0.007	0.012	0.002
PS 1 1	0.005	0.000	0.004	0.004	0.000	0.000	0.000	0.000	0.000	0.000
PS 2 2	0.001	0.000	0.000	0.000	0.000	0.000	0.000	0.000	0.000	0.000
PS 3 3	-0.000	-0.003	-0.012	0.023	0.000	-0.000	-0.000	0.000	-0.000	0.000

CORRELATIONS OF ESTIMATES

	PH 4 1	PH 4 2	PH 4 3	PH 4 4	PS 1 1	PS 2 2	PS 3 3
PH 4 1	1.000						
PH 4 2	0.047	1.000					
PH 4 3	0.017	-0.164	1.000				
PH 4 4	0.181	0.296	0.058	1.000			
PS 1 1	0.000	0.000	0.000	0.000	1.000		
PS 2 2	0.000	0.000	0.000	0.000	0.001	1.000	
PS 3 3	-0.000	0.000	-0.000	0.000	0.000	0.000	1.000

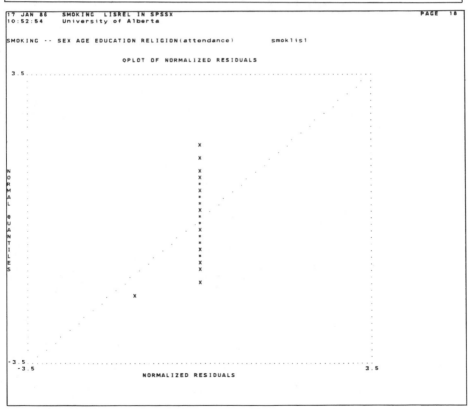

```
17 JAN 86    SMOKING  LISREL IN SPSSX                                    PAGE  17
10:52:53     University of Alberta


SMOKING -- SEX AGE EDUCATION RELIGION(attendance)       smoklis1


FITTED MOMENTS AND RESIDUALS

        FITTED MOMENTS

            smoking1   asviews    asacts     sex        age       educatio   R-attend
smoking1     0.247
asviews     -0.234     3.314
asacts      -0.084     0.290      1.302
sex         -0.010     0.153      0.044      0.250
age         -0.900     4.840     -0.195      0.166    246.898
educatio    -0.213     0.375      0.362      0.016     -7.282     6.909
R-attend    -0.188     0.553     -0.158      0.139      7.269     0.233      4.656

        FITTED RESIDUALS

            smoking1   asviews    asacts     sex        age       educatio   R-attend
smoking1    -0.000
asviews      0.000     -0.000
asacts      -0.000      0.001      0.000
sex          0.000     -0.000     -0.000      0.000
age         -0.007      0.011     -1.171      0.000      0.000
educatio    -0.000      0.000     -0.005      0.000      0.030      0.000
R-attend     0.000     -0.000      0.004     -0.000     -0.031     -0.000      0.000

        NORMALIZED RESIDUALS

            smoking1   asviews    asacts     sex        age       educatio   R-attend
smoking1    -0.000
asviews      0.000     -0.000
asacts      -0.013      0.011      0.000
sex          0.000     -0.000     -0.000      0.000
age         -0.019      0.008     -1.356      0.001      0.000
educatio    -0.000      0.000     -0.036      0.000      0.015      0.001
R-attend     0.000     -0.000      0.033     -0.000     -0.018     -0.001      0.001
```

```
17 JAN 86    SMOKING  LISREL IN SPSSX                                    PAGE  18
10:52:54     University of Alberta


SMOKING -- SEX AGE EDUCATION RELIGION(attendance)       smoklis1

                      QPLOT OF NORMALIZED RESIDUALS

    3.5..........................................................................
        .                                                           .   .  .  .
        .                                                       .  .
        .                                                    .  .
        .                                                 . .
        .                                               .
        .                                            .
        .                            x            .
        .                                      .
        .                            x      .
    N   .                            x    .
    O   .                            x  .
    R   .                            *.
    M   .                            x
    A   .                          . *
    L   .                         .  x
        .                        .   *
    Q   .                      .    .*
    U   .                     .     x
    A   .                    .    . *
    N   .                   .      *
    T   .                 .       x
    I   .                .       x
    L   .              .         x
    E   .            .          x
    S   .           .          x
        .          .         x
        .        .
        .      .       x
        .    .
        .  .
        ..
   -3.5..........................................................................
     -3.5                                                         3.5
                      NORMALIZED RESIDUALS
```

```
17 JAN 86    SMOKING  LISREL IN SPSSX                                           PAGE   19
10:52:55     University of Alberta


SMOKING -- SEX AGE EDUCATION RELIGION(attendance)         smoklis1


TOTAL EFFECTS

          TOTAL EFFECTS OF KSI ON  ETA

                 sex           age        educatio      R-attend
smoking1       -0.015        -0.004       -0.037        -0.036
asviews         0.557         0.020        0.080         0.073
asacts          0.203         0.003        0.063        -0.052

          TOTAL EFFECTS OF KSI ON  Y

                 sex           age        educatio      R-attend
smoking1       -0.015        -0.004       -0.037        -0.036
asviews         0.557         0.020        0.080         0.073
asacts          0.203         0.003        0.063        -0.052

          TOTAL EFFECTS OF ETA ON  ETA

               smoking1      asviews       asacts
smoking1         0.0           0.0           0.0
asviews         -0.879         0.0           0.0
asacts          -0.371         0.068         0.0


LARGEST EIGENVALUE OF (I-BETA)*(I-BETA)-TRANSPOSED (STABILITY INDEX) IS   0.870
```

```
17 JAN 86    SMOKING  LISREL IN SPSSX                                           PAGE   20
10:52:55     University of Alberta


SMOKING -- SEX AGE EDUCATION RELIGION(attendance)         smoklis1


VARIANCES AND COVARIANCES

          ETA - ETA

               smoking1      asviews       asacts
smoking1         0.223
asviews         -0.234        3.149
asacts          -0.084        0.290         1.236

          ETA - KSI

                 sex           age        educatio      R-attend
smoking1       -0.010        -0.900       -0.213        -0.188
asviews         0.153         4.840        0.375         0.553
asacts          0.044        -0.195        0.362        -0.158

          Y - ETA

               smoking1      asviews       asacts
smoking1         0.223        -0.234       -0.084
asviews         -0.234        3.149        0.290
asacts          -0.084        0.290        1.236

          Y - KSI

                 sex           age        educatio      R-attend
smoking1       -0.010        -0.900       -0.213        -0.188
asviews         0.153         4.840        0.375         0.553
asacts          0.044        -0.195        0.362        -0.158

          X - ETA

               smoking1      asviews       asacts
sex            -0.010         0.153         0.044
age            -0.900         4.840        -0.195
educatio       -0.213         0.375         0.362
R-attend       -0.188         0.553        -0.158

          X - KSI

                 sex           age        educatio      R-attend
sex             0.248         0.166         0.016         0.139
age             0.166       234.553        -7.282         7.269
educatio        0.016        -7.282         6.218         0.233
R-attend        0.139         7.269         0.233         4.191
```

```
17 JAN 86    SMOKING  LISREL IN SPSSX                                    PAGE  21
10:52:57     University of Alberta


         KSI - KSI

              sex         age         educatio    R-attend
sex          0.248       0.166         0.016        0.139
age          0.166     234.553        -7.282        7.269
educatio     0.016      -7.282         6.218        0.233
R-attend     0.139       7.269         0.233        4.191
```

```
17 JAN 86    SMOKING  LISREL IN SPSSX                                    PAGE  22
10:52:57     University of Alberta

SMOKING -- SEX AGE EDUCATION RELIGION(attendance)        smoklis1


FIRST ORDER DERIVATIVES

      BETA

           smoking1    asviews     asacts
smoking1   -0.000       0.000       0.020
asviews     0.000       0.000      -0.007
asacts     -0.000      -0.000      -0.000

      GAMMA

           sex         age         educatio    R-attend
smoking1   0.000      -0.000       -0.000      -0.000
asviews    0.000      -0.000        0.000       0.000
asacts     0.000       0.904       -0.000      -0.000

      PHI

           sex         age         educatio    R-attend
sex        -0.000
age        -0.000      -0.000
educatio   -0.000      -0.000       -0.000
R-attend    0.000       0.000        0.000      -0.000

      PSI

           smoking1    asviews     asacts
smoking1   -0.000
asviews     0.000       0.000
asacts      0.017      -0.006       -0.000

      THETA EPS

           smoking1    asviews     asacts
smoking1    0.008
asviews    -0.003       0.001
asacts      0.012      -0.006       -0.000

      THETA DELTA

           sex         age         educatio    R-attend
sex        -0.001
age        -0.001      -0.000
educatio   -0.001      -0.000       -0.001
R-attend    0.001       0.000        0.001      -0.001
```

352   APPENDIX B

SMOKING -- SEX AGE EDUCATION RELIGION(attendance)   smoklis1

FACTOR SCORES REGRESSIONS

        Y

	smoking1	asviews	asacts	sex	age	educatio	R-attend
smoking1	0.886	-0.006	-0.006	0.003	-0.000	-0.003	-0.004
asviews	-0.043	0.943	0.009	0.029	0.001	0.002	0.003
asacts	-0.015	0.003	0.947	0.008	-0.000	0.002	-0.003

        X

	smoking1	asviews	asacts	sex	age	educatio	R-attend
sex	0.000	0.000	0.000	0.990	-0.000	-0.000	0.000
age	-0.142	0.069	-0.002	-0.050	0.944	-0.070	0.079
educatio	-0.085	0.009	0.023	-0.008	-0.004	0.891	0.008
R-attend	-0.067	0.008	-0.023	0.056	0.003	0.005	0.889

SMOKING -- SEX AGE EDUCATION RELIGION(attendance)   smoklis1

STANDARDIZED SOLUTION

        LAMBDA Y

	smoking1	asviews	asacts
smoking1	0.472	0.0	0.0
asviews	0.0	1.774	0.0
asacts	0.0	0.0	1.112

        LAMBDA X

	sex	age	educatio	R-attend
sex	0.498	0.0	0.0	0.0
age	0.0	15.315	0.0	0.0
educatio	0.0	0.0	2.494	0.0
R-attend	0.0	0.0	0.0	2.047

        BETA

	smoking1	asviews	asacts
smoking1	0.0	0.0	0.0
asviews	-0.234	0.0	0.0
asacts	-0.132	0.108	0.0

        GAMMA

	sex	age	educatio	R-attend
smoking1	-0.015	-0.126	-0.197	-0.155
asviews	0.153	0.147	0.066	0.049
asacts	0.072	0.0	0.102	-0.126

        PHI

	sex	age	educatio	R-attend
sex	1.000			
age	0.022	1.000		
educatio	0.013	-0.191	1.000	
R-attend	0.136	0.232	0.046	1.000

        PSI

smoking1	asviews	asacts
0.918	0.869	0.935

        CORRELATION MATRIX FOR ETA

	smoking1	asviews	asacts
smoking1	1.000		
asviews	-0.280	1.000	
asacts	-0.159	0.147	1.000

```
17 JAN 86    SMOKING  LISREL IN SPSSX                                    PAGE  25
10:52:58     University of Alberta


        CORRELATION MATRIX FOR Y

            smoking1    asviews     asacts
smoking1     1.000
asviews     -0.280      1.000
asacts      -0.159      0.147      1.000

        TOTAL EFFECTS     ETA ON KSI (STANDARDIZED)

            sex         age         educatio    R-attend
smoking1    -0.015      -0.126      -0.197      -0.155
asviews      0.156       0.176       0.112       0.085
asacts       0.091       0.036       0.141      -0.096

        TOTAL EFFECTS     Y ON X (STANDARDIZED)

            sex         age         educatio    R-attend
smoking1    -0.015      -0.126      -0.197      -0.155
asviews      0.156       0.176       0.112       0.085
asacts       0.091       0.036       0.141      -0.096


THE PROBLEM REQUIRED   1148 DOUBLE PRECISION WORDS,
THE CPU-TIME WAS       0.97 SECONDS
```

```
17 JAN 86    SMOKING  LISREL IN SPSSX                                    PAGE  26
10:52:59     University of Alberta


PRECEDING TASK REQUIRED        1.17 SECONDS CPU TIME;      58.98 SECONDS ELAPSED.

   32   0           finish

      32 COMMAND LINES READ.
       0 ERRORS DETECTED.
       0 WARNINGS ISSUED.
       1 SECONDS CPU TIME.
      98 SECONDS ELAPSED TIME.
         END OF JOB.
```

Appendix C

LISREL Output with Multiple Indicators

The LISREL program and results for the following model are provided in this appendix.

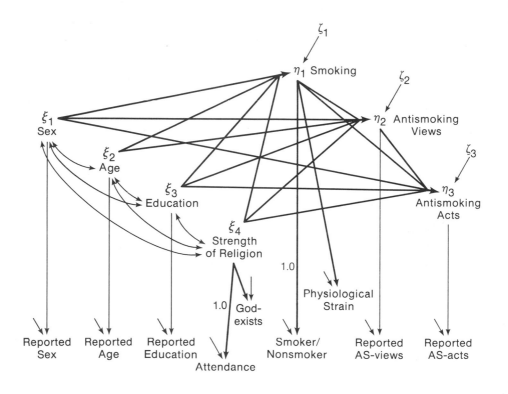

```
21 AUG 86    SPSS-X RELEASE 2.1  FOR IBM VM/MTS                                    PAGE    1
15:30:53     University of Alberta

For                     University of Alberta              License Number 30

    1   O           title 'SMOKING MODEL -- MULTIPLE INDICATORS of SMOKING AND RELIGION'
    2   O           file handle 8/name='smokcov1'
    3   O           INPUT PROGRAM
    4   O           NUMERIC A
    5   O           END FILE
    6   O           END INPUT PROGRAM
    7   O           USERPROC NAME=LISREL
    8   O           *** MULTIPLE INDICATORS S1 + S4, attend + Godexists      Smoklis32a
    9   O           DA NI=13 NO=432 MA=CM
   10   O           CM UN=8 FU FO
   11   O           (8F10.7)
   12   O           LA
   13   O           'smoking1' 'smoking2' 'smoking3' 'smoking4' 'asviews' 'asacts'
   14   O           'sex' 'age' 'education' 'Godexists' 'strongX' 'attend' 'important'
   15   O           SE
   16   O           'smoking1' 'smoking4' 'asviews' 'asacts'
   17   O           'sex' 'age' 'education' 'attend' 'Godexists'/
   18   O           MO NY=4 NX=5 NE=3 NK=4 LY=FU,FI LX=FU,FI BE=FU,FI GA=FU,FR C
   19   O            PH=FU,FR PS=DI,FR TE=DI,FI TD=DI,FI
   20   O           FR BE(2,1) BE(3,1) BE(3,2)
   21   O           FI GA(3,2)
   22   O           VA 1.0 LY(1,1) LY(3,2) LY(4,3)
   23   O           FR LY(2,1)
   24   O           VA 1.0 LX(1,1) LX(2,2) LX(3,3) LX(4,4)
   25   O           FR LX(5,4)
   26   O           VA  .0247  TE(1,1)
   27   O           VA  .2016  TE(2,2)
   28   O           VA  .1657 TE(3,3)
   29   O           VA  .0651 TE(4,4)
   30   O           VA  .0025 TD(1,1)
   31   O           VA  12.3449 TD(2,2)
   32   O           VA  .6909 TD(3,3)
   33   O           VA  .4657 TD(4,4)
   34   O           VA  .4292 TD(5,5)
   35   O           OU ML SE TV RS MI FD SS TM=15
   36   O           end user

THERE ARE   64984 BYTES OF SPSS MEMORY AVAILABLE.
THE LARGEST CONTIGUOUS AREA HAS   64984 BYTES.
SPSS IS USING   65536 BYTES OF SYSTEM MEMORY
THERE ARE AT LEAST  2031616 BYTES OF SYSTEM MEMORY AVAILABLE.
```

```
21 AUG 86    SMOKING MODEL -- MULTIPLE INDICATORS of SMOKING AND RELIGION          PAGE    2
15:30:57     University of Alberta

                         L I S R E L  VI - VERSION 6.6

                                      BY

                   KARL G JORESKOG AND DAG SORBOM

*** MULTIPLE INDICATORS S1 + S4, attend + Godexists      Smoklis32a

THE FOLLOWING LISREL CONTROL LINES HAVE BEEN READ :

DA NI=13 NO=432 MA=CM
CM UN=8 FU FO
(8F10.7)
LA
'smoking1' 'smoking2' 'smoking3' 'smoking4' 'asviews' 'asacts'
'sex' 'age' 'education' 'Godexists' 'strongX' 'attend' 'important'
SE
'smoking1' 'smoking4' 'asviews' 'asacts'
'sex' 'age' 'education' 'attend' 'Godexists'/
MO NY=4 NX=5 NE=3 NK=4 LY=FU,FI LX=FU,FI BE=FU,FI GA=FU,FR C
 PH=FU,FR PS=DI,FR TE=DI,FI TD=DI,FI
FR BE(2,1) BE(3,1) BE(3,2)
FI GA(3,2)
VA 1.0 LY(1,1) LY(3,2) LY(4,3)
FR LY(2,1)
VA 1.0 LX(1,1) LX(2,2) LX(3,3) LX(4,4)
FR LX(5,4)
VA  .0247  TE(1,1)
VA  .2016  TE(2,2)
VA  .1657 TE(3,3)
VA  .0651 TE(4,4)
VA  .0025 TD(1,1)
VA  12.3449 TD(2,2)
VA  .6909 TD(3,3)
VA  .4657 TD(4,4)
VA  .4292 TD(5,5)
OU ML SE TV RS MI FD SS TM=15
```

L I S R E L   VI - VERSION 6.6

*** MULTIPLE INDICATORS S1 + S4, attend + Godexists   Smoklis32a

                        NUMBER OF INPUT VARIABLES 13

                        NUMBER OF Y - VARIABLES   4

                        NUMBER OF X - VARIABLES   5

                        NUMBER OF ETA - VARIABLES   3

                        NUMBER OF KSI - VARIABLES   4

                        NUMBER OF OBSERVATIONS   432

                            OUTPUT REQUESTED

                    TECHNICAL OUTPUT   NO

                    STANDARD ERRORS   YES

                    T - VALUES   YES

                    CORRELATIONS OF ESTIMATES   NO

                    FITTED MOMENTS   YES

                    TOTAL EFFECTS   NO

                    VARIANCES AND COVARIANCES   NO

                    MODIFICATION INDICES   YES

                    FACTOR SCORES REGRESSIONS   NO

                    FIRST ORDER DERIVATIVES   YES

                    STANDARDIZED SOLUTION   YES

                    PARAMETER PLOTS   NO

                    AUTOMATIC MODIFICATION   NO

*** MULTIPLE INDICATORS S1 + S4, attend + Godexists   Smoklis32a

        COVARIANCE MATRIX TO BE ANALYZED

	smoking1	smoking4	asviews	asacts	sex	age	educatio	attend	Godexist
smoking1	0.247								
smoking4	0.636	2.016							
asviews	-0.234	-0.745	3.314						
asacts	-0.084	-0.203	0.291	1.302					
sex	-0.010	-0.042	0.153	0.044	0.250				
age	-0.907	-2.599	4.850	-1.366	0.167	246.898			
educatio	-0.213	-0.667	0.375	0.357	0.016	-7.251	6.909		
attend	-0.188	-0.606	0.553	-0.154	0.139	7.239	0.233	4.857	
Godexist	-0.026	-0.067	0.401	0.141	0.088	1.088	-0.584	1.204	2.146

        DETERMINANT = 0.934258E+03

```
21 AUG 86   SMOKING MODEL -- MULTIPLE INDICATORS of SMOKING AND RELIGION          PAGE   5
15:30:58    University of Alberta


*** MULTIPLE INDICATORS S1 + S4, attend + Godexists      Smoklis32a


PARAMETER SPECIFICATIONS

            LAMBDA Y

                 ETA  1        ETA  2        ETA  3
smoking1            0             0             0
smoking4            1             0             0
asviews             0             0             0
asacts              0             0             0

            LAMBDA X

                 KSI  1        KSI  2        KSI  3        KSI  4
sex                 0             0             0             0
age                 0             0             0             0
educatio            0             0             0             0
attend              0             0             0             0
Godexist            0             0             0             2

            BETA

                 ETA  1        ETA  2        ETA  3
ETA  1              0             0             0
ETA  2              3             0             0
ETA  3              4             5             0

            GAMMA

                 KSI  1        KSI  2        KSI  3        KSI  4
ETA  1              6             7             8             9
ETA  2             10            11            12            13
ETA  3             14             0            15            16

            PHI

                 KSI  1        KSI  2        KSI  3        KSI  4
KSI  1             17
KSI  2             18            19
KSI  3             20            21            22
KSI  4             23            24            25            26

            PSI

                 ETA  1        ETA  2        ETA  3
                   27            28            29

            THETA EPS

               smoking1      smoking4      asviews       asacts
                   0             0             0             0
```

```
21 AUG 86   SMOKING MODEL -- MULTIPLE INDICATORS of SMOKING AND RELIGION          PAGE   6
15:30:58    University of Alberta


            THETA DELTA

               sex           age           educatio      attend        Godexist
                   0             0             0             0             0
```

```
21 AUG 86    SMOKING MODEL -- MULTIPLE INDICATORS of SMOKING AND RELIGION                    PAGE    7
15:30:58    University of Alberta


*** MULTIPLE INDICATORS S1 + S4, attend + Godexists      Smoklis32a


INITIAL ESTIMATES (TSLS)

          LAMBDA Y

               ETA 1         ETA 2         ETA 3
smoking1       1.000         0.0           0.0
smoking4       2.636         0.0           0.0
asviews        0.0           1.000         0.0
asacts         0.0           0.0           1.000

          LAMBDA X

               KSI 1         KSI 2         KSI 3         KSI 4
sex            1.000         0.0           0.0           0.0
age            0.0           1.000         0.0           0.0
educatio       0.0           0.0           1.000         0.0
attend         0.0           0.0           0.0           1.000
Godexist       0.0           0.0           0.0           0.111

          BETA

               ETA 1         ETA 2         ETA 3
    ETA 1      0.0           0.0           0.0
    ETA 2     -0.911         0.0           0.0
    ETA 3     -0.227         0.071         0.0

          GAMMA

               KSI 1         KSI 2         KSI 3         KSI 4
    ETA 1     -0.031        -0.004        -0.044        -0.042
    ETA 2      0.524         0.016         0.041         0.042
    ETA 3      0.154         0.0           0.045        -0.059

          PHI

               KSI 1         KSI 2         KSI 3         KSI 4
    KSI 1      0.248
    KSI 2      0.167       234.553
    KSI 3      0.016        -7.251         6.218
    KSI 4      0.147         7.270         0.166         4.371

          PSI

               ETA 1         ETA 2         ETA 3
               0.231         2.697         1.167

          THETA EPS

            smoking1      smoking4       asviews        asacts
               0.025         0.202         0.166         0.065
```

```
21 AUG 86    SMOKING MODEL -- MULTIPLE INDICATORS of SMOKING AND RELIGION                    PAGE    8
15:30:58    University of Alberta


          THETA DELTA

            sex           age           educatio       attend        Godexist
               0.002        12.345         0.691         0.466         0.429

          SQUARED MULTIPLE CORRELATIONS FOR Y - VARIABLES

            smoking1      smoking4       asviews        asacts
               0.900         0.900         0.950         0.950

          TOTAL COEFFICIENT OF DETERMINATION FOR Y - VARIABLES IS   1.000

          SQUARED MULTIPLE CORRELATIONS FOR X - VARIABLES

            sex           age           educatio       attend        Godexist
               0.990         0.950         0.900         0.900         0.800

          TOTAL COEFFICIENT OF DETERMINATION FOR X - VARIABLES IS   1.000

          SQUARED MULTIPLE CORRELATIONS FOR STRUCTURAL EQUATIONS

            ETA 1         ETA 2         ETA 3
               0.097         0.144         0.056

          TOTAL COEFFICIENT OF DETERMINATION FOR STRUCTURAL EQUATIONS   IS   0.167
```

```
21 AUG 86   SMOKING MODEL -- MULTIPLE INDICATORS of SMOKING AND RELIGION          PAGE    9
15:31:00     University of Alberta


*** MULTIPLE INDICATORS S1 + S4, attend + Godexists        Smoklis32a


LISREL ESTIMATES (MAXIMUM LIKELIHOOD)

        LAMBDA Y

              ETA 1      ETA 2      ETA 3
smoking1      1.000      0.0        0.0
smoking4      2.856      0.0        0.0
asviews       0.0        1.000      0.0
asacts        0.0        0.0        1.000

        LAMBDA X

              KSI 1      KSI 2      KSI 3      KSI 4
sex           1.000      0.0        0.0        0.0
age           0.0        1.000      0.0        0.0
educatio      0.0        0.0        1.000      0.0
attend        0.0        0.0        0.0        1.000
Godexist      0.0        0.0        0.0        0.415

        BETA

              ETA 1      ETA 2      ETA 3
ETA 1         0.0        0.0        0.0
ETA 2        -0.919      0.0        0.0
ETA 3        -0.256      0.069      0.0

        GAMMA

              KSI 1      KSI 2      KSI 3      KSI 4
ETA 1        -0.021     -0.004     -0.041     -0.037
ETA 2         0.517      0.016      0.045      0.068
ETA 3         0.150      0.0        0.044     -0.047

        PHI

              KSI 1      KSI 2      KSI 3      KSI 4
KSI 1         0.248
KSI 2         0.166    234.553
KSI 3         0.016     -7.284      6.218
KSI 4         0.151      6.527     -0.026      3.989

        PSI

              ETA 1      ETA 2      ETA 3
              0.203      2.704      1.170

        THETA EPS

            smoking1   smoking4    asviews     asacts
              0.025      0.202      0.166      0.065
```

```
21 AUG 86   SMOKING MODEL -- MULTIPLE INDICATORS of SMOKING AND RELIGION          PAGE   10
15:31:00     University of Alberta

        THETA DELTA

            sex        age        educatio    attend     Godexist
            0.002     12.345      0.691       0.465      0.429

        SQUARED MULTIPLE CORRELATIONS FOR Y - VARIABLES

            smoking1   smoking4    asviews     asacts
            0.900      0.900       0.950      0.950

        TOTAL COEFFICIENT OF DETERMINATION FOR Y - VARIABLES IS   1.000

        SQUARED MULTIPLE CORRELATIONS FOR X - VARIABLES

            sex        age        educatio    attend     Godexist
            0.990      0.950       0.900      0.900      0.800

        TOTAL COEFFICIENT OF DETERMINATION FOR X - VARIABLES IS   1.000

        SQUARED MULTIPLE CORRELATIONS FOR STRUCTURAL EQUATIONS

            ETA 1      ETA 2      ETA 3
            0.088      0.141      0.054

        TOTAL COEFFICIENT OF DETERMINATION FOR STRUCTURAL EQUATIONS   IS   0.160

            MEASURES OF GOODNESS OF FIT FOR THE WHOLE MODEL :

     CHI-SQUARE WITH  16 DEGREES OF FREEDOM IS    686.35 (PROB. LEVEL = 0.0  )

                 GOODNESS OF FIT INDEX IS 0.627

             ADJUSTED GOODNESS OF FIT INDEX IS-0.048

             ROOT MEAN SQUARE RESIDUAL IS       0.386
```

```
21 AUG 86    SMOKING MODEL -- MULTIPLE INDICATORS of SMOKING AND RELIGION                PAGE  11
15:31:00     University of Alberta


*** MULTIPLE INDICATORS S1 + S4, attend + Godexists      Smoklis32a


MODIFICATION INDICES

          LAMBDA Y

                 ETA 1        ETA 2        ETA 3
smoking1          0.0         2.038        1.183
smoking4          0.0         2.038        1.027
asviews           0.0          0.0         2.816
asacts            0.0          0.0          0.0

          LAMBDA X

                 KSI 1        KSI 2        KSI 3        KSI 4
sex               0.0         2.816         0.0          0.0
age               0.0          0.0          0.0          0.0
educatio          0.0         2.817         0.0          0.0
attend           3.210       11.812       54.165         0.0
Godexist         3.210       12.671       54.164         0.0

          BETA

                 ETA 1        ETA 2        ETA 3
ETA  1            0.0          0.0         2.816
ETA  2            0.0          0.0         2.816
ETA  3            0.0          0.0          0.0

          GAMMA

                 KSI 1        KSI 2        KSI 3        KSI 4
ETA  1            0.0          0.0          0.0          0.0
ETA  2            0.0          0.0          0.0          0.0
ETA  3            0.0         2.816         0.0          0.0

          PHI

                 KSI 1        KSI 2        KSI 3        KSI 4
KSI  1            0.0
KSI  2            0.0          0.0
KSI  3            0.0          0.0          0.0
KSI  4            0.0          0.0          0.0          0.0

          PSI

                 ETA 1        ETA 2        ETA 3
                  0.0          0.0          0.0

          THETA EPS

               smoking1     smoking4      asviews      asacts
                 0.034        0.022        2.816         0.0
```

```
21 AUG 86    SMOKING MODEL -- MULTIPLE INDICATORS of SMOKING AND RELIGION                PAGE  12
15:31:00     University of Alberta


          THETA DELTA

                 sex          age         educatio      attend       Godexist
                2.821         0.0          2.820       1456.447      1732.063

       MAXIMUM MODIFICATION INDEX IS 1732.06 FOR ELEMENT ( 5, 5) OF THETA DELTA
```

```
21 AUG 86    SMOKING MODEL -- MULTIPLE INDICATORS of SMOKING AND RELIGION              PAGE  13
15:31:00     University of Alberta


*** MULTIPLE INDICATORS S1 + S4, attend + Godexists        Smoklis32a


STANDARD ERRORS

         LAMBDA Y

              ETA 1        ETA 2        ETA 3
smoking1       0.0          0.0          0.0
smoking4       0.067        0.0          0.0
asviews        0.0          0.0          0.0
asacts         0.0          0.0          0.0

         LAMBDA X

              KSI 1        KSI 2        KSI 3        KSI 4
sex            0.0          0.0          0.0          0.0
age            0.0          0.0          0.0          0.0
educatio       0.0          0.0          0.0          0.0
attend         0.0          0.0          0.0          0.0
Godexist       0.0          0.0          0.0          0.018

         BETA

              ETA 1        ETA 2        ETA 3
ETA 1          0.0          0.0          0.0
ETA 2          0.188        0.0          0.0
ETA 3          0.127        0.033        0.0

         GAMMA

              KSI 1        KSI 2        KSI 3        KSI 4
ETA 1          0.046        0.002        0.010        0.012
ETA 2          0.167        0.006        0.036        0.045
ETA 3          0.111        0.0          0.023        0.029

         PHI

              KSI 1        KSI 2        KSI 3        KSI 4
KSI 1          0.017
KSI 2          0.379       16.819
KSI 3          0.063        2.020        0.471
KSI 4          0.051        1.616        0.265        0.303

         PSI

              ETA 1        ETA 2        ETA 3
               0.016        0.197        0.084

         THETA EPS

             smoking1     smoking4     asviews      asacts
               0.0          0.0          0.0          0.0
```

```
21 AUG 86    SMOKING MODEL -- MULTIPLE INDICATORS of SMOKING AND RELIGION              PAGE  14
15:31:01     University of Alberta


         THETA DELTA

             sex          age          educatio     attend       Godexist
              0.0          0.0          0.0          0.0          0.0
```

*** MULTIPLE INDICATORS S1 + S4, attend + Godexists   Smoklis32a

T-VALUES

          LAMBDA Y

                  ETA 1   ETA 2   ETA 3
smoking1   0.0   0.0   0.0
smoking4   42.900   0.0   0.0
asviews   0.0   0.0   0.0
asacts   0.0   0.0   0.0

          LAMBDA X

                  KSI 1   KSI 2   KSI 3   KSI 4
sex   0.0   0.0   0.0   0.0
age   0.0   0.0   0.0   0.0
educatio   0.0   0.0   0.0   0.0
attend   0.0   0.0   0.0   0.0
Godexist   0.0   0.0   0.0   23.066

          BETA

                  ETA 1   ETA 2   ETA 3
ETA 1   0.0   0.0   0.0
ETA 2   -4.879   0.0   0.0
ETA 3   -2.023   2.056   0.0

          GAMMA

                  KSI 1   KSI 2   KSI 3   KSI 4
ETA 1   -0.462   -2.575   -4.194   -3.048
ETA 2   3.089   2.790   1.244   1.502
ETA 3   1.355   0.0   1.914   -1.614

          PHI

                  KSI 1   KSI 2   KSI 3   KSI 4
KSI 1   14.533
KSI 2   0.439   13.946
KSI 3   0.246   -3.606   13.212
KSI 4   2.954   4.040   -0.098   13.147

          PSI

                  ETA 1   ETA 2   ETA 3
                  13.065   13.748   13.863

          THETA EPS

                  smoking1   smoking4   asviews   asacts
                  0.0   0.0   0.0   0.0

          THETA DELTA

                  sex   age   educatio   attend   Godexist
                  0.0   0.0   0.0   0.0   0.0

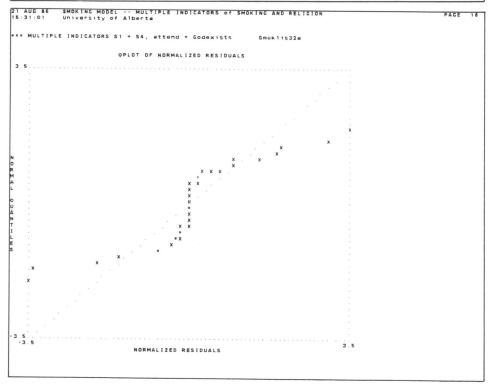

```
21 AUG 86    SMOKING MODEL -- MULTIPLE INDICATORS of SMOKING AND RELIGION                    PAGE   17
15:31:01     University of Alberta

*** MULTIPLE INDICATORS S1 + S4, attend + Godexists       Smoklis32a

FITTED MOMENTS AND RESIDUALS

        FITTED MOMENTS

             smoking1    smoking4    asviews     asacts      sex        age       educatio    attend     Godexist
smoking1      0.247
smoking4      0.636       2.017
asviews      -0.248      -0.707       3.314
asacts       -0.077      -0.221       0.290      1.302
sex          -0.012      -0.035       0.153      0.044      0.250
age          -0.905      -2.585       4.838     -0.043      0.166    246.898
educatio     -0.223      -0.637       0.375     -0.362      0.016     -7.284      6.909
attend       -0.178      -0.508       0.618     -0.079      0.151      6.527     -0.026      4.454
Godexist     -0.074      -0.211       0.257     -0.033      0.063      2.716     -0.011      1.660      1.120

        FITTED RESIDUALS

             smoking1    smoking4    asviews     asacts      sex        age       educatio    attend     Godexist
smoking1      0.000
smoking4     -0.000      -0.001
asviews       0.013      -0.038      -0.000
asacts       -0.007       0.018       0.001     -0.000
sex           0.002      -0.007      -0.000     -0.000      0.000
age          -0.002      -0.013       0.012     -1.323      0.000      0.000
educatio      0.010      -0.030       0.000     -0.006      0.000      0.033      0.000
attend       -0.011      -0.097      -0.065     -0.075     -0.012      0.711      0.258      0.202
Godexist      0.049       0.144       0.144      0.174      0.025     -1.828     -0.573     -0.455      1.026

        NORMALIZED RESIDUALS

             smoking1    smoking4    asviews     asacts      sex        age       educatio    attend     Godexist
smoking1      0.009
smoking4     -0.001      -0.007
asviews       0.293      -0.293      -0.000
asacts       -0.241       0.225       0.012     -0.000
sex           0.201      -0.201      -0.000     -0.001      0.000
age          -0.006      -0.012       0.000     -1.532      0.001      0.000
educatio      0.163      -0.164       0.000     -0.038      0.000      0.016      0.001
attend       -0.206      -0.666      -0.346     -0.643     -0.224      0.437      0.967      0.666
Godexist      1.895       1.971       1.536      2.983      0.992     -2.006     -4.275     -3.398     13.457


21 AUG 86    SMOKING MODEL -- MULTIPLE INDICATORS of SMOKING AND RELIGION                    PAGE   18
15:31:01     University of Alberta

*** MULTIPLE INDICATORS S1 + S4, attend + Godexists       Smoklis32a

                 QPLOT OF NORMALIZED RESIDUALS

  3.5.........................................................................
     .                                                          .
     .                                                       .
     .                                                    .
     .                                                 .
     .                                              .
     .                                           .
     .                                        .
     .                                     .         X
     .                                  .
     .                               .            X
     .                            .         X
     .                         .        X
     .                      .       X  X
N    .                   .      X X X
O    .                .    X X X .
R    .             .     X X
M    .          .      X X
A    .       .       X
L    .            .   X
     .         .   X
Q    .      .    N
U    .   .     X
A    . .     X
N    .     X  X
T    .   X X
I    .  X
L    . * X
E    .  X
S    .    .
     .  X .
     . X.
   .X
  .X
  X
     .
     .
     .
     .
 -3.5.........................................................................
   -3.5                                                      3.5
                 NORMALIZED RESIDUALS
```

```
21 AUG 86    SMOKING MODEL -- MULTIPLE INDICATORS of SMOKING AND RELIGION                    PAGE  19
15:31:01    University of Alberta


*** MULTIPLE INDICATORS S1 + S4, attend + Godexists        Smoklis32a


FIRST ORDER DERIVATIVES

        LAMBDA Y

               ETA 1        ETA 2        ETA 3
smoking1      -0.000       -0.509        0.252
smoking4      -0.000        0.178       -0.082
asviews       -0.000        0.000       -0.007
asacts        -0.000        0.000        0.000

        LAMBDA X

               KSI 1        KSI 2        KSI 3        KSI 4
sex           -0.000       -0.152        0.000        0.000
age            0.000       -0.000       -0.000       -0.000
educatio       0.000       -0.045       -0.000       -0.000
attend         0.025       -1.405       -0.489        0.000
Godexist      -0.059        3.491        1.175        0.000

        BETA

               ETA 1        ETA 2        ETA 3
   ETA 1      -0.000       -0.000        0.024
   ETA 2      -0.000        0.000       -0.007
   ETA 3      -0.000        0.000        0.000

        GAMMA

               KSI 1        KSI 2        KSI 3        KSI 4
   ETA 1      -0.000        0.000       -0.000        0.000
   ETA 2       0.000       -0.000       -0.000        0.000
   ETA 3       0.000        1.011       -0.000        0.000

        PHI

               KSI 1        KSI 2        KSI 3        KSI 4
   KSI 1      -0.000
   KSI 2       0.000       -0.000
   KSI 3       0.000       -0.000       -0.000
   KSI 4       0.000       -0.000       -0.000        0.000

        PSI

               ETA 1        ETA 2        ETA 3
   ETA 1      -0.000
   ETA 2      -0.000        0.000
   ETA 3       0.020       -0.006       -0.000
```

```
21 AUG 86    SMOKING MODEL -- MULTIPLE INDICATORS of SMOKING AND RELIGION                    PAGE  20
15:31:01    University of Alberta


        THETA EPS

              smoking1     smoking4     asviews      asacts
smoking1      -0.257
smoking4       0.010        0.025
asviews       -0.172        0.059        0.001
asacts         0.257       -0.085       -0.006       -0.000

        THETA DELTA

               sex          age         educatio     attend       Godexist
sex           -0.001
age           -0.001       -0.000
educatio      -0.001       -0.000       -0.001
attend         0.046       -0.010       -0.095       -0.932
Godexist      -0.108        0.024        0.231        2.279       -5.572
```

21 AUG 86   SMOKING MODEL -- MULTIPLE INDICATORS of SMOKING AND RELIGION   PAGE 21
15:31:02   University of Alberta

*** MULTIPLE INDICATORS S1 + S4, attend + Godexists   Smoklis32a

STANDARDIZED SOLUTION

        LAMBDA Y

                ETA 1   ETA 2   ETA 3
smoking1   0.472   0.0   0.0
smoking4   1.347   0.0   0.0
asviews   0.0   1.774   0.0
asacts   0.0   0.0   1.112

        LAMBDA X

                KSI 1   KSI 2   KSI 3   KSI 4
sex   0.498   0.0   0.0   0.0
age   0.0   15.315   0.0   0.0
educatio   0.0   0.0   2.494   0.0
attend   0.0   0.0   0.0   1.997
Godexist   0.0   0.0   0.0   0.831

        BETA

                ETA 1   ETA 2   ETA 3
ETA 1   0.0   0.0   0.0
ETA 2   -0.244   0.0   0.0
ETA 3   -0.109   0.109   0.0

        GAMMA

                KSI 1   KSI 2   KSI 3   KSI 4
ETA 1   -0.022   -0.132   -0.215   -0.158
ETA 2   0.145   0.140   0.064   0.077
ETA 3   0.067   0.0   0.100   -0.085

        PHI

                KSI 1   KSI 2   KSI 3   KSI 4
KSI 1   1.000
KSI 2   0.022   1.000
KSI 3   0.013   -0.191   1.000
KSI 4   0.151   0.213   -0.005   1.000

        PSI

                ETA 1   ETA 2   ETA 3
                0.912   0.859   0.946

21 AUG 86   SMOKING MODEL -- MULTIPLE INDICATORS of SMOKING AND RELIGION   PAGE 22
15:31:02   University of Alberta

        CORRELATION MATRIX FOR ETA

                ETA 1   ETA 2   ETA 3
ETA 1   1.000
ETA 2   -0.296   1.000
ETA 3   -0.148   0.147   1.000

        TOTAL EFFECTS   ETA ON KSI (STANDARDIZED)

                KSI 1   KSI 2   KSI 3   KSI 4
ETA 1   -0.022   -0.132   -0.215   -0.158
ETA 2   0.151   0.172   0.116   0.115
ETA 3   0.086   0.033   0.136   -0.055

THE PROBLEM REQUIRED 1468 DOUBLE PRECISION WORDS,
THE CPU-TIME WAS   1.43 SECONDS

Appendix D

The Moment Matrix Fit Function

This appendix shows that maximum likelihood estimates of intercepts and means requires only relatively minor changes to the input data and fit function discussed in Chapter 5. Three fit functions must be clearly distinguished. *First* there is the fit function providing maximum likelihood estimates of all the coefficients in a model (collectively denoted by the vector Θ in this appendix) if both the means and covariances are functions of the model coefficients (Eqs. 9.16–9.18 with κ). That is, the estimates of the model coefficients, when placed in the model's equations, predict a specific covariance matrix among the observed variables (as discussed in Section 4.4) and a specific vector of observed means (as discussed in Section 9.2.2). We select as estimates of the model coefficients those coefficient values that maximize the likelihood that the observed means and covariances could arise as mere random sampling fluctuations if the model-implied vector of means μ and covariance matrix Σ described the population from which the sample of observations was selected. This fit function will be denoted $F(\mu(\Theta), \Sigma(\Theta))$, and it is a rewritten version of the likelihood ratio appearing in Theorem 10.9.1 of Anderson (1958). In a LISREL context, the formula is provided by Joreskog and Sorbom (1980: Eq. B9; and 1978a: LISREL IV manual Appendix B, Eq. B16), and both of these follow Sorbom (1974: Eq. 5a). The formula is

$$F(\mu(\Theta),\Sigma(\Theta)) = \log|\Sigma| + \text{tr}(T\Sigma^{-1}) - \log|S| - (p + q) \qquad \text{D.1}$$

where

$$T = S + (\bar{z} - \mu)(\bar{z} - \mu)' \qquad \text{D.2}$$

We take this formula as given and show how it can be transformed into the second and third fit functions discussed below.

The *second* fit function $F(\Omega(\Theta))$ describes the fit between the observed moment matrix $\mathbf{M}$ and the model-implied moment matrix Ω. This fit function was developed "by analogy" in Section 9.2.3.2. The formula for calculating its value is

$$F(\Omega(\Theta)) = \log|\Omega| + \text{tr}(\Omega^{-1}\mathbf{M}) - \log|\mathbf{M}| - (p + q + 1) \qquad \text{D.3}$$

The moment maximum likelihood fit function (Eq. D.3) is not reported in either the LISREL V or VI manuals. It is presented in Appendix B of later versions of the LISREL IV manual and in Joreskog and Sorbom (1980). Most of this appendix is devoted to showing that optimizing the fit of $\mathbf{M}$ to Ω is identically equivalent to obtaining maximum likelihood estimates via the fit function $F(\mu(\Theta), \Sigma(\Theta))$, where the means and covariances are presented as separate entities. The equivalence of these fit functions lets us capitalize on the parallel between the moment fit function and the basic maximum likelihood fit function (Eq. 5-9), which is the *third* fit function.

$$F(\Sigma(\Theta)) = \log|\Sigma| + \text{tr}(\mathbf{S}\Sigma^{-1}) - \log|\mathbf{S}| - (p + q) \qquad \text{D.4}$$

The parallel between the moment matrix and covariance matrix fit functions has two payoffs. First, the maximum likelihood estimates of all the model coefficients, including intercepts and means, are obtainable with the same program steps appropriate for covariance matrices (namely LISREL's iterative procedure). We merely enter the moment matrix as data, in which case the extra row/column of the matrix required by the variable ONE automatically increments the number of input variables from $p + q$ to $p + q + 1$, and proceed with estimation as before.

Second, the similarity in structural form "implies that LISREL . . . will use the correct first order derivatives and the correct information matrix (matrix of expected second order derivatives), and hence also that the correct standard errors will be obtained as well" (Joreskog and Sorbom, 1980:B6; LISREL IV manual: B9).

We now begin the proof that the fit function describing the fit of the means and covariances separately (Eqs. D.1 and D.2) is identical to the fit function for the moment matrix (Eq. D.3). Our presentation relies heavily on Joreskog (1971a), Sorbom (1974), Joreskog and Sorbom (1980), and Appendix B to the LISREL IV manual. First, recall from the definitions in Chapter 9 that the model-implied moment matrix Ω can be written as a partitioned matrix:

$$\Omega = \begin{bmatrix} \Sigma + \mu\mu' & \mu \\ \mu' & 1 \end{bmatrix} \qquad \text{D.5}$$

The partitioning of Ω allows us to apply the following well-known rule about determinants of partitioned matrices (cf. Searle 1966:96, 1982:258):

$$\begin{vmatrix} A & B \\ C & D \end{vmatrix} = |D||A - BD^{-1}C| = |A||D - CA^{-1}B| \qquad \text{D.6}$$

By direct application of this rule

$$|\Omega| = |1|(\Sigma + \mu\mu') - \mu 1^{-1}\mu'| \qquad \text{D.7}$$

$$= 1|\Sigma + \mu\mu' - \mu\mu'| \qquad \text{D.8}$$

$$= |\Sigma| \qquad \text{D.9}$$

This last step proves that the first terms to the right of the equals signs in Eq. D.1 and D.3 are identical.

The rule giving the determinant of partitioned matrices also applies to the observed moment matrix **M**.

$$|M| = \begin{vmatrix} S + \bar{z}\bar{z}' & \bar{z} \\ \bar{z}' & 1 \end{vmatrix} = |1|(S + \bar{z}\bar{z}') - \bar{z}1^{-1}\bar{z}'| \qquad \text{D.10}$$

$$= |(S + \bar{z}\bar{z}' - \bar{z}\bar{z}')| \qquad \text{D.11}$$

$$= |S| \qquad \text{D.12}$$

Showing $|M| = |S|$ proves that the third terms to the right of the equals signs in Eq. D.1 and D.3 are identical.

Proving the equality of the sums of the remaining two terms in D.1 and D.3 is more complex. As a first step we note that we need the inverse of the partitioned matrix Ω; hence we return to Searle (1966:210–211, 1982:260, Eq. 14), where we find that the inverse of any partitioned matrix can be written as follows:

$$\begin{bmatrix} A & B \\ C & D \end{bmatrix}^{-1} = \begin{bmatrix} (A - BD^{-1}C)^{-1} & -(A - BD^{-1}C)^{-1}BD^{-1} \\ -D^{-1}C(A - BD^{-1}C)^{-1} & D^{-1} + D^{-1}C(A - BD^{-1}C)^{-1}BD^{-1} \end{bmatrix} \quad \text{D.13}$$

Direct substitution of the corresponding portions of Ω gives the inverse of Ω. This substitution is simplified considerably if we first substitute into the term $(A - BD^{-1}C)^{-1}$, which appears in each of the four portions of the partition.

$$(A - BD^{-1}C)^{-1} = ((\Sigma + \mu\mu') - \mu 1^{-1}\mu')^{-1} \qquad \text{D.14}$$

$$= (\Sigma + \mu\mu' - \mu\mu')^{-1} \qquad \text{D.15}$$

$$= \Sigma^{-1} \qquad \text{D.16}$$

With this simplification in hand, we now substitute in all the remaining terms to obtain

$$\Omega^{-1} = \begin{bmatrix} \Sigma + \mu\mu' & \mu \\ \mu' & 1 \end{bmatrix}^{-1} = \begin{bmatrix} \Sigma^{-1} & -\Sigma^{-1}\mu 1^{-1} \\ -1^{-1}\mu'\Sigma^{-1} & 1^{-1} + 1^{-1}\mu'\Sigma^{-1}\mu 1^{-1} \end{bmatrix} \qquad \text{D.17}$$

This simplifies to

$$\Omega^{-1} = \begin{bmatrix} \Sigma^{-1} & -\Sigma^{-1}\mu \\ -\mu'\Sigma^{-1} & 1 + \mu'\Sigma^{-1}\mu \end{bmatrix} \qquad \text{D.18}$$

With Ω^{-1} in a relatively simple form, we are now prepared to reexpress the quantity $\text{tr}(\Omega^{-1}M)$, which is the second term on the right of Eq. D.3.

$$\text{tr}(\Omega^{-1}M) = \text{tr}\left(\begin{bmatrix} \Sigma^{-1} & -\Sigma^{-1}\mu \\ -\mu'\Sigma^{-1} & 1 + \mu'\Sigma^{-1}\mu \end{bmatrix} \begin{bmatrix} S + \bar{z}\bar{z}' & \bar{z} \\ \bar{z}' & 1 \end{bmatrix} \right) \qquad \text{D.19}$$

Since the trace is the sum of the diagonal elements, only the top left and lower right of the partitioned product are required. Since the sum of all the diagonal elements must equal the sum of the diagonals of the top left partition plus the diagonals of the bottom right partition,

$$\text{tr}(\Omega^{-1}M) = \text{tr}[\Sigma^{-1}(S + \bar{z}\bar{z}') - \Sigma^{-1}\mu\bar{z}'] + \text{tr}[-\mu'\Sigma^{-1}\bar{z} + 1 + \mu'\Sigma^{-1}\mu] \qquad \text{D.20}$$

Since the trace of a sum of matrices equals the sum of the traces,

$$\text{tr}(\Omega^{-1}M) = \text{tr}[\Sigma^{-1}(S + \bar{z}\bar{z}') - \Sigma^{-1}\mu\bar{z}']$$
$$+ \text{tr}[-\mu'\Sigma^{-1}\bar{z}] + \text{tr}[\mu'\Sigma^{-1}\mu] + \text{tr}[1] \qquad \text{D.21}$$

Using the fact that $\text{tr}(ABC) = \text{tr}(CAB)$ (Searle, 1982:45), we have

$$\text{tr}(\Omega^{-1}M) = \text{tr}[\Sigma^{-1}(S + \bar{z}\bar{z}') - \Sigma^{-1}\mu\bar{z}']$$
$$+ \text{tr}[-\Sigma^{-1}\bar{z}\mu'] + \text{tr}[\Sigma^{-1}\mu\mu'] + 1 \qquad \text{D.22}$$

Using the sum of traces and traces of a sum, again, produces

$$\text{tr}(\Omega^{-1}M) = \text{tr}[\Sigma^{-1}(S + \bar{z}\bar{z}') - \Sigma^{-1}\mu\bar{z}' - \Sigma^{-1}\bar{z}\mu' + \Sigma^{-1}\mu\mu'] + 1 \qquad \text{D.23}$$

from which Σ^{-1} can be factored out of many of the internal components.

$$\text{tr}(\Omega^{-1}M) = \text{tr}[\Sigma^{-1}(S + \bar{z}\bar{z}' - \mu\bar{z}' - \bar{z}\mu' + \mu\mu')] + 1 \qquad \text{D.24}$$

The matrix T, as defined in Eq. D.2, implies that

$$T = S + \bar{z}\bar{z}' - \bar{z}\mu' - \mu\bar{z}' + \mu\mu' \qquad \text{D.25}$$

Recognizing this as the matrix in the inner parentheses in Eq. D.24 allows us to rewrite $\text{tr}(\Omega^{-1}M)$ as

$$\text{tr}(\Omega^{-1}M) = \text{tr}(\Sigma^{-1}T) + 1 \qquad \text{D.26}$$

Since the $\text{tr}(AB) = \text{tr}(BA)$ (Searle, 1982:45),

$$\text{tr}(\Omega^{-1}M) = \text{tr}(T\Sigma^{-1}) + 1 \qquad \text{D.27}$$

Cumulating the various pieces, we have, from Eq. D.9, that $\log|\Sigma| = \log|\Omega|$, from Eq. D.12 that $\log|S| = \log|M|$, and from Eq. D.27 that $\text{tr}(T\Sigma^{-1}) = \text{tr}(\Omega^{-1}M) - 1$. Replacing the parts of D.1 corresponding to the left sides of these equations, with the right sides, results in exactly D.3 and completes the proof. Thus, the fit function specifying simultaneous MLE of the vector of means and the covariance matrix (D.1) is identical to the fit function for MLE of the moment matrix (D.3).

The parallel of both these equations to the fit function for a covariance matrix (Eq. D.4 or 5.9) is direct, once we note that each matrix on the right of D.3 has one more variable than previously (the variable ONE), and hence the total number of input variables has increased from $p + q$ to $p + q + 1$.

Bibliography

Acock, Alan, and Theodore D. Fuller. 1985. "Standardized solutions using LISREL on multiple populations." *Sociological Methods and Research* 13:551–557.

Allison, Paul D. 1977. "Testing for interaction in multiple regression." *American Journal of Sociology* 83:144–153.

Alwin, Duane F. 1985. "Addendum: The analysis of experimental data using structural equation models." Pp. 82–88 in H. M. Blalock (ed.), *Causal Models in Panel and Experimental Designs.* New York: Aldine. This addendum follows a reprint of the 1974 article by Alwin and Tessler.

Alwin, D., and R. Hauser. 1975. "The decomposition of effects in path analysis." *American Sociological Review* 40:37–47.

Alwin, Duane F., and David J. Jackson. 1980. "Measurement models for response errors in surveys: Issues and applications." Pp. 68–119 in Karl F. Schuessler (ed.), *Sociological Methodology 1980.* San Francisco: Jossey-Bass.

Alwin, Duane F., and David J. Jackson. 1981. "Applications of simultaneous factor analysis to issues of factorial invariance." Pp. 249–279 in David Jackson and Edgar Borgatta (eds.), *Factor Analysis and Measurement in Sociological Research: A Multi-Dimensional Perspective.* Beverly Hills: Sage.

Alwin, Duane F., and Richard C. Tessler. 1974. "Causal models, unobserved variables and experimental data." *American Journal of Sociology* 80:59–86. A six page addendum follows the reprint of this article in H. M. Blalock Jr. (ed.), 1985. *Causal Models in Panel and Experimental Designs.* New York: Aldine.

Alwin, Duane F., and Arland Thornton. 1984. "Family origins and the schooling process: Early versus late influence of parental characteristics." *American Sociological Review* 49:784–802.

Anderson, James C. 1985. "A measurement model to assess measure-specific factors in multiple-informant research." *Journal of Marketing Research* 22:86–92.

Anderson, James C., and David W. Gerbing. 1984. "The effect of sampling error on convergence, improper solutions and goodness-of-fit indices for maximum likelihood confirmatory factor analysis." *Psychometrika* 49:155–173.

Anderson, James G. 1987. "Structural equation models in the social and behavioral sciences: Model building." *Child Development* 58:49–64.

Anderson, T. W. 1958. *An Introduction to Multivariate Statistical Analysis.* New York: Wiley.

Andrews, D. F., R. Gnanadesikan, and J. L. Warner. 1973. "Methods of assessing multivariate normality." Pp. 95–116 in P. R. Krishnaiah (ed.), *Multivariate Analysis III.* New York: Academic Press.

Arminger, Gerhard. 1986. "Linear stocastic differential equation models for panel data with unobserved variables." Pp. 187–212 in Nancy Brandon Tuma (ed.), *Sociological Methodology 1986.* San Francisco: Jossey-Bass.

Arminger, Gerhard. 1987. "Misspecification, asymptotic stability, and ordinal variables in the analysis of panel data." *Sociological Methods and Research* 15:336–348.

Baer, Douglas E., and James E. Curtis. 1984. "French Canadian–English Canadian differences in values: National survey findings." *Canadian Journal of Sociology* 9:405–427.

Bagozzi, Richard P. 1980. *Causal Models in Marketing.* New York: Wiley.

Bagozzi, Richard P. 1981. "Attitudes, intentions, and behavior: A test of some key hypotheses." *Journal of Personality and Social Psychology* 41:607–627.

Bagozzi, Richard P., and Lynn W. Phillips. 1982. "Representing and testing organizational theories: A holistic construal." *Administrative Science Quarterly* 27:459–489.

Baker, R. L., B. Mednick, and W. Brock. 1984. "An application of causal modeling techniques to prospective longitudinal data bases." Pp. 106–132 in S. A. Mednick, M. Harway, and K. M. Finello (eds.), *Handbook of Longitudinal Research*, Vol. 1. New York: Praeger.

Bartholomew, D. J. 1985. "Foundations of factor analysis: Some practical implications." *British Journal of Mathematical and Statistical Psychology* 38:1–10.

Beckman, Linda L., Rhonda Aizenberg, Alan B. Forsythe, and Tom Day. 1983. "A theoretical analysis of antecedents of young couples' fertility decisions and outcomes." *Demography* 20:519–533.

Belsley, David A., Edwin Kuh, and Roy E. Welsch. 1980. *Regression Diagnostics: Identifying Infuential Data and Sources of Collinearity.* New York: Wiley.

Bentler, P. M. 1980. "Multivariate analysis with latent variables: causal modeling." *Annual Review of Psychology* 31:419–456.

Bentler, P. M. 1982. "Linear systems with multiple levels and types of latent variables." Pp. 101–130 in K. G. Joreskog and H. Wold (eds.), *Systems under Indirect Observation*, Part I. Amsterdam: North-Holland.

Bentler, P. M. 1983a. "Some contributions to efficient statistics in structural models: Specification and estimation of moment structures." *Psychometrika* 48:493–517.

Bentler, P. M. 1983b. "Theory and implementation of EQS, a structural equations program." Los Angeles: BMDP Statistical Software.

Bentler, P. M. 1983c. "Simultaneous equation systems as moment structure models." *Journal of Econometrics* 22:13–42.

Bentler, P. M. 1984. "Structural equation models in longitudinal research." Pp. 88–105 in S. A. Mednick, M. Harway, and K. M. Finello (eds.), *Handbook of Longitudinal Research*, Vol. 1. New York: Praeger.

Bentler, P. M. 1986. "Structural modeling and Psychometrika: An historical perspective on growth and achievements." *Psychometrika* 51:35–51.

Bentler, P. M. 1987. "Drug use and personality in adolescence and young adulthood: Structural equation models with nonnormal variables." *Child Development* 58:65–79.

Bentler, P. M., and Douglas G. Bonett. 1980. "Significance tests and goodness of fit in the analysis of covariance structures." *Psychological Bulletin* 88:588–606.

Bentler, P. M., and Edward H. Freeman. 1983. "Tests for stability in linear structural equation systems." *Psychometrika* 48:143–145.

Bentler, P. M., and Sik-Yum Lee. 1983. "Structural models under polynomial constraints: Applications to correlation and alpha-type structural models." *Journal of Educational Statistics* 8:207–222.

Bentler, P. M., and Michael D. Newcomb. 1986. "Personality, sexual behavior, and drug use revealed through latent variable methods." *Clinical Psychology Review* 6:363–385.

Bentler, P. M., and Jeffrey S. Tanaka. 1983. "Problems with EM algorithms for ML factor analysis." *Psychometrika* 48:247–251.

Bentler, P. M., and David G. Weeks. 1979. "Interrelations among models for the analysis of moment structures." *Multivariate Behavioral Research* 14:169–186.

Bentler, P. M., and David G. Weeks. 1980. "Linear structural equations with latent variables." *Psychometrika* 45:289–308.

Bentler, P. M., and David G. Weeks. 1982. "Multivariate analysis with latent variables." Pp. 747–771 in P. R. Krishnaiah and L. N. Kanal (eds.), *Handbook of Statistics*, Vol. 2. Amsterdam: North-Holland.

Bentler, P. M., and David G. Weeks. 1985. "Some comments on structural equation models." *British Journal of Mathematical and Statistical Psychology* 38:120–121.

Bentler, P. M., and J. Arthur Woodward. 1978. "A Head Start reevaluation: Positive effects are not yet demonstrable." *Evaluation Quarterly* 2:493–510.

Berk, Richard A. 1983. "An introduction to sample selection bias in sociological data." *American Sociological Review* 48:386–398.

Berry, William D. 1984. *Nonrecursive causal models*. Beverly Hills: Sage series on Quantitative Applications in the Social Sciences 07-037.

Biddle, Bruce J., and Marjorie M. Marlin. 1987. "Causality, confirmation, credulity, and structural equation modeling." *Child Development* 58:4–17.

Bielby, Denise Del Vento, and William T. Bielby. 1984. "Work commitment, sex-role attitudes, and women's employment." *American Sociological Review* 49:234–247.

Bielby, William T. 1981. "Neighborhood effects: A LISREL model for clustered samples." *Sociological Methods and Research* 10:82–111.

Bielby, William T. 1986a. "Arbitrary metrics in multiple-indicator models of latent variables." *Sociological Methods and Research.* 15:3–23.

Bielby, William T. 1986b. "Arbitrary normalizations: Comments on issues raised by Sobel, Arminger, and Henry." *Sociological Methods and Research.* 15:62–63.

Bielby, William T. and Robert M. Hauser. 1977. "Structural equation models." *Annual Review of Sociology* 3:137–161.

Blalock, Hubert M., Jr. 1979a. "Measurement and conceptualization problems: The major obstacle to integrating theory and research." *American Sociological Review* 44:881–894.

Blalock, Hubert M., Jr. 1979b. *Social Statistics*, 2nd. rev. ed. New York: McGraw-Hill.

Blalock, Hubert M., Jr. 1982. *Conceptualization and Measurement in the Social Sciences*. Beverly Hills: Sage.

Blalock, Hubert M., Jr. 1986. "Multiple causation, indirect measurement and generalizability in the social sciences." *Synthese* 68:13–36.

Bohrnstedt, George W. 1983. "Measurement." Pp. 69–121 in Peter Rossi, James Wright, and Andy Anderson (eds.), *Handbook of Survey Research* New York: Academic Press.

Bohrnstedt, George W., and Edgar F. Borgatta (eds.) 1981. *Social Measurement: Current Issues*. Beverly Hills: Sage.

Bohrnstedt, George W., and Arthur S. Goldberger. 1969. "On the exact covariance of products of random variables." *Journal of the American Statistical Association* 64:1439–1442.

Bohrnstedt, George W., and Gerald Marwell. 1978. "The reliability of products of two random variables." Pp. 254–273 in K. F. Schuessler (ed.), *Sociological Methodology*. San Francisco: Jossey-Bass.

Bohrnstedt, George W., Peter P. Mohler, and Walter Muller. 1987. "Editors' introduction" (to a special issue on the reliability of survey research items). *Sociological Methods and Research* 15:171–176.

Bollen, Kenneth A. 1984. "Multiple indicators: internal consistency or no necessary relationship?" *Quality and Quantity* 18:377–385.

Bollen, Kenneth A. 1986. "Sample size and Bentler and Bonnet's nonnormed fit index." *Psychometrika* 51:375–377.

Bollen, Kenneth A. 1987. *Structural Equation Modeling with Latent Variables*. New York: Wiley.

Bollen, Kenneth A., and Robert W. Jackman. 1985. "Regression diagnostics: An expository treatment of outliers and influential cases." *Sociological Methods and Research* 13:510–542.

Bollen, Kenneth A., and Karl G. Joreskog. 1985. "Uniqueness does not imply identification: A note on confirmatory factor analysis." *Sociological Methods and Research* 14:155–163.

Bookstein, F. L. 1982. "Panel discussion: Modeling and Method." Pp. 317–321 in K. Joreskog and H. Wold (eds.), Systems under Indirect Observation, Part II. Amsterdam: North-Holland.

Boomsma, Anne. 1982. "The robustness of LISREL against small sample size in factor analysis models." Pp. 149–173 in K. Joreskog and H. Wold (eds.), *Systems under Indirect Observation*, Part I. Amsterdam: North-Holland.

Boomsma, Anne. 1983. "On the robustness of LISREL (maximum likelihood estimation) against small sample size and non-normality." Doctoral dissertation, Department of Statistics and Measurement, State University of Groningen, Holland.

Boomsma, Anne. 1985. "Nonconvergence, improper solutions, and starting values in LISREL maximum likelihood estimation." *Psychometrika* 50:229–242.

Borgatta, Edgar F., and George W. Bohrnstedt. 1981. "Level of measurement: Once over again." *Sociological Methods and Research* 9:147–160. Reprinted as Pp. 23–37 in George W. Bohrnstedt and Edgar F. Borgatta (eds.), *Social Measurement: Current Issues*. Beverly Hills: Sage.

Brown, R. L. 1986a. "A comparison of the LISREL and EQS programs for obtaining parameter estimates in confirmatory factor analysis studies." *Behavior Research Methods, Instruments, and Computers* 18:382–388.

Brown, R. L. 1986b. "A cautionary note on the use of LISREL's automatic start values in confirmatory factor analysis studies." *Applied Psychological Measurement* 10:239–245.

Browne, Michael W. 1974. "Generalized least squares estimators in the analysis of covariance structures." *South African Statistical Journal* 8:1–24.

Browne, Michael W. 1982. "Covariance structures." Pp. 72–141 in Douglas M. Hawkins (ed.), *Topics in Applied Multivariate Analysis.* Cambridge: Cambridge University Press.

Browne, Michael W. 1984. "Asymptotically distribution-free methods for the analysis of covariance structures." *British Journal of Mathematical and Statistical Psychology* 37:62–83.

Burt, Ronald S. 1973. "Confirmatory factor-analytic structures and the theory construction process (plus corrigenda)." *Sociological Methods and Research* 2:131–190.

Burt, Ronald S. 1976. "Interpretational confounding of unobserved variables in structural equation models." *Sociological Methods and Research* 5:3–51.

Burt, Ronald S. 1981. "A note on interpretational confounding of unobserved variables in structural equation models." Pp. 299–318 in P. V. Marsden (ed.), *Linear Models in Social Research.* Beverly Hills: Sage.

Burt, Ronald S., Michael G. Fischer, and Kenneth P. Christman. 1979. "Structures of well-being: Sufficient conditions for identification as restricted covariance models." *Sociological Methods and Research* 8:111–120.

Busemeyer, Jerome R., and Lawrence E. Jones. 1983. "Analysis of multiplicative combination rules when the causal variables are measured with error." *Psychological Bulletin* 93:549–562.

Bye, Barry V., Salvatore J. Gallicchio, and Janice M. Dykacz. 1985. "Multiple-indicator, multiple-cause models for a single latent variable with ordinal indicators." *Sociological Methods and Research* 13:487–509.

Bynner, J. M., and D. M. Romney. 1985. "LISREL for beginners." *Canadian Psychology* 26:43–49.

Bynner, J. M. and Romney, D. M. 1986. "Intelligence, fact or artefact: Alternative structures for cognitive abilities." British Journal of Educational Psychology 56:13-23.

Campbell, Richard T. 1983. "Status attainment research: End of the beginning or beginning of the end." *Sociology of Education* 56:47–62.

Carmines, Edward G. 1986. "The analysis of covariance structure models." Pp. 23–55 in William Berry and Michael Lewis-Beck (eds.), *New Tools for Social Scientists: Advances and Applications in Research Methods*. Beverly Hills: Sage Publications.

Carmines, Edward, and John McIver. 1981. "Analyzing models with unobserved variables: Analysis of covariance structures." Pp. 65–115 in G. Bohrnstedt and E. Borgatta (eds.), *Social Measurement: Current Issues*. Beverly Hills: Sage.

Chen, Huey-Tsyh. 1983. "Flowgraph analysis of effect decompositon: Use in recursive and nonrecursive models." *Sociological Methods and Research* 12:3–29.

Chen, Meei-Shia, and Kenneth C. Land. 1986. "Testing the health belief model: LISREL analysis of alternative models of causal relationships between health beliefs and preventive dental behavior." *Social Psychology Quarterly* 49:45–60.

Cliff, Norman. 1983. "Some cautions concerning the application of causal modeling methods." *Multivariate Behavioral Research* 18:115–126.

Cole, David A., and Scott E. Maxwell. 1985. "Multitrait–multimethod comparisions across populations: a confirmatory factor analytic approach." *Multivariate Behavioral Research* 20:389–417.

Comrey, Andrew L. 1985. "A method for removing outliers to improve factor analytic results." *Multivariate Behavioral Research* 20:273–281.

Cooley, William W. 1979. "Structural equations and explanatory observational studies." Introduction to K. G. Joreskog and D. Sorbom. *Advances in Factor Analysis and Structural Equation Models*. Cambridge, Mass.: Abt Books.

Corcoran, Mary. 1980. "Sex differences in measurement error in status attainment models." *Sociological Methods and Research* 9:199–217.

Costner, Herbert L., and Ronald Schoenberg. 1973. "Diagnosing indicator ills in multiple indicator models." Pp. 167–199 in Arthur S. Goldberger and Otis Dudley Duncan (eds.), *Structural Equation Models in the Social Sciences*. New York: Seminar Press.

Cramer, H. 1946. *Mathematical Methods of Statistics*. Princeton, N.J.: Princeton University Press.

Crano, William D., and Jorge L. Mendoza. 1987. "Maternal factors that influence children's positive behavior: Demonstration of a structural equation analysis of selected data from the Berkeley Growth Study." *Child Development* 58:38–48.

Cudeck, Robert, and Michael W. Browne. 1983. "Cross-validation of covariance structures." *Multivariate Behavioral Research* 18:147–167.

Cullen, Charles G. 1972. *Matrices and Linear Transformations*, 2nd ed. Reading, Mass.: Addison-Wesley.

Cuttance, Peter. 1985. "A general structural equation modeling framework for the social and behavioral sciences." Pp. 408–463 in Robert B. Smith (ed.), *A Handbook of Social Science Methods*, Vol. 3. New York: Praeger.

Davidon, W. C. 1959. "Variable metric method for minimization." A. E. C. Research and Development Report, ANL5990, Argonne National Laboratory.

Davis, Philip J. 1973. *The Mathematics of Matrices*, 2nd ed. Lexington, Mass.: Xerox.

Davis, James A. 1985. *The Logic of Causal Order*. Beverly Hills: Sage.

DeLeeuw, Jan. 1983. "Models and methods for the analysis of correlation coefficients." *Journal of Econometrics* 22:113–137.

DeLeeuw, Jan, Wouter Keller, and Tom Wansbeek. 1983. "Editors Introduction." *Journal of Econometrics* 22:1–12.

DeLeeuw, Jan. 1985. "Review of four recent texts." *Psychometrika* 50:371–375.

Devlin, Susan J., R. Gnanadesikan, and J. R. Kettenring. 1975. "Robust estimation and outlier detection with correlation coefficients." *Biometrika* 62:531–545.

Dijkstra, Theo. 1983. "Some comments on maximum likelihood and partial least squares methods." *Journal of Econometrics* 22:67–90.

Dijkstra, Taeke Klass. 1985. *Latent Variables in Linear Stochastic Models: Reflections on "Maximum Likelihood" and "Partial Least Squares" Methods*, 2nd ed. Amsterdam: Sociometric Research Foundation.

Dillon, William R., Ajith Kumar, and Narendra Mulani. 1987. "Offending estimates in covariance structure analysis: Comments on the causes and solutions to Heywood cases." *Psychological Bulletin* 101:126–135.

Driel, Otto P. van. 1978. "On various causes of improper solutions in maximum likelihood factor analysis." *Psychometrika* 43:225–243.

Duncan, Otis D. 1975. *Introduction to Structural Equation Models*. New York: Academic Press.

Duncan, Otis D., Douglas M. Sloane, and Charles Brody. 1982. "Latent classes inferred from response-consistency effects." Pp. 19–64 in K. G. Joreskog and H. Wold (eds.), *Systems under Indirect Observation*, Part I. Amsterdam: North-Holland.

Dupacova, J., and H. Wold. 1982. "On some identification problems in ML modeling of systems with indirect observations." Pp. 293–315 in K. G. Joreskog and H. Wold (eds.), *Systems under Indirect Observation: Causality, Structure, Prediction*, Part II. Amsterdam: North-Holland.

Entwisle, Doris R., and Leslie A. Hayduk, with the collaboration of Thomas W. Reilly. 1982. *Early Schooling: Cognitive and Affective Outcomes*. Baltimore: Johns Hopkins University Press.

Everitt, B. S. 1984. *An Introduction to Latent Variable Models*. London: Chapman and Hall.

Faulbaum, Frank. 1987. "Intergroup comparisons of latent means across waves." *Sociological Methods and Research* 15:317–335.

Fink, Edward L., and Timothy I. Mabee. 1978. "Linear equations and nonlinear estimation: A lesson from a nonrecursive example." *Sociological Methods and Research* 7:107–120.

Fisher, Franklin M. 1966. *The Identification Problem in Econometrics*. New York: McGraw-Hill.

Fisher, Franklin M. 1970. "A correspondence principle for simultaneous equation models." *Econometrica* 38:73–92.

Fletcher, R., and M. J. D. Powell. 1963. "A rapidly convergent decent method for minimization." *Computer Journal* 6:163–168.

Fornell, Claes (ed.) 1982. *A Second Generation of Multivariate Analysis*, Vol. 2. New York: Praeger. Fornell, Claes. 1983. "Issues in the application of covariance structure analysis: A comment." *Journal of Consumer Research* 9:443–448.

Fornell, Claes, and David F. Larcker. 1981. "Structural equation models with unobservable variables and measurement error: Algebra and statistics." *Journal of Marketing Research* 18:382–388.

Fox, John. 1980. "Effect analysis in structural equation models: Extensions and simplified methods of computation." *Sociological Methods and Research* 9:3–28 (Errata 10:119). Reprinted in 1980 as Pp. 178–202 in P. V. Marsden (ed.), *Linear Models in Social Research*. Beverly Hills: Sage.

Fox, John. 1984. *Linear Statistical Models and Related Methods: With Applications to Social Science Research*. New York: Wiley.

Fox, John. 1985. "Effect analysis in structural-equation models II: Calculation of specific indirect effects." *Sociological Methods and Research* 14:81–95.

Fraser, C. 1980. *COSAN User's Guide*. Toronto, Ontario: The Ontario Institute for Studies in Education.

Fulker, David W., Laura A. Baker, and R. Darrell Bock. 1983. "Estimating components of covariance using LISREL." *Data Analyst* (Scientific Software, Inc.) 1:5–8.

Fuller, Leonard E. 1962. *Basic Matrix Theory.* Englewood Cliffs, N.J.: Prentice-Hall.

Gallini, Joan K., and Garrett K. Mandeville. 1984. "An investigation of the effect of sample size and specification error on the fit of structural equation models." *Journal of Experimental Education* 53:9–19.

Geraci, Vincent J. 1977. "Identification of simultaneous equation models with measurement error." Pp. 163–185 in D. Aigner and A. S. Goldberger (eds.), *Latent Variables in Socioeconomic Models.* Amsterdam: North-Holland.

Gerbing, David W., and James C. Anderson. 1985. "The effects of sampling error and model characteristics on parameter estimation for maximum likelihood confirmatory factor analysis." *Multivariate Behavioral Research* 20:255–271.

Geweke, John F., and Kenneth J. Singleton. 1980. "Interpreting the likelihood ratio statistic in factor models when sample size is small." *Journal of the American Statistical Association: Theory and Methods* 75:133–137.

Glymour, Clark, and Richard Scheines. 1986. "Causal modeling with the TETRAD program." *Synthese* 68:37–63.

Gnanadesikan, R. 1977. *Methods for Statistical Data Analysis of Multivariate Observations.* New York: Wiley.

Gnanadesikan, R., and J. R. Kettenring. 1972. "Robust estimates, residuals, and outlier detection with multiresponse data." *Biometrics* 28:81–124.

Gollob, Harry F., and Charles S. Reichardt. 1987. "Taking account of time lags in causal models." *Child Development* 58:80–92.

Graff, J., and P. Schmidt. 1982. "A general model for decomposition of effects." Pp. 131–148 in K. G. Joreskog and H. Wold (eds.), *Systems under Indirect Observation,* Part I. Amsterdam: North-Holland.

Greenberg, David F., and Ronald C. Kessler. 1982. "Equilibrium and identification in linear panel models." *Sociological Methodology and Research* 10:435–451.

Greene, Vernon L. 1977. "An algorithm for total and indirect causal effects." *Political Methodology* 44:369–381.

Guilford, J. P., and Benjamin Fruchter. 1978. *Fundamental Statistics in Psychology and Education.* New York: McGraw-Hill.

Hammarling, S. J. 1970. *Latent Roots and Latent Vectors.* Toronto: University of Toronto Press.

Hauser, Robert M., Shu-Ling Tsai, and William H. Sewell. 1983. "A model of stratification with response error in social and psychological variables." *Sociology of Education* 56:20–46.

Hayduk, Leslie A. 1985. "Personal space: the conceptual and measurement implications of structural equation models." *Canadian Journal of Behavioural Science* 17:141–149.

Hayduk, Leslie A., and Tom Wonnacott. 1980. " 'Effect equations' or 'effect coefficients': A note on the visual and verbal presentation of multiple regression interactions." *Canadian Journal of Sociology* 5:399–404.

Hays, William L. 1963. *Statistics.* New York: Holt, Rinehart and Winston.

Heise, David. 1972. "Employing nominal variables, induced variables, and block variables in path analysis." *Sociological Methods and Research* 1:147–173.

Heise, David. 1975. *Causal Analysis.* New York: Wiley.

Heise, David. 1986. "Estimating nonlinear models: Correcting for measurement error." *Sociological Methods and Research* 14:447–472.

Heise, David R., and Roberta G. Simmons. 1985. "Some computer-based developments in sociology." *Science* 228:428–433.

Henry, Neil W. 1986. "On 'arbitrary metrics' and 'normalization issues.'" *Sociological Methods and Research.* 15:59–61.

Hertel, Bradley R. 1976. "Minimizing error variance introduced by missing data routines in survey analysis." *Sociological Methods and Research* 4:459–474.

Herting, Jerald R. 1985. "Multiple indicator models using LISREL." Pp. 263–319 in H. M. Blalock (ed.), *Causal Models in the Social Sciences,* 2nd ed. New York: Aldine.

Herting, Jerald R., and Herbert L. Costner. 1985. "Respecification in multiple indicator models." Pp. 321–393 in H. M. Blalock (ed.), *Causal Models in the Social Sciences,* 2nd ed. New York: Aldine.

Hertzog, Christopher, and John R. Nesselroade. 1987. "Beyond autoregressive models: Some implications of the trait-state distinction for the structural modeling of developmental change." *Child Development* 58:93–109.

Hoelter, Jon W. 1983a. "The analysis of covariance structures: Goodness-of-fit indices." *Sociological Methods and Research* 11:325–344.

Hoelter, Jon W. 1983b. "Factorial invariance and self-esteem: Reassessing race and sex differences." *Social Forces* 61:834–846.

Hoppe, Hans-Hermann. 1980. "On how not to make inferences about measurement error." *Quality and Quantity* 14:503–510.

Horn, J. L., and J. Jack McArdle. 1980. "Perspectives on mathematical/statistical model building (MASMOB) in aging research." Pp. 503–541 in L. W. Poon (ed.), *Aging in the 1980's.* Washington, D.C.: American Psychological Association.

Huba, George J., and Lisa L. Harlow. 1987. "Robust structural equation models: Implications for developmental psychology." *Child Development* 58:147–166.

Huggins, William H., and Doris R. Entwisle. 1968. *Introductory Systems and Design.* Waltham, Mass.: Blaisdell.

Hultsch, David F., Christopher Hertzog, and Roger A. Dixon. 1984. "Text recall in adulthood: The role of intellectual abilities." *Developmental Psychology* 20:1193–1209.

Jackson, Douglas N., and David W. Chan 1980. "Maximum-likelihood estimation in common factor analysis: A cautionary note." *Psychological Bulletin* 88:502–508.

Jagodzinski, Wolfgang. 1984. "The overestimation of stability coefficients in LISREL applications: Some problems in the correction for attenuation." (In German) *Zeitschrift fur Soziologie* 13:225–242.

James, Lawrence R., Stanley A. Mulaik, and Jeanne M. Brett. 1982. *Causal Analysis: Assumptions, Models and Data.* Beverly Hills: Sage.

Johnson, David R., and James C. Creech. 1983. "Ordinal measures in multiple indicator models: A simulation study of categorization errors." *American Sociological Review* 48:398–407.

Joreskog, Karl G. 1963. *Statistical Estimation in Factor Analysis: A New Technique and Its Foundations.* Stockholm: Almqvist and Wiksell.

Joreskog, Karl G. 1966 "Testing simple structure hypotheses in factor analysis." *Psychometrika* 31:165–178.

Joreskog, Karl G. 1967. "Some contributions to maximum likelihood factor analysis." *Psychometrika* 32:443–482.

Joreskog, Karl G. 1968. "Statistical models for congeneric test scores." Pp. 213–214 in *The Proceedings of the 76th Annual Convention of the American Psychological Association.*

Joreskog, Karl G. 1969. "A general approach to confirmatory maximum likelihood factor analysis." *Psychometrika* 34:183–202. (Reprinted with an important addendum as Chapter 2 in K. G. Joreskog and D. Sorbom, 1979).

Joreskog, Karl G. 1970a. "A general method for analysis of covariance structures." *Biometrika* 57:239–251.

Joreskog, Karl G. 1970b. "Estimation and testing of simplex models." *British Journal of Mathematical and Statistical Psychology* 23:121–145.

Joreskog, Karl G. 1971a. "Simultaneous factor analysis in several populations." *Psychometrika* 36:409–426. (Reprinted as Chapter 7 in K. G. Joreskog and D. Sorbom, 1979.)

Joreskog, Karl G. 1971b. "Statistical analysis of sets of congeneric tests." *Psychometrika* 32:109–133.

Joreskog, Karl G. 1973a. "A general method for estimating a linear structural equation system." Pp. 85–112 in A. S. Goldberger and O. D. Duncan (eds.), *Structural Equation Models in the Social Sciences*. New York: Seminar Press.

Joreskog, Karl G. 1973b. "Analysis of covariance structures." Pp. 263–285 in P. R. Krishnaiah (ed.), *Multivariate Analysis III*. New York: Academic Press.

Joreskog, Karl G. 1977a. "Structural equation models in the social sciences: Specification, estimation, and testing." Pp. 265 287 in P. R. Krishnaiah (ed.), *Applications of Statistics*. Amsterdam: North-Holland. (Reprinted as Chapter 4 in K. G. Joreskog and D. Sorbom, 1979.)

Joreskog, Karl G. 1977b. "Statistical models and methods for analysis of longitudinal data." Chapter 16 in D. V. Aigner and A. S. Goldberger (eds.), *Latent Variables in Socioeconomic Models*. Amsterdam: North-Holland. (Reprinted as Chapter 5 in K. G. Joreskog and D. Sorbom, 1979.)

Joreskog, Karl G. 1977c. "Factor analysis by least-squares and maximum-likelihood methods." Pp. 125–153 in Kurt Enslein, A. Ralston, and H. Wilf (eds.), *Statistical Methods for Digital Computers*. New York: Wiley-Interscience.

Joreskog, Karl G. 1978a. "Structural analysis of covariance and correlation matrices." *Psychometrika* 43:443–477.

Joreskog, Karl G. 1978b. "An econometric model for multivariate panel data." *Annales de l'INSEE* 30–31:355–366.

Joreskog, Karl G. 1979a. "Basic ideas of factor and component analysis." Pp. 5 20 in K. G. Joreskog and D. Sorbom (eds.), *Advances in Factor Analysis and Structural Equation Models*. Cambridge, Mass.: Abt Books.

Joreskog, Karl G. 1979b. "Analyzing psychological data by structural analysis of covariance matrices." Pp. 45–100 in K. G. Joreskog and D. Sorbom (eds.), *Advances in Factor Analysis and Structural Equation Models*. Cambridge, Mass.: Abt Books.

Joreskog, Karl G. 1979c. "Statistical estimation of structural models in longitudinal-developmental investigations." Pp. 303–351 in John Nesselroade and Paul Baltes (eds.), *Longitudinal Research in the Study of Behavior and Development*. New York: Academic Press.

Joreskog, Karl G. 1981a. "Analysis of covariance structures." *Scandinavian Journal of Statistics* 8:65–92.

Joreskog, Karl G. 1981b. "Basic issues in the application of LISREL." *Data* (later *Data Analyst*) 1:1–6.

Joreskog, Karl G. 1982a. "The LISREL approach to causal model building in the social sciences." Pp. 81–99 in K. G. Joreskog and H. Wold (eds.), *Systems under Indirect Observation*: Part I. Amsterdam: North-Holland.

Joreskog, Karl G. 1982b. "Selected references on LISREL and related topics." Mimeograph. Department of Statistics, University of Uppsala.

Joreskog, Karl G. 1983. "Factor analysis as an errors-in-variables model." Pp. 185–196 in Howard Wainer and Samuel Messick (eds.), *Principals of Modern Psychological Measurement*. Hillsdale, N.J.: Lawrence Erlbaum Associates.

Joreskog, Karl G. 1986. "Estimation of the polyserial correlation from summary statistics." Mimeograph, Department of Statistics, University of Uppsala, Sweden.

Joreskog, Karl G., and Arthur S. Goldberger. 1972. "Factor analysis by generalized least squares." *Psychometrika* 37:243–260.

Joreskog, Karl G., and Arthur S. Goldberger. 1975. "Estimation of a model with multiple indicators and multiple causes of a single latent variable." *Journal of the American Statistical Association* 70:631–639.

Joreskog, Karl G., Gunnar T. Gruvaeus, and Marielle van Thillo. 1970. *ACOVS: A General Computer Program for Analysis of Covariance Structures*. Princeton, N.J.: Educational Testing Services.

Joreskog, Karl G., and D. N. Lawley. 1968. "New methods in maximum likelihood factor analysis." *British Journal of Mathematical and Statistical Psychology* 21:85–96.

Joreskog, Karl, and Dag Sorbom. 1976a. *LISREL III: Estimation of Linear Structural Equation Systems by Maximum Likelihood Methods*. Chicago: National Educational Resources, Inc.

Joreskog, Karl, and Dag Sorbom. 1976b. "Statistical models and methods for test-retest situations." Pp. 135–157 in D. N. M. De Gruijter and L. J. T. van der Kamp (eds.), *Advances in Psychological and Educational Measurement*. London: Wiley.

Joreskog, Karl, and Dag Sorbom. 1977. "Statistical models and methods for analysis of longitudinal data." Pp. 285–325 in D. Aigner and A. S. Goldberger (eds.), *Latent Variables in Socioeconomic Models*. Amsterdam: North-Holland.

Joreskog, Karl, and Dag Sorbom. 1978a. *LISREL IV: Analysis of Linear Structural Relationships by the Method of Maximum Likelihood (Release 2)*. Chicago: National Educational Resources, Inc.

Joreskog, Karl, and Dag Sorbom. 1978b. "EFAP II Users Guide: Exploratory Factor Analysis Program." Mimeograph, Department of Statistics, University of Uppsala, Sweden.

Joreskog, Karl, and Dag Sorbom. 1979. *Advances in Factor Analysis and Structural Equation Models*. Cambridge, Mass.: Abt Books. (J. Magidson, ed.).

Joreskog, Karl, and Dag Sorbom. 1980. "Simultaneous Analysis of Longitudial Data from Several Cohorts." Research Report 80-5 (ISSN 0348-2987), Department of Statistics, University of Uppsala.

Joreskog, Karl, and Dag Sorbom. 1981. *LISREL V: Analysis of Linear Structural Relationships by Maximum Likelihood and Least Squares Methods*. Chicago: National Educational Resources. Distributed by International Educational Services, Chicago.

Joreskog, Karl, and Dag Sorbom. 1982. "Recent developments in structural equation modeling." *Journal of Marketing Research* 19:404–416.

Joreskog, Karl, and Dag Sorbom. 1983. "Supplement to the LISREL V manual." Mooresville, Ind.: Scientific Software, Inc.

Joreskog, Karl, and Dag Sorbom. 1984. *LISREL VI: Analysis of Linear Structural Relationships by Maximum Likelihood, Instrumental Variables, and Least Squares Methods*. Mooresville, Ind.: Scientific Software, Inc.

Joreskog, Karl, and Dag Sorbom. 1985a. "PRELIS: A Program for Multivariate Data Screening and Data Summarization: A Preprocessor for LISREL." Department of Statistics, University of Uppsala.

Joreskog, Karl, and Dag Sorbom. 1985b. "Simultaneous analysis of longitudinal data from several cohorts." (with a discussion and reply) Pp. 323–370 in William Mason and Stephen Fienberg (eds.), *Cohort Analysis in Social Research*. New York: Springer-Verlag.

Joreskog, Karl, and Dag Sorbom. 1986. "SIMPLIS: A Fast and Simple Version of LISREL, preliminary version." Department of Statistics, University of Uppsala.

Joreskog, Karl, G., and Dag Sorbom. 1987. "New developments in LISREL." *Data Analyst* 4:1–22.

Joreskog, Karl G., and H. Wold (eds.) 1982a. *Systems under Indirect Observation: Causality, Structure, Prediction*, Part I. New York: North-Holland.

Joreskog, Karl G., and H. Wold (eds.) 1982b. *Systems under Indirect Observation: Causality, Structure, Prediction*, Part II. New York: North-Holland.

Joreskog, Karl G., and H. Wold. 1982c. "The ML and PLS techniques for modeling with latent variables: Historical and comparative aspects." Pp. 263–270 in K. G. Joreskog and H. Wold (eds.), *Systems under Indirect Observation*, Part I. Amsterdam: North-Holland.

Kendall, M. G., and A. Stuart. 1958. *The Advanced Theory of Statistics*. Vol. 1. London: Charles Griffin and Company.

Kennedy, Leslie, and Clifford Kinzel. 1984. "Edmonton Area Study Sampling Report 1984." Edmonton Arca Series Report Number 33. Edmonton, Alberta: Population Research Laboratory, Department of Sociology, University of Alberta.

Kennedy, Patricia H., Susan L. Starrfield, and Charles Baffi. 1983. "Using LISREL analysis for drug research." *Journal of School Health* 53:277–281.

Kenny, David A. 1979. *Correlation and Causality*. New York: Wiley-Interscience.

Kenny, David A., and Charles M. Judd. 1984. "Estimating the nonlinear and interactive effects of latent variables." *Psychological Bulletin* 96:201–210.

Kessler, Ronald C., and David F. Greenberg. 1981. *Linear Panel Analysis: Models of Quantitative Change*. New York: Academic Press.

Kim, Jae-On. 1984. "An approach to sensitivity analysis in sociological research. *American Sociological Review* 49:272–282.

Kohn, Melvin L., and Carmi L. Schooler. 1981. "Job conditions and intellectual flexibility: a longitudinal assessment of their reciprocal effects." Pp. 281–313 in D. J. Jackson and E. F. Borgatta (eds.), *Factor Analysis and Measurement in Sociological Research*. Beverly Hills: Sage.

Kohn, Melvin L., and Carmi L. Schooler. 1983. *Work and Personality: An Inquiry into the Impact of Social Stratification*. Norwood, N.J.: Ablex.

Land, Kenneth C., and Marcus Felson. 1978. "Sensitivity analysis of arbitrarily identified simultaneous-equation models." *Sociological Methods and Research* 6:283–307.

Laumann, Edward O., David Knoke, and Yong-Hak Kim. 1985. "An organizational approach to state policy formation: A comparative study of energy and health domains." *American Sociological Review* 50:1–19.

Lawley, D. N., and A. E. Maxwell. 1971. *Factor Analysis As a Statistical Method*. New York: American Elsevier.

Lee, S. Y. 1980. "Estimation of covariance structure models with parameters subject to functional restraints". *Psychometrika* 45:309–324.

Lee, S. Y., and R. I Jennrich. 1979. "A study of algorithms for covariance structure analysis with specific comparisons using factor analysis. *Psychometrika* 44:99–113.

Lee, Sik-Yum, and Robert I. Jennrich. 1984. "The analysis of structural equation models by means of derivative free nonlinear least squares." *Psychometrika* 49:521–528.

Lee, Sik-Yum, and Kowk-Leung Tsui. 1982. "Covariance structure analysis in several populations." *Psychometrika* 47:297–308.

Lee, Sik-Yum. 1986. "Estimation for structural equation models with missing data." *Psychometrika* 51:93–99.

Lewis-Beck, Michael S., and Lawrence B. Mohr. 1976. "Evaluating effects of independent variables." *Political Methodology* 3:27–47.

Lincoln, James R., and Arne L. Kalleberg. 1985. "Work organization and workforce commitment: A study of plants and employees in the U.S. and Japan." *American Sociological Review* 50:738–760.

Lindsay, Paul, and William E. Knox. 1984. "Continuity and change in work values among young adults: A longitudinal study." *American Journal of Sociology* 89:918–931.

Liska, Allen E., and Mark D. Reed. 1985. "Ties to conventional institutions and delinquency: Estimating reciprocal effects." *American Sociological Review* 50:547–560.

Lomax, Richard G. 1982. "A guide to LISREL-type structural equation modeling." *Behavior Research Methods and Instrumentation* 14:1–8.

Lomax, Richard G. 1983. "A guide to multiple-sample structural equation modeling." *Behavior Research Methods and Instrumentation* 15:580–584.

Long, J. Scott. 1976. "Estimation and hypothesis testing in linear models containing measurement error: A review of Joreskog's model for the analysis of covariance structures." *Sociological Methods and Research* 5:157–206.

Long, J. Scott. 1981. "Estimation and hypothesis testing in linear models containing measurement error: A review of Joreskog's model for the analysis of covariance structures." Pp. 209–256 in P. V. Marsden (ed.), *Linear Models in Social Research*. Beverly Hills: Sage.

Long, J. Scott. 1983a. *Confirmatory Factor Analysis: A Preface to LISREL*. Beverly Hills: Sage.

Long, J. Scott. 1983b. *Covariance Structure Models: An Introduction to LISREL*. Beverly Hills: Sage.

Lord, F. M., and M. R. Novick, with contributions by A. Brinbaum. 1968. *Statistical Theories of Mental Test Scores.* Reading, Mass.: Addison-Wesley.

Lorence, Jon, and Jeylan T. Mortimer. 1985. "Job involvement through the life course: A panel study of three age groups." *American Sociological Review* 50:618–638.

Lorens, Charles S. 1964. *Flowgraphs for the Modeling and Analysis of Linear Systems.* New York: McGraw-Hill.

MacCallum, Robert. 1986. "Specification searches in covariance structure modeling." *Psychological Bulletin* 100:107–120.

Magidson, Jay. 1977. "Toward a causal model approach for adjusting for preexisting differences in the nonequivalent control group situation: A general alternative to ANCOVA." *Evaluation Quarterly* 1:399–420.

Magidson, Jay. 1978. "Reply to Bentler and Woodward." *Evaluation Quarterly* 2:511–520.

Mare, Robert D., and William M. Mason. 1981. "Children's reports of parental socioeconomic status." Pp. 187–207 in G. Bohrnstedt and E. Borgatta (eds.), *Social Measurement: Current Issues.* Beverly Hills: Sage.

Marsden, Peter V. 1983. "On interaction effects involving block variables." *Sociological Methods and Research* 11:305–232.

Marsh, Herbert W., and Dennis Hocevar. 1983. "Confirmatory factor analysis of multitrait-multimethod matrices." *Journal of Educational Measurement* 20:231–248.

Marsh, Herbert W., and Dennis Hocevar. 1985. "Application of confirmatory factor analysis to the study of self-concept: First- and higher-order factor models and their invariance across groups." *Psychological Bulletin* 97:562–582.

Mason, Samuel J. 1953. "Feedback theory: Some properties of signal flow graphs." *Proceedings of the IRE* 41:1144–1156 (now *Proceedings of the Institute of Electrical and Electronics Engineers*).

Mason, Samuel J. 1956. "Feedback theory: further properties of signal flow graphs." *Proceedings of the IRE* 44:920–926 (now *Proceedings of the Institute of Electrical and Electronics Engineers*).

Mason, Samuel J., and Henry J. Zimmermann. 1960. *Electronic Circuits, Signals, and Systems.* New York: Wiley.

Matsueda, Ross L., and William T. Bielby. 1986. "Statistical power in covariance structure models." Pp. 120–158 in Nancy Brandon Tuma (ed.), *Sociological Methodology 1986.* San Francisco: Jossey-Bass.

McArdle, J. Jack, 1980. "Causal modeling applied to psychonomic systems simulation." *Behavior Research Methods and Instrumentation* 12:193–209.

McArdle, J. J., and David Epstein. 1987. "Latent growth curves within developmental structural equation models." *Child Development* 58:110–133.

McArdle, J. Jack, and Roderick P. McDonald. 1984. "Some algebraic properties of the reticular action model for moment structures." *British Journal of Mathematical and Statistical Psychology* 37:234–251.

McCarthy, John D., and Dean R. Hoge. 1984. "The dynamics of self-esteem and delinquency." *American Journal of Sociology* 90:396–410.

McDonald, Roderick P. 1978. "A simple comprehensive model for the analysis of covariance structures." *British Journal of Mathematical and Statistical Psychology* 31:59–72.

McDonald, Roderick P. 1979. "The structural analysis of multivariate data: a sketch of a general theory." *Multivariate Behavioral Research* 14:21–28.

McDonald, Roderick P. 1980. "A simple comprehensive model for the analysis of covariance structures: some remarks on applications." *British Journal of Mathematical and Statistical Psychology* 33:161–183.

McDonald, Roderick P. 1982. "A note on the investigation of local and global identifiability." *Psychometrika* 47:101–103.

McDonald, Roderick P. 1985. *Factor Analysis and Related Methods.* Hillsdale, N.J.: Lawrence Erlbaum Associates.

McDonald, Roderick P. 1986. "Describing the elephant: structure and function in multivariate data." *Psychometrika* 51:513–534.

McDonald, Roderick P., and William R. Krane. 1977. "A note on local identifiability and degrees of freedom in the asymptotic likelihood ratio test." *British Journal of Mathematical and Statistical Psychology* 30:198–203.

McDonald, Roderick P., and William R. Krane. 1979. "A Monte Carlo study of local identifiability and degrees of freedom in the asymptotic likelihood ratio test." *British Journal of Mathematical and Statistical Psychology* 32:121–132.

McGaw, Barry, and Karl G. Joreskog 1971. "Factorial invariance of ability measures in groups differing in intelligence and socioeconomic status." *British Journal of Mathematical and Statistical Psychology* 24:154–168.

Miller, Karen A., Melvin L. Kohn, and Carmi Schooler. 1985. "Educational self-direction and the cognitive functioning of students." *Social Forces* 63:923–944.

Molenaar, Peter C. 1985. "A dynamic factor model for the analysis of multivariate time series." *Psychometrika* 50:181–202.

Monfort, A. 1978. "First-order identification in linear models." *Journal of Economics* 7:333–350.

Mood, Alexander M., Franklin A. Graybill, and Duane C. Boes. 1974. *Introduction to the Theory of Statistics,* 3rd ed. New York: McGraw-Hill.

Morrison, Donald F. 1976. *Multivariate Statistical Methods,* 2nd ed. New York: McGraw-Hill.

Mosteller, Frederick, Robert E. Rourke, and George B. Thomas. 1961. *Probability and Statistics.* Reading, Mass.: Addison-Wesley.

Mulaik, Stanley A. 1972. *The Foundations of Factor Analysis.* New York: McGraw-Hill.

Mulaik, Stanley A. 1987. "Toward a conception of causality applicable to experimentation and causal modeling." *Child Development* 58:18–32.

Mulaik, Stanley A. Forthcoming. "Toward a synthesis of deterministic and probabilistic formulations of causal relations by the functional relation concept." *Philosophy of Science.*

Muthen, Bengt. 1983. "Latent variable structural equation modeling with categorical data." *Journal of Econometrics* 22:43–65.

Muthen, Bengt. 1984. "A general structural equation model with dichotomous, ordered categorical, and continuous latent variable indicators." *Psychometrika* 49:115–132.

Muthen, Bengt, and Karl G. Joreskog. 1983. "Selectivity problems in quasi-experimental studies." *Evaluation Review* 7:139–174.

Muthen, Bengt, and David Kaplan. 1985. "A comparison of some methodologies for the factor analysis of non-normal Likert variables." *British Journal of Mathematical and Statistical Psychology.* 38:171–189.

Nel, D. G. 1980. "On matrix differentiation in statistics." *South African Statistical Journal* 14:137–193.

Nelson, F. Howard, Richard G. Lomax, and Ronald Perlman. 1984. "A structural equation model of second language acquisition for adult learners." *Journal of Experimental Education* 53:29–39.

Nesselroade, John R., and J. Jack McArdle. 1985. "Multivariate causal modeling in alcohol use research." *Social Biology* 32:272–296.

Newton, Rae R., Velma A. Komaeoka, Jon W. Hoelter, and Junko Tanaka-Matsumi. 1984. "Maximum likelihood estimation of factor structures of anxiety measures: A multiple group comparison." *Educational and Psychological Measurement* 44:179–193.

Neyman, J., and E. S. Pearson. 1933. "On the problem of the most efficient tests of statistical hypotheses." *Philosophical Transactions of the Royal Society of London, Series A* 231:289–337.

O'Grady, Kevin E. 1983. "A confirmatory maximum likelihood factor analysis of the WAIS-R." *Journal of Consulting and Clinical Psychology* 51:826–831.

Olsson, Ulf. 1979a. "Maximum likelihood estimation of the polychoric correlation coefficient." *Psychometrika* 44:443–460.

Olsson, Ulf. 1979b. "On the robustness of factor analysis against crude classification of the observations." *Multivariate Behavioral Research* 14:485–500.

Olsson, Ulf, Fritz Drasgow, and Neil J Dorans. 1982. "The polyserial correlation coefficient." *Psychometrika* 47:337–347.

Otter, Pieter W. 1986. "Dynamic structural systems under indirect observation: Identification and estimation aspects from a system theoretic perspective." *Psychometrika* 51:415–428.

Parkerson, Jo Ann, Richard G. Lomax, Diane P. Schiller, and Herbert J. Walberg. 1984. "Exploring causal models of educational achievement." *Journal of Educational Psychology* 76:638–646.

Piliavin, Irving, Craig Thornton, Rosemary Gartner, and Ross Matsueda. 1986. "Crime, deterrence and rational choice." *American Sociological Review* 51:101–119.

Protter, M. H., and C. B. Morrey, Jr. 1964. *Modern Mathematical Analysis*. Reading, Mass.: Addison-Wesley.

Reilly, Thomas W. 1981. "Social class background and women's socialization into parenthood." Ph.D. Dissertation, Department of Sociology, Johns Hopkins University.

Rindskopf, David. 1983a. "Using inequality constraints to prevent Heywoood cases: The LISREL parameterization." *Data Analyst* 1:1–3.

Rindskopf, David. 1983b. "Parameterizing inequality constraints on unique variances in linear structural equation models." *Psychometrika* 48:73–83.

Rindskopf, David. 1984a. "Structural equation models: Empirical identification, Heywood cases, and related problems." *Sociological Methods and Research* 13:109–119.

Rindskopf, David. 1984b. "Using phantom and imaginary latent variables to parameterize constraints in linear structural models." *Psychometrika* 49:37–47.

Rock, D. A., C. E. Werts, R. L. Linn, and K. G. Joreskog. 1977. "A maximum likelihood solution to the errors in variables and errors in equations model." *The Journal of Multivariate Behavioral Research* 12:187–197.

Rubin, Donald B., and Dorothy T. Thayer. 1982. "EM algotithms for ML factor analysis." *Psychometrika* 47:69–76.

Rubin, Donald B., and Dorothy T. Thayer. 1983. "More on EM for ML factor analysis." *Psychometrika* 48:253–257.

Saris, W. E., W. M. de Pijper, and P. Zegwaart. 1979. "Detection of specification errors in linear structural equation models." Pp. 151–171 in Karl F. Schuessler (ed.), *Sociological Methodology 1979*. San Francisco: Jossey-Bass.

Saris, Willem E., and L. Henk Stronkhorst. 1984. *Causal Modelling in Nonexperimental Research: An Introduction to the LISREL Approach.* Amsterdam: Sociometric Research Foundation.

Satorra, Albert, and Willem E. Saris. 1985. "Power of the likelihood ratio test in covariance structure analysis." *Psychometrika* 50:83–90.

Schoenberg, Ronald. 1972. "Strategies for meaningful comparison." Pp. 1–35 in Herbert L. Costner (ed.), *Sociological Methodology 1972*. San Francisco: Jossey-Bass.

Schoenberg, Ronald. 1982. "Multiple indicator models: Estimation of unconstrained construct means and their standard errors." *Sociological Methods and Research* 10:421–433.

Schoenberg, Ronald, and Carol Richtand. 1984. "Application of the EM method: A study of maximum likelihood estimation of multiple indicator and factor analysis models." *Sociological Methods and Research* 13:127–150.

Schonemann, Peter H. 1985. "On the formal differentiation of traces and determinants." *Multivariate Behavioral Research* 20:113–139.

Searle, S. R. 1966. *Matrix Algebra for the Biological Sciences.* New York: Wiley.

Searle, Shayle R. 1982. *Matrix Algebra Useful for Statistics.* New York: Wiley.

Silverman, Robert A., and Leslie W. Kennedy. 1985. "Loneliness, satisfaction and fear of crime: A test for nonrecursive effects." *Canadian Journal of Criminology* 27:1–12.

Skinner, C. J. 1986. "Regression estimation and post-stratification in factor analysis." *Psychometrika* 51:347–356.

Smith, Douglas A., and E. Britt Patterson. 1984. "Applications and generalization of MIMIC models to criminological research." *Journal of Research in Crime and Delinquency* 21:333–352.

Sobel, Michael E. 1982. "Asymptotic confidence intervals for indirect effects in structural equation models." Pp. 290–312 in S. Leinhardt (ed.), *Sociological Methodolgy 1982*, San Francisco: Jossey-Bass.

Sobel, Michael E. 1986. "Some new results on indirect effects and their standard errors in covariance structure models." Pp. 159–186 in Nancy Brandon Tuma (ed.), *Sociological Methodology 1986*. San Francisco: Jossey-Bass.

Sobel, Michael E., and Gerhard Arminger. 1986. "Platonic and operational true scores in covariance structure analyses: An invited comment on Bielby's 'Arbitrary metrics in multiple indicator models of latent variables.'" *Sociological Methods and Research* 15:44–58.

Sobel, Michael E., and George W. Bohrnstedt. 1985. "Use of null models in evaluating the fit of covariance structure models." Pp. 152–178 in Nancy Brandon Tuma (ed.), *Sociological Methodology 1985*. San Francisco: Jossey-Bass.

Sorbom, Dag. 1974. "A general method for studying differences in factor means and factor structure between groups." *British Journal of Mathematical and Statistical Psychology* 27:229–239. (Reprinted as Chapter 8 in K. G. Joreskog and D. Sorbom, 1979.)

Sorbom, Dag. 1975. "Detection of correlated errors in longitudinal data." *British Journal of Mathematical and Statistical Psychology* 28:138–151. (Reprinted as Chapter 6 in K. G. Joreskog and D. Sorbom, 1979.)

Sorbom, Dag. 1976. "A statistical model for the measurement of change in true scores." Pp. 159–170 in D. N. M. de Gruijter and Leo J. Th. van der Kamp (eds.), *Advances in Psychological and Educational Measurement*. New York: Wiley.

Sorbom, Dag. 1978. "An alternative to the methodology for analysis of covariance." *Psychometrika* 43:381–396. (Reprinted as Chapter 9 in K. G. Joreskog and D. Sorbom, 1979.)

Sorbom, Dag. 1982. "Structural equation models with structured means." Pp. 183–195 in K. G. Joreskog and H. Wold (eds.), *Systems under Indirect Observation*, Part 1. Amsterdam: North-Holland.

Sorbom, Dag. 1986. "Model modification." Mimeograph, Department of Statistics, University of Uppsala, Sweden.

Sorbom, Dag, and Karl G. Joreskog. 1981. "The use of LISREL in sociological model building." Pp. 179–199 in D. J. Jackson and E. F. Borgatta (eds.), *Factor Analysis and Measurement in Sociological Research*. Beverly Hills: Sage.

Sorbom, Dag, and Karl G. Joreskog. 1982. "The use of structural equation models in evaluation research." Pp. 381–418 in Claes Fornell (ed.), *A Second Generation of Multivariate Analysis*, Vol. 2. New York: Praeger.

SPSS Inc. 1984. *USERPROC LISREL: Using LISREL VI within SPSSX*. Chicago.

Stahura, John M. 1986. "Suburban development, black suburbanization and the civil rights movement since World War II." *American Sociological Review* 51:131–144.

Steiger, James H. 1979. "Factor indeterminacy in the 1930's and the 1970's: Some interesting parallels." *Psychometrika* 44:157–167.

Stelzl, Ingeborg. 1986. "Changing a causal hypothesis without changing the fit: Some rules for generating equivalent path models." *Multivariate Behavioral Research* 21:309–331.

Stolzenberg, Ross M. 1980. "The measurement and decomposition of causal effects in nonlinear and nonadditive models." Pp. 459–488 in Karl F. Schueller (ed.), *Sociological Methodology 1980.* San Francisco: Jossey-Bass.

Stolzenberg, Ross M., and Kenneth C. Land. 1983. "Causal modeling and survey research." Pp. 613–675 in Peter Rossi, James Wright, and Andy Anderson (eds.), *Handbook of Survey Research.* New York: Academic Press.

Tanaka, J. S. 1987. "How big is big enough?: Sample size and goodness of fit in structural equation models with latent variables." *Child Development* 58:134–146.

Thomson, Elizabeth. 1983. "Individual and couple utility of children." *Demography* 20:507–518.

Thomson, Elizabeth, and Richard Williams. 1982. "Beyond wives' family sociology: A method for analyzing couple data." *Journal of Marriage and the Family* 44:999–1008.

Thornberry, Terence P., and R. L. Christenson. 1984. "Unemployment an criminal involvement: An investigation of reciprocal causal structures." *American Sociological Review* 49:389–411.

Thornton, Arland, Duane F. Alwin, and Donald Camburn. 1983. "Causes and consequences of sex-role attitudes and attitude change." *American Sociological Review* 48:211–227.

Tucker, L. R., and C. Lewis. 1973. "A reliability coefficient for maximum likelihood factor analysis." *Psychometrika* 38:1–10.

Van de Geer, John P. 1971. *Introduction to Multivariate Analysis for the Social Sciences.* San Francisco: W. H. Freeman.

Varga, Richard S. 1962. *Matrix Iterative Analysis.* Englewood Cliffs, N.J.: Prentice-Hall.

Walling, Derald, H. Lawrence Hotchkiss, and Evans W. Curry. 1984. "Power models and error terms." *Sociological Methods and Research* 13:121–126.

Werts, Charles E., Karl G. Joreskog, and Robert L. Linn. 1973. "Identification and estimation in path analysis with unmeasured variables." *American Journal of Sociology* 78:1469–1484.

Werts, Charles E., Karl G. Joreskog, and Robert L. Linn. 1985. "Comment on 'The estimation of measurement error in panel data.'" Pp. 145–150 in H. M. Blalock, Jr. (ed.), *Causal Models in Panel and Experimental Designs.* New York: Aldine.

Werts, Charles E., Robert L. Linn, and Karl G. Joreskog. 1971. "Estimating the parameters of path models involving unmeasured variables." Pp. 400–409 in H. M. Blalock, Jr. (ed.), *Causal Models in the Social Sciences.* Chicago: Aldine.

Werts, Charles E., R. L. Linn, and K. G. Joreskog. 1974. "Intraclass reliability estimates: Testing structural assumptions." *Educational and Psychological Measurement* 34:25–33.

Werts, Charles E., R. L. Linn, and K. G. Joreskog. 1977. "A simplex model for analyzing academic growth." *Educational and Psychological Measurement* 37:745–756.

Werts, Charles E., R. L. Linn, and K. G. Joreskog. 1978. "Reliability of college grades from longitudinal data." *Educational and Psychological Measurement* 38:89–95.

Werts, C. E., D. A. Rock, R. L. Linn, and K. G. Joreskog. 1976. "Comparison of correlations, variances, covariances, and regression weights with or without measurement error." *Psychological Bulletin* 83:1007–1013.

Werts, C. E., D. A. Rock, R. L. Linn, and K. G. Joreskog. 1977. "Validating psychometric assumptions within and between several populations." *Educational and Psychological Measurement* 37:863–872.

Wheaton, Blair, Bengt Muthen, Duane Alwin, and Gene Summers. 1977. "Assessing reliability and stability in panel models." Pp. 84–136 in D. Heise (ed.), *Sociological Methodology 1977.* San Francisco: Jossey-Bass.

Wilk, M. B., and R. Gnanadesikan. 1968. "Probability plotting methods for the analysis of data." *Biometrika* 55:1–17.

Wilks, Samuel S. 1938. "The large-sample distribution of the likelihood ratio for testing composite hypotheses." *Annals of Mathematical Statistics* 9:60–62.

Wilks, Samuel S. 1962. *Mathematical Statistics.* New York: Wiley.

Williams, Richard, and Elizabeth Thompson. 1986a. "Normalization issues in latent variable modeling." *Sociological Methods and Research* 15:24–43.

Williams, Richard, and Elizabeth Thompson. 1986b. "Problems needing solutions or solutions needing problems?: Final thoughts on the normalization controversy." *Sociological Methods and Research* 15:64–68.

Winship, Christopher, and Robert D. Mare. 1983. "Structural equations and path analysis for discrete data." *American Journal of Sociology* 89:54–110.

Wishart, John. 1928. "The generalised product moment distribution in samples from a normal multivariate population." *Biometrika* 20 A:32–52.

Wolfle, Lee M. 1982. "Causal models with unmeasured variables: An introduction to LISREL." *Multiple Linear Regression Viewpoints* 11:9–54.

Wolfle, Lee M., and Corinna A. Ethington. 1985. "SEINE: Standard errors of indirect effects." *Educational and Psychological Measurement* 45:161–166.

Wolfle, Lee M., and Corinna A. Ethington. 1986. "Within-variable, between-occasion error covariance in models of educational achievement." *Educational and Psychological Measurement* 46:571–583.

Wolkowicz, H., and G. P. H. Styan. 1980. "Bounds for eigenvalues using traces." *Linear Algebra and Its Applications* 29:471–506.

Wonnacott, Ronald J., and Thomas H. Wonnacott. 1970. *Econometrics.* New York: Wiley.

Index